Parasitology Research Monographs

More information about this series at http://www.springer.com/series/8816

Sven Klimpel • Thomas Kuhn • Heinz Mehlhorn
Editors

Biodiversity and Evolution of Parasitic Life in the Southern Ocean

Springer

Editors
Sven Klimpel
Goethe University
Institute for Ecology, Evolution, and Diversity
Senckenberg Biodversity and Climate
Research Centre
Frankfurt/Main
Germany

Heinz Mehlhorn
Heinrich Heine University
Department of Zoomorphology
Düsseldorf
Germany

Thomas Kuhn
Goethe University
Institute for Ecology, Evolution, and Diversity
Senckenberg Biodversity and Climate
Research Centre
Frankfurt/Main
Germany

ISSN 2192-3671 ISSN 2192-368X (electronic)
Parasitology Research Monographs
ISBN 978-3-319-46342-1 ISBN 978-3-319-46343-8 (eBook)
DOI 10.1007/978-3-319-46343-8

Library of Congress Control Number: 2017931024

Printed on acid-free paper

This Springer imprint is published by Springer Nature
The registered company is Springer International Publishing AG
The registered company address is: Gewerbestrasse 11, 6330 Cham, Switzerland

Preface

Antarctica is the most Southern continent on earth and had millions of years time to adapt its environments from tropical ones of the giant continent *Gondwana* to most cold ones in our times. This led to an enormous reduction of species and for the survivors strict specialization and adaption to the new environment was needed. Parasites, which lived in or on these animals (migrating with their continents), had the same problems like their hosts. They had to adapt their life cycles and their body properties to the new conditions.

This book reports from the sometimes sophisticated adaptions of some of these survivors of the struggle for life in and around Antarctica.

Düsseldorf, Germany — Heinz Mehlhorn

Acknowledgment

The quick and attractive publication of so many data is not possible without the help of many persons. Each group of authors has to thank heartily a broad spectrum of "helping ghosts" in their institutes, who gave text and figures their final shape. Our thanks are also directed to Dr. Lars Koerner and Dr. Martina Himberger and the production team of SPi Global. Their duly and competent efforts made it possible to present these comprehensive insights into the present knowledge of some of the host-parasite problems in the Antarctica region.

Düsseldorf, Frankfurt a.M., August 2016
Sven Klimpel
Thomas Kuhn
Heinz Mehlhorn

Contents

Contributors

Andrés Barbosa Departamento de Ecología Evolutiva, Museo Nacional de Ciencias Naturales – CSIC, Madrid, Spain

Michaela Blyton Western Sydney University, Richmond, Australia

Rodney A. Bray Department of Life Sciences, Natural History Museum, London, UK

Paolo Cipriani Department of Public Health and Infectious Diseases, Section of Parasitology, Sapienza - University of Rome, Rome, Italy

Department of Ecological and Biological Sciences, Tuscia University, Viterbo, Italy

Erli Schneider Costa Universidade Estadual do Rio Grande do Sul and APECS-Brazil, Porto Alegre, Brazil

Meagan Dewar Deakin University, Waurn Ponds, Australia

Julia I. Diaz Centro de Estudios Parasitológicos y de Vectores (CEPAVE), FCNyM, UNLP, CONICET, La Plata, Argentina

Anna Faltýnková Institute of Parasitology, Biology Centre of the Czech Academy of Sciences, České Budějovice, Czech Republic

Bruno Fusaro Instituto Antártico Argentino (IIA), Plan Anual, Argentina

Simona Georgieva Institute of Parasitology, Biology Centre of the Czech Academy of Sciences, České Budějovice, Czech Republic

Daniel González-Acuña Facultad de Ciencias Veterinarias, Universidad de Concepción, Concepción, Chile

Rachael Gray Faculty of Veterinary Science, School of Life and Environmental Sciences, The University of Sydney, Sydney, Australia

Regina Klapper Goethe-University, Institute for Ecology, Evolution and Diversity, Senckenberg Gesellschaft für Naturforschung, Senckenberg Biodiversity and Climate Research Centre, Frankfurt/Main, Germany

Sven Klimpel Goethe-University, Institute for Ecology, Evolution and Diversity, Senckenberg Gesellschaft für Naturforschung, Senckenberg Biodiversity and Climate Research Centre, Frankfurt/Main, Germany

Judith Kochmann Goethe-University, Institute for Ecology, Evolution and Diversity, Senckenberg Gesellschaft für Naturforschung, Senckenberg Biodiversity and Climate Research Centre, Frankfurt/Main, Germany

Aneta Kostadinova Institute of Parasitology, Biology Centre of the Czech Academy of Sciences, České Budějovice, Czech Republic

Thomas Kuhn Goethe-University, Institute for Ecology, Evolution and Diversity, Senckenberg Gesellschaft für Naturforschung, Senckenberg Biodiversity and Climate Research Centre, Frankfurt/Main, Germany

Zdzisław Laskowski W. Stefanski Institute of Parasitology of the Polish Academy of Sciences, Warsaw, Poland

Ken MacKenzie School of Biological Sciences (Zoology), The University of Aberdeen, Aberdeen, Scotland, UK

Simonetta Mattiucci Department of Public Health and Infectious Diseases, Section of Parasitology, Sapienza - University of Rome, Rome, Italy

Heinz Mehlhorn Institute for Parasitology, Heinrich-Heine-University, Düsseldorf, Germany

Birgit Mehlhorn Institute for Parasitology, Heinrich-Heine-University, Düsseldorf, Germany

Gary Miller Institute for Marine and Antarctic Studies, University of Tasmania, Sandy Bay, TAS, Hobart, Australia

Julian Münster Goethe-University, Institute for Ecology, Evolution and Diversity, Senckenberg Gesellschaft für Naturforschung, Senckenberg Biodiversity and Climate Research Centre, Frankfurt/Main, Germany

Giuseppe Nascetti Department of Ecological and Biological Sciences, Tuscia University, Viterbo, Italy

Michela Paoletti Department of Public Health and Infectious Diseases, Section of Parasitology, Sapienza - University of Rome, Rome, Italy

Department of Ecological and Biological Sciences, Tuscia University, Viterbo, Italy

Michelle Power Department of Biological Sciences, Macquarie University, Sydney, Australia

Anna Rocka W. Stefański Institute of Parasitology, Polish Academy of Sciences, Warsaw, Poland

Juan T. Timi Laboratorio de Ictioparasitología, Instituto de Investigaciones Marinas y Costeras (IIMyC), Consejo Nacional de Investigaciones Científicas y Técnicas (CONICET)-Facultad de Ciencias Exactas y Naturales, Universidad Nacional de Mar del Plata, Buenos Aires, Argentina

Ralph Vanstreels Department of Pathology, Faculty of Veterinary Medicine and Animal Science, University of São Paulo, São Paulo, Brazil

Virginia Vidal Departamento de Ecología Evolutiva, Museo Nacional de Ciencias Naturales – CSIC, Madrid, Spain

Stephen C. Webb Cawthron Institute, Nelson, New Zealand

Krzysztof Zdzitowiecki W. Stefanski Institute of Parasitology of the Polish Academy of Sciences, Warsaw, Poland

About the Editors

Sven Klimpel studied Biology at the Christian-Albrechts-University Kiel/IfM-Geomar (now Helmholtz Centre for Ocean Research) and then completed his doctorate and his habilitation at the Institute of Zoomorphology, Cell Biology and Parasitology of the Heinrich-Heine-University Düsseldorf. Since 2010 he is full professor and head of the department "Integrative Parasitology and Zoophysiology (IPZ)" of the Goethe-University Frankfurt am Main in cooperation with the Senckenberg Biodiversity and Climate Research Centre/Senckenberg Society for Nature Research and the director of the Institute for Ecology, Evolution and Diversity. His primary research interests are the ecology, evolution, life-cycle strategies and host-parasite co-evolution of aquatic and terrestrial protozoan/metazoan parasites, pathogens and their invertebrate/vertebrate intermediate hosts and vectors. In his laboratory, he and his coworkers combine traditional morphological methods with up-to-date molecular techniques. He participates in numerous scientific research cruises, including some to the Southern Ocean (Antarctica).

Thomas Kuhn studied biology at the Heinrich-Heine-University (HHU) Düsseldorf, majoring in parasitology, zoology, and genetics. In 2013, he obtained his doctorate at the Institute for Ecology, Evolution and Diversity of the Goethe-University and the Biodiversity and Climate Research Centre (BiK-F) in Frankfurt/Main, Germany. Since 2013, he works as a junior research group leader (Molecular and Aquatic Parasitology) at the department "Integrative Parasitology and Zoophysiology" at the Goethe-University, Frankfurt/Main, Germany. His research aims to understand the complex interrelationships between aquatic metazoan pathogens and their respective vertebrate and invertebrate hosts. He is particularly interested in the morphological and molecular identification, (co-)evolution and ecology as well as the zoogeography of aquatic, zoonotic parasites and their implications on food safety.

Heinz Mehlhorn, Düsseldorf, Germany. He has investigated the transmission pathways of human and animal parasites for over 40 years at German and international universities and he and his university spin-off company Alpha-Biocare have developed many antiparasitic medical products based on more than 20 patents – several in cooperation with big international companies. He is editor and author of the Springer *Encyclopedia of Parasitology* and has published 25 books, more than 250 original papers, and has served as Managing Editor of the journal *Parasitology Research* since 1981. A long list of renown international scientists did their PhD work in his laboratory and remain still today interconnected as a large group of lovers of parasitology.

Chapter 1
Introduction: Biodiversity and Evolution of Parasitic Life in the Southern Ocean

Sven Klimpel, Thomas Kuhn, and Heinz Mehlhorn

Researchers of various disciplines, including taxonomy, ecology, and physiology, have long been attracted to the Southern Ocean environment that lies at the limits of the physical conditions capable of supporting life and thus constitutes an exceptional ecosystem for undertaking fundamental research on the relationship between the climate and evolutionary processes (Clarke et al. 2007a and ref. therein; Ducklow et al. 2007). The establishment of the Antarctic Circumpolar Current (ACC) and its associated oceanographic regime in the Early Cenozoic fostered unique adaptations of both, marine and terrestrial organisms, relatively unaffected by biotic exchange (Clarke et al. 2007a). Low air and water temperatures, lack of coastal zones due to a thick shelf-ice cover, and drifting and stranding of icebergs are only some of those unique environmental features that necessitate special adaptations of terrestrial and marine floral and faunal species to extreme environmental conditions (Klimpel et al. 2010). A particular characteristic in the marine environment is the missing of a strict separation between the continental shelf and the deep-sea, enabling deep-sea species to occur also in shallower waters and especially benthodemersal shallow water species to extend their range into the deep-sea (Klimpel et al. 2010).

To date, many endemic species have been recorded from Antarctica, illustrating the unique history and environment of the region. However, dramatic climatic

---

S. Klimpel (✉) • T. Kuhn
Goethe-University, Institute for Ecology, Evolution and Diversity, Senckenberg Gesellschaft für Naturforschung, Senckenberg Biodiversity and Climate Research Centre, Max-von-Laue-Str. 13, D-60439 Frankfurt/Main, Germany
e-mail: klimpel@bio.uni-frankfurt.de; t.kuhn@bio.uni-frankfurt.de

H. Mehlhorn
Institute for Parasitology, Heinrich-Heine-University, Universitätsstr. 1, D-40225 Düsseldorf, Germany
e-mail: mehlhorn@uni-duesseldorf.de

S. Klimpel et al. (eds.), *Biodiversity and Evolution of Parasitic Life in the Southern Ocean*, Parasitology Research Monographs, Vol. 9,
DOI 10.1007/978-3-319-46343-8_1

changes have caused major shifts in the species composition (e.g., Clarke and Crame 1992; Clarke et al. 2007a, b). Recent climate change and rising temperatures will likely intensify this effect on the endemic biota in the high Antarctic and Southern Ocean, possibly leading to another shift in species composition and distribution in the future.

The Antarctic Peninsula with its surrounding islands (e.g., Elephant and King George Island, both South Shetlands), for example, is one of those areas on the globe which is currently experiencing rapid regional climatic changes, with more than 1.5 °C rise in mean annual temperature since 1950 (compared with a global mean increase of appr. 0.6 °C) (e.g., Clarke et al. 2007a; Vaughan et al. 2003). The loss of seven larger ice fields during the past 60 years, including the collapse of the Wordie Ice Shelf in the 1980s and the middle section of the Larsen Ice Shelf (Larsen B) in 2002 are only the most evident consequences of climatic impacts on the local environment (Vaughan and Doake 1996; Doake and Vaughan 1991; Clarke et al. 2007a; Domack et al. 2005).

Krill, cephalopods, and Antarctic fish species are considered the key species of the Antarctic marine food web (e.g., Loeb et al. 1997). The fish species composition, biomass, zoogeographical distribution, feeding ecology, and reproduction are comparatively well known (e.g., Kock 1992; Kock and Stransky 2000; Flores et al. 2004; Bushula et al. 2005; Eastman 2005; Kock 2005a, b). With currently 283 recognized species, it is generally dominated by the perciform suborder Notothenioidei (Kock 2005a, b; Froese and Pauly 2016), which comprises the majority of species in shelf waters down to 500 m water depth (Flores et al. 2004; Kock and Stransky 2000).

Being a species-rich but often well-hidden component of the Southern Ocean fauna, fish parasites have been studied by various research groups. Earlier works focused on new species descriptions and the faunistic description, especially of parasitic helminths (e.g., Digenea: Zdzitowiecki 1991a, 1996, 1997; Laskowski et al. (2014); Cestoda: Rocka and Zdzitowiecki 1998; Wojciechowska 1991; Wojciechowska et al. 1994; Nematoda: Klöser et al. 1992; Palm et al. 1994, 1998; Acanthocephala: Zdzitowiecki 1990, 1991b, 1996). Most research activities have been carried out on Antarctic notothenioids and also channichtyids from shallow coastal waters or the open sea shelf (e.g., Zdzitowiecki 1991a, 1997; Santoro et al. 2014), where species are easy to catch and, therefore, more available for such studies. Investigations along the Antarctic continental slope and the deep-sea are limited (e.g., Walter et al. 2002). Most parasitological studies from the Southern Ocean, especially from the Antarctic Peninsula and the eastern Weddell-Sea, revealed a species-rich fish parasite fauna, including mainly endemic and noncosmopolitan species (e.g., Palm et al. 1998; Zdzitowiecki and Laskowski 2004; Brickle et al. 2005; Rocka 2006).

Other examples of fish parasitological investigations were published mainly by scientists such as Rocka (Rocka 2002, 2003, 2004); Rocka and Zdzitowiecki (1998); Wojciechowska (1991); Wojciechowska et al. (1994); Zdzitowiecki (1990, 1991b, 1996); Zdzitowiecki and Laskowski (2004) and Zdzitowiecki and Pisano (1996). Rocka (2006) summarized the available information about the life cycle biology, specificity, and geographical distribution of the parasitic helminth groups Digenea,

Cestoda, Nematoda, and Acanthocephala of Antarctic bony fishes and elasmobranchs. The author stated that almost all of the helminth species maturing in Antarctic bony fishes are endemic, whereas only extremely few parasite species are cosmopolitan or bipolar. Specificity to the intermediate or paratenic hosts of the majority of Antarctic helminths is low, whereas that for the definitive host is often higher (Rocka 2006).

During first investigations on the life cycle biology and the zoogeography of, e.g., anisakid nematodes in the Weddell Sea and around the South Shetland Islands, different benthic and pelagic life cycles could be identified for the anisakid nematodes *Contracaecum radiatum*, *C. osculatum* (e.g., Klöser et al. 1992; Klöser and Plötz 1992)), *Pseudoterranova decipiens* (e.g., Palm et al. 1994; Palm 1999), and specimens of the genus *Anisakis* (e.g., Klimpel et al. 2010; Kuhn et al. 2011)) (Fig. 1.1). Although these anisakids have explored the extreme Antarctic environment, they have maintained the principal life cycle biology that is known for their relatives from non-Antarctic waters such as in the North Atlantic.

Generally, the biodiversity of fish parasites in benthodemersal fish from shallow waters and from deep water fish is species rich, but demonstrates low host specificity for most of the collected species (Palm et al. 1998, 2007; Walter et al. 2002;

Fig. 1.1 Research vessel (*RV*) and scientific equipment (**a**, **b**), scientific investigation of Antarctic fish material, and typical ectoparasites (**c**, **d**). (**a**) *Polarstern* during the research cruise ANT XXIII/8. The research vessel *Polarstern* is the most important tool for German polar research. (**b**) Fish trawl from the waters around Elephant Island. (**c**) Scientist during data collection in the wet lab of the RV *Polarstern*. (**d**) Mackerel icefish *Champsocephalus gunnari*, Channichthyidae) with parasitic leeches

Klimpel et al. 2009). Mammalian parasites, for example, seem to use mainly the nototheniids and channichthyids as common transmission routes into their seal final hosts; however, some have also explored parallel host systems that utilize different combinations of final and intermediate hosts (e.g., Palm et al. 2007). Until now, comparative investigations that could indicate long-term changes in the parasite fauna are still missing, and many, often more rare fish species, have not or only sporadically been examined. The present monograph should cover some of the still missing aspects on fish parasitological research in Antarctic waters.

References

Brickle P, MacKenzie K, Pike A (2005) Parasites of the Patagonian toothfish, *Dissostichus eleginoides* Smitt 1898, in different parts of the Subantarctic. Polar Biol 28:663–671

Bushula T, Pakhomov EA, Kaehler S, Davis S, Kalin RM (2005) Diet and daily ration of two nototheniid fish on the shelf of the sub-Antarctic Prince Edward Islands. Polar Biol 28:585–593

Clarke A, Crame JA (1992) The Southern Ocean benthic fauna and climate change: a historical perspective. Philos Trans R Soc B 338:299–309

Clarke A, Johnston NM, Murphy EJ, Rogers AD (2007a) Antarctic ecology from genes to ecosystems: the impact of climate change and the importance of scale. Philos Trans R Soc B 362:5–9

Clarke A, Murphy EJ, Meredith MP, King JC, Peck LS, Barnes DKA, Smith RC (2007b) Climate change and the marine ecosystem of the western Antarctic Peninsula. Philos Trans R Soc B 362:149–166

Doake CSM, Vaughan DG (1991) Rapid disintegration of the Wordie Ice Shelf in response to atmospheric warming. Nature 350:328–330

Domack E, Duran D, Leventer A, Ishman S, Doane S, McCallum S, Amblas D, Ring J, Gilbert R, Prentice M (2005) Stability of the Larsen B ice shelf on the Antarctic Peninsula during the Holocene epoch. Nature 436:681–685

Ducklow HW et al (2007) Marine pelagic ecosystems: the west Antarctic Peninsula. Philos Trans R Soc Lond B Biol Sci 362(1477):67–94

Eastman JT (2005) The nature of the diversity of Antarctic fishes. Polar Biol 28:93–107

Flores H, Kock KH, Wilhelms S, Jones CD (2004) Diet of two icefish species from the South Shetland Islands and Elephant Island, *Champsocephalus gunnari* and *Chaenocephalus aceratus*. Polar Biol 27:119–129

Froese R, Pauly D (2016) FishBase. World Wide Web electronic publication. www.fishbase.org, version (01/2016)

Klimpel S, Busch MW, Kellermanns E, Kleinertz S, Palm HW (2009) Metazoan deep-sea fish parasites. Acta biologica benrodis, Supplementband 11. Verlag Natur & Wissenschaft, Solingen, pp 384

Klimpel S, Busch MW, Kuhn T, Rohde A, Palm HW (2010) The *Anisakis simplex* complex off the South Shetland Islands (Antarctica): endemic populations versus introduction through migratory hosts. Mar Ecol Prog Ser 403:1–11

Klöser H, Plötz J, Palm HW, Bartsch A, Hubold G (1992) Adjustment of anisakid nematode life cycles to the high Antarctic food web as shown by *Contracaecum radiatum* and *C. osculatum* in the Weddell Sea. Antarct Sci 4:171–178

Klöser H, Plötz J (1992) Morphological distinction between adult *Contracaecum radiatum* and *Contracaecum osculatum* (Nematoda, Anisakidae) from the Weddell seal (*Leptonychotes weddelli*). Zool Scr 21:129–132

Kock KH (1992) Antarctic fish and fisheries. Cambridge University Press, Cambridge

Kock KH, Stransky C (2000) The composition of the coastal fish fauna around Elephant Island (South Shetland Islands, Antarctica). Polar Biol 23:825–832

Kock KH (2005a) Antarctic icefishes (Channichthyidae): a unique family of fishes. A review, part I. Polar Biol 28:862–895
Kock KH (2005b) Antarctic icefishes (Channichthyidae): a unique family of fishes. A review, part II. Polar Biol 28:897–909
Kuhn T, García-Màrquez J, Klimpel S (2011) Adaptive radiation within marine anisakid nematodes: a zoogeographical modeling of cosmopolitan, zoonotic parasites. PLoS One 6(12):e28642
Laskowski Z, Jezewski W, Zdzitowiecki K (2014) Changes in digenean infection of the Antarctic fish *Notothenia coriiceps* in Admiralty Bay, King George Island, over three decades. Polish Polar Research 35(3):513–520
Loeb V, Siegel V, Holm-Hansen O, Hewitt R, Fraser W, Trivelpiece W, Trivelpiece S (1997) Effects of sea-ice extent and krill or salp dominance on the Antarctic food web. Nature 387:897–900
Palm HW (1999) Ecology of *Pseudoterranova decipiens* (Krabbe, 1878) (Nematoda: Anisakidae) from Antarctic waters. Parasitol Res 85:638–646
Palm HW, Andersen K, Klöser H, Plötz J (1994) Occurrence of *Pseudoterranova decipiens* (Nematoda) in fish from the south-eastern Weddell Sea (Antarctic). Polar Biol 14:539–544
Palm HW, Reimann N, Spindler M, Plötz J (1998) The role of the rock cod *Notothenia coriiceps* Richardson, 1844 in the life-cycle of Antarctic parasites. Polar Biol 19:399–406
Palm HW, Klimpel S, Walter T (2007) Demersal fish parasite fauna around the South Shetland Islands: High species richness and low host specificity in deep Antarctic waters. Polar Biol 30:1513–1522
Rocka A (2002) Nematodes of fishes in the Weddell Sea (Antarctica). Acta Parasitol 47:294–299
Rocka A (2003) Cestodes of the Antarctic fishes. Polish Polar Research 24:261–276
Rocka A (2004) Nematodes of the Antarctic fishes. Polish Polar Research 25:135–152
Rocka A (2006) Helminths of Antarctic fishes: life cycle biology, specificity and geographical distribution. Acta Parasitol 51:26–35
Rocka A, Zdzitowiecki K (1998) Cestodes in fishes of the Weddell Sea. Acta Parasitol 43:64–70
Santoro M, Mattiucci S, Cipriani P, Bellisario B, Romanelli F, Cimmaruta R, Nascetti G (2014) Parasite communities of icefish (*Chionodraco hamatus*) in the Ross Sea (Antarctica): influence of the host sex on the helminth infracommunity structure. PLoS One 9(2):e88876
Vaughan DG, Doake CSM (1996) Recent atmospheric warming and retreat of ice shelves on the Antarctic Peninsula. Nature 379:328–331
Vaughan DG, Marshall GJ, Connolley WM, Parkinson CL, Mulvaney R, Hodgson DA, King JC, Pudsey CJ, Turner J (2003) Recent rapid regional climate warming on the Antarctic Peninsula. Clim Change 60:243–274
Walter T, Palm HW, Piepiorka S, Rückert S (2002) Parasites of the Antarctic rattail *Macrourus whitsoni* (Regan, 1913) (Macrouridae, Gadiformes). Polar Biol 25:633–640
Wojciechowska A (1991) Some tetraphyllidean and diphyllidean cestodes from Antarctic batoid fishes. Acta Parasitol Pol 36:69–74
Wojciechowska A, Zdzitowiecki K, Pisano E, Vacchi M (1994) The tetraphyllidean cercoids from bony fishes accruing in the Ross Sea (Antarctic). Acta Parasitol 39:13–15
Zdzitowiecki K (1990) Occurrence of acanthocephalans in fishes of the open sea off the South Shetlands and South Georgia (Antarctic). Acta Parasitol Pol 35:131–142
Zdzitowiecki K (1991a) Occurrence of digeneans in open sea fishes off the South Shetland Islands and South Georgia, and a list of fish digeneans in the Antarctic. Polish Polar Research 12:55–72
Zdzitowiecki K (1991b) Antarctic Acanthocephala. Koeltz Scientific Books, Königstein
Zdzitowiecki K (1996) Acanthocephala in fish in the Weddell Sea (Antarctic). Acta Parasitol 41:199–203
Zdzitowiecki K (1997) Antarctic Digenea, parasites of fish. Koeltz Scientific Books, Königstein
Zdzitowiecki K, Laskowski Z (2004) Helminths of an Antarctic fish, *Notothenia coriiceps*, from the Vernadsky Station (Western Antarctic) in comparison with Admirality Bay (South Shetland Islands). Helminthologia 41:201–207
Zdzitowiecki K, Pisano E (1996) New records of Digenea infecting elasmobranches and teleost fish off Heard Island (Kerguelen sub-region, sub-Antarctic). Arch Fish Mar Res 43:265–272

Chapter 2
Antarctica: The Peculiar World

Birgit Mehlhorn and Heinz Mehlhorn

The continent Antarctica, which was officially discovered in the year 1820, obtained its name from the Greek term *antarktikos* = being situated opposite to the arctic, which comprises the Northern ice region on the globe. The Antarctica of our days represents a land area of about 13–14 million square kilometers being covered by very thick ice layers (up to 4700 m), which laterally overlap often considerably for many miles the icy waters around this fifth continent. This antarctical ice represents around 80–90 % of the ice on earth (= respectively about 70 % of the freshwater). The land mass of this continent includes numerous large lakes (up to 2500 m deep), which are all covered by this enormous ice shield. However, besides its outer icy aspect Antarctica includes a belt of active volcanos, which stretches over the continent from Victorialand to the Antarctic peninsula. The biggest volcano is the 3800 m high Mount Erebus on Ross Island.

The development of the continent Antarctica has a long history. Its oldest regions (e.g., Enderbyland) contains material that has an age of 3 billion years. About 170 million years ago Antarctica was a part of the large continent *Gondwana*. This region of earth was free of ice and gave room to fruitful soil, plants and a rich spectrum of animals including dinosaurs as is proven by the finding of their fossils dated 145–100 million years before our times.

During the period of the late Jura period, the supercontinent Gondwana started to become divided into precursors of the continents of our times and a land mass comprising Antarctica/Australia. Both were later separated from each other. Since this separation proceeded very slowly, animals and plants had sufficient time to become adapted at the changing temperatures and/or to develop sophisticated survival strategies in changing climates. Thus the species living today on the continent and in the surroundings of Antarctica are completely different from those at the beginning of

B. Mehlhorn • H. Mehlhorn (✉)
Institute for Parasitology, Heinrich-Heine-University, Düsseldorf, Germany
e-mail: mehlhorn@uni-duesseldorf.de

S. Klimpel et al. (eds.), *Biodiversity and Evolution of Parasitic Life in the Southern Ocean*, Parasitology Research Monographs, Vol. 9,
DOI 10.1007/978-3-319-46343-8_2

the continental drift. However, although the recent living conditions seem bad with respect to human needs, the fauna in the sea and around Antarctica is extremely large, but is in many aspects not yet really known.

Animals like seals, fishes, crustaceans, penguins, whales etc. within the water or birds with regular short contacts to the water had time enough during the continental shift to adapt their body needs to the high salinity of the ocean in Antarctica, where 35 per mille salt contents in the sea water are reached in contrast to only 3–5 per mille salt measured in the ice. However, the high saline content decreases the freezing point of the Antarctic sea water to –1.9 degrees Celsius. This temperature is "rather warm" compared to Antarctic air temperatures of –40 °C, which often occur and have to become survived e.g., by penguins.

The fauna in the sea around Antarctica is very rich and shows many sophisticated adaptations to peculiar conditions. Giant masses of typical Antarctic crustaceans (krill, *Euphausia superba*, Fig. 2.1) and related species are the basic food of fishes (~200 species) and whales (Fig. 2.2), which spend their time there in the Antarctic summer and ingest in addition to the krill also giant amounts (~50 million tons) of cephalopods (= squids). Penguins (Fig. 2.3) and seals breed on shore and feed fish, which are attacked by squads of ecto- and endoparasites like copepods of the families Ergasilidae and Lernaeidae. The latter appear worm-like and penetrate from outside with their anchor-like anterior ends into the body cavity of fishes. They can be easily recognized by their two egg-sacks, which may reach often a length of 4–5 cm (Figs. 2.4 and 2.5).

All these animals belonging to practically all tribes of the animal phylum had developed their skills to survive in about 40–45 million years, when the first ice development started reaching a full coverage about 3 million years ago. Humans would not have the chance to survive under the present conditions, if they would not be transported by ice-breaking ships and wear warm-holding suits (Figs. 2.6, 2.7, and 2.8).

Thus the authors of the present book want to report on the adaptations of several Antarctic parasites, which have learnt to escape the attacks of their hosts and had become able to survive the extreme low temperatures in their icy biotopes (Tables 2.1 and 2.2).

Fig. 2.1 Macrophoto of an adult crustacean (*Euphausia superba*) belonging to the so-called krill

Fig. 2.2 Photo of a jumping humpback whale, which are not shy and come close to boats. They stay in Antarctica in summer, but in winter in the Australian and South American sea, where they give birth to their progeny

Fig. 2.3 Two gentoo penguins (*Pygoscelis papua*) at the Antarctic shore close to the German Dallmann summer station

Fig. 2.4 Macrophoto of the surface of an Antarctic fish with an attached female lernaeid copepod (Crustacea). Note the two very long egg sacks

Fig. 2.5 Macrophoto of the opened inner side of the same fish depicted in Fig. 2.4 showing the deep anchoring system of the copepod's anterior end

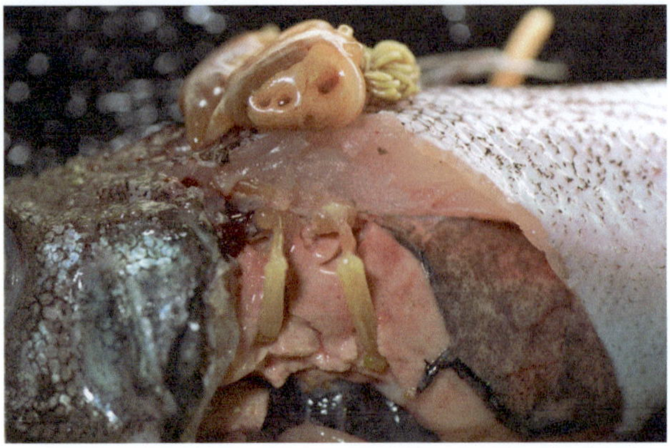

Fig. 2.6 Photo of the German research vessel *Polarstern* (Polar star) during the "Century Antarctic Expedition" in February until April of the year 2000. It was photographed during a helicopter flight to the German All-Year research station Neumayer showing also the high borders of the shelf ice

Fig. 2.7 Photo of the researchers Heinz and Birgit Mehlhorn on board of the *Polarstern* vessel in the year 2000 during the "Century Antarctic Expedition"

Fig. 2.8 Photo of the author and a co-worker of the Paul Wegener Institute Bremerhaven, Germany, close to the German Dallmann Station looking for body lice on Southern elephant seals

Table 2.1 Seals on and around Antarctica

Southern elephant seal	*Mirounga leonina*
Crabeater seal	*Lobodon carcinophaga*
Ross seal	*Ommatophoca rossii*
Leopard seal	*Hydrurga leptonyx*
Weddel seal	*Leptonychotes weddellii*

Table 2.2 Some penguins of Antarctica

Emperor penguin *(Aptenodytes forsteri)* These animals reach as adults a length of 100–130 cm and body weights between 20 and 38 kg. Both sexes do not show morphological differences in their outer appearance. This is the only species found in circumpolar region that and breeds on ice. Their feathers and a fat layer protects them from the cold. Females lay a single egg, which is kept warm by the male being placed on its feet and covered by a belly fold.
Adélie penguin *(Pygoscelis adeliae)* Besides the Emperor penguin this species is the other one which occurs in the mainland of Antarctica. It is named honoring the wife of the French researcher J.D. d'Urville. Females are smaller than males reaching weights between 3.9 and 5.5 kg.
Gentoo penguin *(Pygoscelis papua)* This species (Fig. 2.3) occurs at the Antarctic peninsula and subantarctic islands and is characterized by a white spot at the lateral sides of the head. Gentoo penguins reach a height of 51–90 cm. Males weight between 4.9 and 8.5 kg, females reach a weight of 4.5–8.2 kg.

Further Reading

Arndt CE, Swadling KM (2006) Crustacea in Arctic and Antarctic sea ice: distribution, diet and life history strategies. Adv Mar Biol 51:197–315

Black C et al (2016) Why Huddle? Ecological drivers of chick aggregations in Gentoo penguins, *Pygoscelis papua*, across latitudes. PLoS One 11:e0145676

Chambert T et al (2015) Female Weddell seals show flexible strategies of colony attendance related to varying environmental conditions. Ecology 96:479–488

Corrigan LJ et al (2016) Population differentiation in the context of Holocene climate change for a migratory marine species, the southern elephant seal. J Evol Biol. doi:10.1111/jeb.12870

Fütterer D et al (2006) Antarctica – contributions to global earth science. Springer, Berlin

Herr H et al (2016) Collection and analysis of data on occurrence, distribution and abundance of cetaceans in the Southern Ocean following international standards. Proceedings of the Umweltbundesamt, Berlin

Kleinertz S et al (2014) Gastrointestinal parasite fauna of Emperor penguins (*Aptenodytes forsteri*) at the Atka Bay, Antarctica. Parasitol Res 113:4133–4139

Laakmann S et al (2012) Evolution in the deep sea: biological traits, ecology and phylogenetics of pelagic copepods. Mol Phylogenet Evol 65:535–546

Montero E et al (2016) First record of *Babesia* sp. in Antarctic penguins. Ticks Tick Borne Dis 7:498–501

Peter T et al (2012) An Emperor penguin population estimate. PLoS One 7:e33751

Roland NW (2009) Antarctica. Research in persistent ice. Springer Spektrum, Heidelberg

Tuck G, Heinzel H (1980) Sea birds of the world. Parey, Hamburg/Berlin

Wandrey R (1997) Whales and seals of the world. Franck Kosmos Verlag, Stuttgart

Chapter 3
The History of Antarctic Parasitological Research

Ken MacKenzie

3.1 Introduction

To begin with, it is necessary to define the northern limit of the Antarctic. The Antarctic Circle at 70° South is a non-starter because it excludes most of the marine environment considered to be Antarctic in nature. For the purposes of this review, I have therefore accepted the Antarctic Convergence as the geographical limit. Also known as the Polar Front, this is an irregular line circling the Antarctic continent where the cold northward-flowing Antarctic waters sink beneath the relatively warmer waters of the sub-Antarctic. The Antarctic Convergence lies south of the southernmost tip of South America and between the Falkland Islands and South Georgia, so the considerable body of parasitological research carried out around Patagonia and the Falkland Islands is necessarily excluded from this review.

The history of Antarctic parasitological research is relatively recent. Reports and descriptions of parasites inevitably appear after those of their hosts. It is no surprise therefore that there are so few published reports on parasites of Antarctic organisms until well into the twentieth century. Most of the earliest reports of Antarctic parasites resulted from the examination of host specimens collected during voyages of exploration which included the collection of large amounts of biological material from this hitherto little-explored environment. These expeditions included the British "Challenger" expedition (1873–1876), the German expedition to South Georgia (1882–1883), the Danish "Ingolf" Expedition (1885–1886), the Belgian Antarctic Expedition (1897–1899), the Scottish National Antarctic Expedition (1902–1904), the Russian Polar Expedition (1900–1903), the two French Antarctic

K. MacKenzie
School of Biological Sciences (Zoology), The University of Aberdeen,
Tillydrone Avenue, Aberdeen AB24 2TZ, Scotland, UK
e-mail: k.mackenzie@abdn.ac.uk

S. Klimpel et al. (eds.), *Biodiversity and Evolution of Parasitic Life in the Southern Ocean*, Parasitology Research Monographs, Vol. 9,
DOI 10.1007/978-3-319-46343-8_3

Expeditions (1903–1905 and 1908–1910), the British Antarctic "Terra Nova" Expedition (1910–1912), and the Australian Antarctic Expedition (1911–1914). Many more national and international expeditions followed after the end of the First World War in 1918.

It is not possible in a review such as this to refer to every publication on parasites of Antarctic organisms. I have attempted therefore to select the most significant and important publications from the huge body of literature on this subject, but it is inevitable that some readers will disagree with my selection. If my omissions include any important publications I apologise in advance.

3.2 The Early Years (1853–1920)

Possibly the earliest published description of a parasite from the Antarctic is that of Baird (1855), who described the nematode *Ascaris similis*, claiming that the host was a seal brought back from an Antarctic expedition. The identity of the host has been disputed, but Johnston (1938) presented evidence that an elephant seal from the Antarctic may have been the original host. Whatever the true identity of the host of this nematode, the next report of a parasite from the Antarctic appears to be that of Graff (1884), who described a new genus of Myzostomida parasitic in Antarctic echinoderms, collected during the voyage of HMS *Challenger*. Linstow (1892) reported a number of helminths collected from mammalian and fish hosts during a German expedition to South Georgia, including descriptions of six new species. The following year, Burger (1893) reported gregarine parasites from nemertinean hosts in South Georgia, collected during the Belgian Antarctic Expedition.

The first report of the twentieth century was that of Shipley (1901) who referred to and commented on the new myzostomid described by Graff (1884). Ludwig (1903) and Mortensen (1903) reported nematodes and a parasitic copepod from echinoderms collected during the Belgian Antarctic Expedition and the Danish "Ingolf" Expedition respectively. Linstow (1905, 1907, 1911) described a new species of nematode collected from an Antarctic seal during the Russian Polar Expedition and two new species, a cestode and a nematode, collected from fish during the Scottish National Antarctic Expedition. The latter expedition also led to the description by Rennie (1907) of the first new acanthocephalan parasite from Antarctic fishes. The two French Antarctic Expeditions contributed a wealth of new parasite material and an impressive number of publications in the period 1907 to 1914. Railliet and Henry (1907) described three new nematode parasites collected during the first expedition, then a series of reports of crustacean parasites, collected during the second expedition, were published by Gravier (1912a, b, c, d, e, f, g, 1913, 1914) on invertebrate hosts, and by Quidor (1913) on fish and marine mammals.

The British Antarctic "Terra Nova" Expedition led to three publications on Antarctic parasites: Leiper and Atkinson (1914, 1915) described a large number of helminth parasites collected from fishes and marine mammals, Boulenger (1916) described a new species of myzostomid parasitic from an echinoderm, Woodcock and Lodge (1921) described parasitic protistans from fish, mammals and polychaetes, and Baylis (1923) described a new species of nematode from a sperm whale. The ill-fated "Terra Nova" expedition later prompted two publications of more general interest: one (Campbell 1988) a beautifully written and entertaining account of the characters and achievements of the two parasitologists on board, Leiper and Atkinson; the other describing the historical basis of the binomials assigned to the parasites collected and described (Campbell and Overstreet 1994).

3.3 1929–1950

As with many other fields of scientific research, the First World War and its aftermath effectively halted the flow of literature on Antarctic parasites. It took until 1929 for the first publications of this era to appear. The literature drought ended with the publication of a report by Baylis (1929) on nematodes and acanthocephalans from fishes and marine mammals, some from Antarctica, collected during the *Discovery* cruises of the 1920s. From 1930 to 1938 an important series of papers were published on parasites of Antarctic fishes, mammals and birds, based on material collected during the Australasian Antarctic Expedition of 1911 to 1914 (Johnston 1930, 1931, 1937a, b, 1938; Johnston and Best 1937). This was followed by the first paper (Johnston and Mawson 1945) on parasitic nematodes from fishes, mammals and birds, from material collected during another Antarctic expedition – the British-Australia-New Zealand (BANZ) Antarctic Expedition. Nineteen forty-eight saw the first paper on Antarctic leeches by Brinkmann (1948), who described two new genera and species from Antarctic fishes. Finally, Eichler (1949) described some Mallophaga (lice) from birds in Antarctica.

3.4 1951–1960

It was during this period that parasitological research in the Antarctic began to gain momentum (Table 3.1). The first paper published in this period was that of Heegard (1951) on parasites and commensals of echinoderms. Three papers by Mawson (1953) and Edmonds (1955, 1957) reported on nematodes and acanthocephalans from fishes, birds and mammals collected during two more Antarctic expeditions – the Australian National Expedition to Heard and Macquarie Islands in 1948–1951 and the BANZ Expedition. Laird (1956) reported on myxosporeans from Antarctic fishes, Arsen'ev and Gusev (1958) reported on biological data collected during a Russian Antarctic expedition in 1957 which included examinations of Antarctic fishes for parasites, while Gusev (1958, 1960) published

Table 3.1 Numbers of publications on parasites of Antarctic hosts according to time period and host taxonomic group

	Host group						
Period	Fish	Mammals	Birds	Crustaceans	Molluscs	Other invertebrates	All
Pre 1920	6	7	1	2	0	12	23
1921–1930	2	1	1	0	0	0	2
1931–1940	4	3	2	0	0	0	6
1941–1950	3	1	0	2	0	0	5
1951–1960	5	3	1	0	0	1	7
1961–1970	19	7	5	3	2	1	32
1971–1980	30	17	6	6	0	3	60
1981–1990	63	15	24	7	3	4	114
1991–2000	80	15	19	2	4	8	119
2001–2010	44	15	20	9	3	7	95
2011–2016	23	6	12	1	1	4	45

Note that the figures in the last column (All) do not necessarily coincide with the sum of the figures in the other columns of each row because the same publication may deal with parasites of more than one host group

accounts of Russian parasitological studies on Antarctic fishes. Fell (1961) reported on ophiuroids of the Ross Sea, including reports of parasites and commensals, and Laird (1961) commented on the lack of haematozoa in Antarctic birds and mammals.

3.5 1961–1970

From 1961 to 1970 we see the beginnings of the coming flood of Antarctic parasitological literature, especially papers on helminth parasites of Antarctic fishes. Particularly prominent among the authors of these papers were the Americans H.L. Holloway and W.J. Hargis and the Argentinian L. Szidat, together with their co-authors (Byrd 1963; Dollfus and Euzet 1964; Bychovsky et al. 1965; Dollfus 1965; Szidat 1965; Gusev 1967; Holloway 1967; Holloway et al. 1967; Szidat and Graefe 1967; Becker and Holloway 1968; Dillon and Hargis 1968; Hargis and Dillon 1968a, b; Hargis and Zwerner 1968; Holloway 1968; Holloway and Bier 1968; Holloway and Klewer 1969; Prudhoe 1969; Szidat and Graefe 1969). Publications also appeared on parasites of Antarctic birds and mammals (Fain and Hyland 1963; Murray and Nicholls 1965; Skrjabin 1967, 1969; Graefe 1968; Jones and Williams 1969; Kagei and Kureha 1970; Skrjabin and Muravieva 1970), an acanthocephalan and a leech were described from Antarctic crustaceans (Holloway and Bier 1967; Sawyer and White 1969), and a dicyemid was described from an Antarctic cephalopod (Short and Hochberg 1970).

3.6 1971–1980

From 1971 to 2000 the numbers of publications on parasites of Antarctic hosts increased with each successive decade (Table 3.1). This was especially true of fish parasites, and was most marked during the periods 1971–1980 and 1981–1990. Much of this was due to publications authored by the Polish parasitologist Professor K. Zdzitowiecki and his co-authors. Publications from this source began with three papers on Antarctic fish parasites published in 1978 based on material collected during the Polish Academy of Sciences' Antarctic Expedition of 1977 (Zdzitowiecki 1978a, b, c). Other important publications on Antarctic fish parasites during this decade are those of Gibson (1976) on monogeneans and digeneans; Kovaleva and Gaevskaya (1977) on monogeneans; Kovaleva and Gaevskaya (1974), Prudhoe and Bray (1973) and Parukhin and Lyadov (1979) on digeneans; and Pois (1975) on cestodes. Papers listing or describing the general parasite faunas of Antarctic fishes, birds and mammals were published in this period by Markowski (1971), and Kagei and Watanuki (1977) and Siegel (1980a). Siegel (1980b) also used parasites as biological tags to identify stocks of channichthyid fishes.

Antarctic seals were the subjects of papers on their helminth parasites by Beverley-Burton (1971, 1972), Kurochkin and Nikol'skii (1972) and Nikol'skii (1974), and whales the subjects of a series of papers by the Russian Professor Skrjabin and his co-authors (Skrjabin 1971a, b, 1974; Skrjabin and Muravieva 1971, 1972; Skryabin and Nikol'skii 1971). The first report of parasites of an introduced terrestrial mammal (reindeer) was published by Leaderwilliams (1980), while Williams et al. (1974), Szelenbaum-Cielecka and Zdzitowiecki (1979) and Zdzitowiecki and Drozdz (1980) reported on helminth parasites from Antarctic birds. Among the papers on parasites of Antarctic marine invertebrates (crustaceans, polychaetes and echinoderms) published in this period were those of Platonova and Potin (1972), Lutzen and Jones (1976), Rubtsov (1977), Kagei et al. (1978), Michajlow (1978) and Schultz (1980).

3.7 1981–2000

During the decade 1981–1990, publications from Professor Zdzitowiecki's group accounted for about 35 % of publications on parasites of Antarctic fish, and about 25 % of publications on parasites of all Antarctic hosts. This trend continued into the period 1991–2000, during which they accounted for about two-thirds of publications on parasites of fish and about one-third of those on parasites of all Antarctic hosts. The contribution made by Professor Zdzitowiecki and his co-authors to Antarctic parasitology cannot be overstated. Most of the publications are on acanthocephalans and digeneans, with descriptions of many new species. They are too numerous to list in full in this review, but the following are suggested as probably the most useful references for parasitologists researching Antarctic fish parasites.

Acanthocephala: Zdzitowiecki (1986a, b, c, 1987, 1990, 1991a; Zdzitowiecki and Rokosz 1986).
Digenea: Zdzitowiecki (1988, 1991b, 1997a, b, c, d); Zdzitowiecki and Cielecka (1997a, b, c).
Cestoda: Rocka and Zdzitowiecki (1998), Wojciechowska (1991, 1993a, b, c), (Wojciechowska et al. 1994),(Rocka 1999).

Apart from the major contribution of the Polish group, a number of important papers on fish parasites were also produced during this period by researchers from other countries, notably Russia and Germany. Those considered to be the most significant are listed below.

Myxosporea: Kovaleva and Gaevskaya (1984), Noble (1984)
Monogenea: Rodyuk (1986a), Timofeeva et al. (1987).
Digenea: Santoro et al. (1990)
Acanthocephala: Rodyuk (1986b).
Nematoda: Klöser and Plötz (1992), Klöser et al. (1992), Orecchia et al. (1994), Palm et al. (1994), Arduino et al. (1995), Bullini et al. (1997), Nascetti et al. (1997), Palm (1999), Paggi et al. (2000).
Hirudinea: Yang (1987), Utevsky (1993, 1995, 1997), Epshtein and Utevsky (1994).
Crustacea: Sosinski and Janusz (1986), Rokicki and Skora (1987), Wägele and Brandt (1988), Brandt and Wagele (1991), Janusz and Sosinski (2000).

Publications listing or describing the more general parasite faunas of Antarctic fishes were published in this period by Hoogesteger and White (1981), Parukhin and Lyadov (1981), Beumer et al. (1983), Lyadov (1985), Rodyuk (1985), Reimer (1987), Gaevskaya et al. (1990), Palm et al. (1998), Walter (1998) and Rohde (2000). Moser and Cowen (1991) described the effects of environmental change on some fish parasites in McMurdo Sound and suggested the use of these parasites as biological tags for fish stock identification.

During the period 1971–2000, the Polish group also contributed papers on helminth parasites of Antarctic mammals and birds. These included Zdzitowiecki (1984, 1985; Zdzitowiecki 1991a, b) on acanthocephalans, Zdzitowiecki et al. (1989) on digeneans, and Wojciechowska and Zdzitowiecki (1995), and Cieleska and Zdzitowiecki (1989) and Cieleska et al. (1992) on cestodes. In addition, Odening (1986) and Drozdz (1987) reported on coccidian parasites of Antarctic mammals and birds, and a series of papers by Yurakhno culminated in a paper by Yurakhno and Maltsev (1997) on cestode infections of Antarctic seals. Dailey and Vogelbein (1991) described the parasite faunas of three species of Antarctic whales and discussed the possible use of parasites as biological tags for whale stock identification. Reports of parasites from Antarctic birds included those of Horne and Rounsevell (1982), Zlotorzyska and Modrzejewska (1992), Pugh (1993), Murray et al. (1993, 1999) and Mironov (1991) on lice, Odening (1982) on cestodes, Feiler (1986) on trematodes, Hoberg (1984, 1985, 1986, 1987) on various helminths and a pentastomid and Jones (1988) and Clarke and Kerry (1993) on parasites of penguins.

Parasites of Antarctic crustaceans described in this period included trematode metacercariae from mysids described by Gaevskaya (1982), new parasitic copepods from amphipods and isopods described by Boxshall and Harrison (1988) and gregarines from planktonic crustaceans described by Avdeev and Avdeeva (1989). New harpacticoid copepods parasitic in octopuses were described by Avdeev (1983) and Bresciani and Lutzen (1994), and Palm (1997) reported on parasites of Antarctic molluscs and annelids. Chesunov and Spiridonov (1985) and Lopez-Gonzalez et al. (2000) described new species of nematodes and parasitic copepods from polychaete hosts. A new species of Cirripedia parasitic in an Antarctic starfish was described by Grygier (1981) and the same author (Grygier 1987) published records of Cirripedia infecting Antarctic asteroids. New species of parasitic copepods were also described from Antarctic echinoderms by Bartsch (1994, 1996). Among the more unusual reports were those of Oresland and Pleijel (1991) of an ectoparasitic polychaete infecting a chaetognath and Czaker (1997) of a microsporidian hyperparasitic on a dicyemid mesozoan.

3.8 2001–2016

By the start of the new millennium, most of the common parasites of Antarctic hosts had been described. This is particularly true of the parasites of fish, thanks mainly to the efforts of the aforementioned Polish group led by Professor Zdzitowiecki. New species continued to be described, but at a slower rate, and the emphasis began to change towards more ecological aspects such as analyses of parasite communities, pathology and the place of parasites in the Antarctic food web.

In fish parasitology the literature continued to be dominated by the Polish group, with their contribution in this period still accounting for more than one-third of publications on parasites of Antarctic fish and about 20 % of those on all Antarctic hosts. The following is a selection of some of their most important contributions in this period.

Digenea: Zdzitowiecki (2002a, b, c, d, 2003); Laskowski et al. (2014), Jezewski et al. (2014)

Nematoda: Rocka (2002, 2004), (Rokicki et al. 2009).

Cestoda: Rocka (2003).

Acanthocephala: Laskowski et al. (2012).

Helminths in general: Zdzitowiecki (2001a), Laskowski and Zdzitowiecki (2005), Rocka (2006).

Other important publications on Antarctic fish parasites in this period were by Kovaleva et al. (2002) on myxosporeans, Utevsky (2005, 2007) on leeches, Rokicka (2009) and Rokicka et al. (2009) on gyrodactylid monogeneans, Sokolov and Gordeev (2013, 2015) on digeneans and Evans (2014) on X-cell disease. Sures and Reimann (2003) demonstrated how acanthocephalan parasites can be used as extremely sensitive indicators of heavy metal pollution even in relatively pristine environments, and Palm et al. (2007) and Klimpel and Busch (2008) reported on the species richness, life cycles and molecular identification of Antarctic fish parasites. Rokicki (2009) studied the effects of climate change on anisakid nematode infections in polar regions. Santoro et al. (2013) investigated pathological changes and the effects of infection by larval helminths on fish body condition, and Mattiucci et al. (2015) reported on the genetic variability of some anisakid larvae in fishes of the Ross Sea. The helminth parasite communities of icefish were investigated in relation to the life cycles of the parasites and the effects of host sex by Santoro et al. (2014). Finally, Oguz et al. (2015) published a comprehensive and invaluable list of all metazoan parasites described from Antarctic fish up to 2010.

Among the publications on parasites of Antarctic mammals were several on acanthocephalan parasites of seals (Stryukov 2002, 2004; Yurakhno and Stryukov 2004; Silveira et al. (2014). Penguins were well-represented in the parasitological literature on Antarctic birds, with papers on their coccidian parasites (Golemansky 2003, 2008, 2011) and gastrointestinal parasite faunas (Fredes et al. 2007, 2008; Vidal et al. 2012; Diaz et al. 2013; Kleinertz et al. 2014).

Finally, for invertebrate hosts, Zdzitowiecki (2001b), Zdzitowiecki and Presler (2001) and Laskowski et al. (2010) reported on juvenile acanthocephalans infecting Antarctic amphipods, and Takahashi et al. (2004, 2008, 2011) published studies on gregarine parasites of Antarctic krill. A new species of dicyemid was described from an Antarctic octopus by Furuya and Hochberg (2002), and new genera and species of parasitic copepods were described from Antarctic sponges and polychaetes by Bandera et al. (2005), Lopez-Gonzalez et al. (2006) and Suarez-Morales and Boxshall (2012).

3.9 Antarctic Parasitology Research in the Future

While many more new parasite species await description from Antarctic hosts, the emphasis in the future is likely to be much more on studies of parasite communities. Climate change is much more pronounced in polar regions than elsewhere, with corresponding changes in the distribution of hosts and their parasites. Parasites can be used as particularly sensitive indicators of these environmental changes. With commercial fisheries ever searching for new resources to exploit, populations of fish and invertebrates in the Antarctic will come under increasing pressure. To enable efficient management of these resources, it is essential that we learn more about the population biology and stock structure of the populations being exploited. The use of parasites as biological tags can make an important contribution in this area. There is also more awareness now of zoonotic parasites in marine food products, so the effects of muscle-infecting parasites on the quality of Antarctic marine food products will have to be studied. Climate change will alter the distribution of these zoonotic parasites in marine organisms and these changes will have to be monitored. Parasitological research in the Antarctic is entering a new phase.

References

Arduino P, Nascetti G, Cianchi R, Plotz J, Mattiucci S, D'Amelio S, Paggi L, Orrechia P, Bullini L (1995) Isozyme variation and taxonomic rank of *Contracaecum radiatum* (V. Linstow, 1907) from the Antarctic Ocean (Nematoda, Ascaridoidea). Syst Parasitol 30:1–9

Arsen'ev VA, Gusev AV (1958) Studies on the flora and fauna of the Southern Ocean. The second marine Antarctic expedition of the Academy of Sciences of the USSR. Referat. Zhur. Biol., 1960, No. 18961 (Translation). Priroda 11:56–63

Avdeev GV (1983) New harpacticoid copepods (Tisbidae), parasites of octopuses in the Ross Sea. Zool Zhurnal 62:1775–1785

Avdeev VV, Avdeeva NV (1989) On gregarine fauna from planktonic crustaceans from the Antarctic. In: Lebedev BI (ed) Parasites of animals and plants: collection of papers, DVO AN SSSR. pp 40–44 [In Russian]

Baird W (1855) Descriptions of some new species of entozoa from the collection of the British Museum. Ann Mag Nat Hist 15(2):69–76.

Bandera ME, Conradi M, Lopez-Gonzalez PJ (2005) *Asterocheres hirsutus*, a new species of parasitic copepod (Siphonostomatoida: Asterocheridae) associated with an Antarctic hexatinellid sponge. Helgol Mar Res 59:315–322

Bartsch I (1994) Siphonostomatoid copepods on ophiacanthids (Ophiuroidea): description of three new species of *Cancerilla* (Cancerillidae, Copepoda) from Antarctica. Mitt Zool StInst Hamb 91:71–76

Bartsch I (1996) Parasites of the Antarctic brittle star *Ophiacantha disjuncta* (Ophiacanthidae, Ophiuroidea): redescription of the copepod *Lernaeosaccus ophiacanthae* Heegard, 1951. Mitt Zool StInst Hamb 93:63–72

Baylis HA (1923) An ascarid from the sperm whale. Ann Mag Nat Hist Ser 9(11):211–217

Baylis HA (1929) Nematode & Acanthocephala collected in 1923–1927. Discovery Rep 1:541–560

Becker CD, Holloway HL (1968) A survey for haematozoa in Antarctic vertebrates. Trans Am Microsc Soc 87:354–360

Beumer JP, Ashburner LD, Burbury ME, Jette E, Latham DJ (1983) A checklist of the parasites of fishes from Australia and its adjacent Antarctic territories. CAB Tech Commun Commonwealth Inst Parasitol 48(Suppl):1–99

Beverley-Burton M (1971) Helminths from Weddell seal, *Leptonychotes weddelli* (Lesson, 1826) in Antarctica. Can J Zool 49:75–83

Beverley-Burton M (1972) Trematodes from the Weddell seal, *Leptonychotes weddelli* (Lesson, 1826) with a description of *Orthosplanchus weddelli* n. sp. (Trematoda: Campulidae). Zool Anz 189:49–60

Boulenger CL (1916) Myzostomida. British Antarctic Terra Nova Expedition 1910 Nat Hist Rep Zool 2:135–140

Boxshall GA, Harrison K (1988) New nicothoid copepods (Copepoda, Siphonostomatoida) from an amphipod and from deep-sea isopods. Bull Br Mus Nat Hist (Zool) 54:285–299

Brinkmann A (1948) Some new and remarkable leeches from the Antarctic seas. Norske Vidensk Akad Oslo Sci Results Norwegian Antarct Exped 29:1–16

Brandt A, Wagele W (1991) Parasitic Isopoda of the family Gnathiidae from the Atlantic sector of the Southern Ocean. Redescriptions and remarks on the synonymy (Crustacea, Isopoda). Senck Marit 21:233–261

Bresciani J, Lutzen J (1994) Morphology and anatomy of *Avdeevia antarctica* new genus, new species (Copepoda, Harpacticoida, Tisbidae) parasitic on an Antarctic cephalopod. J Crust Biol 14:744–751

Bullini L, Arduino P, Cianchi R, Nascetti G, D'Amelio S, Mattiucci S, Paggi L, Orecchia P, Ploetz J, Berland B, Smith JW, Brattey J (1997) Genetic and ecological research on Anisakid endoparasites of fish and marine mammals in the Antarctic and Arctic-Boreal regions. In: Battaglia

B, Valencia J, Walton DWH (eds) Antarctic communities: species, structure and survival. Cambridge University Press, Cambridge, pp 39–44
Burger O (1893) Sudgeorgische und andere exotische nemartinen. Zool Jb vii:207–240
Bychovsky BE, Gusev AV, Nagibina LF (1965) Monogenetic trematodes belonging to the family Tetraonchoididae Bychowsky, 1951. Trudy Zool Inst leningr 35:140–166
Byrd MA (1963) Helminth parasites of Antarctic vertebrates. Part I. Digenetic trematodes of marine fishes. J Helminthol Soc Wash 30:129–148
Campbell WC (1988) Heather and ice: an excursion historical parasitology. J Parasitol 74:2–12
Campbell WC, Overstreet RM (1994) Historical basis of binomials assigned to helminths collected on Scott's last Antarctic expedition. J Helminthol Soc Wash 61:1–11
Chesunov AV, Spiridonov SE (1985) *Australonema eulagiscae* new genus new species (Nematoda, Marimemithidae), a parasite of a polychaete from Antarctica. Vest Zool 2:16–21
Cieleska D, Wojciechowka A, Zdzitowiecki K (1992) Cestodes in penguins from King George Island (South Shetlands, Antarctica). Acta Parasitol Pol 37:65–72
Cieleska D, Zdzitowiecki K (1989) Tapeworms from the sheathbill *Chionis alba* on King George Island, West Antarctica. Acta Parasitol Pol 34:15–26
Clarke JR, Kerry JR (1993) Diseases and parasites of penguins. Korean J Polar Res 4:79–96
Czaker R (1997) *Wittmannia antarctica* N. G. N. Sp. (Nosematidae), a new hyperparasite in the Antarctic dicyemid mesozoan *Kantharella antarctica*. J Eukaryot Microbiol 44:438–446
Dailey MD, Vogelbein WK (1991) Parasite fauna of 3 species of Antarctic whales with reference to their use as potential stock indicators. Fish B-NOAA 89:355–365
Diaz JL, Fusaro B, Longarzo L, Coria NR, Vidal V, Jerez S, Ortiz J, Barbosa A (2013) Gastrointestinal helminths of Gentoo penguins (*Pygoscelis papua*) from Stranger Point, 25 de Mayo/King George Island, Antarctica. Parasitol Res 112:1877–1881
Dillon WA, Hargis JW Jr (1968) Helminth parasites of Antarctic vertebrates. Part 3. Monogenetic trematodes from Antarctic fishes: the superfamily Tetraonchoidea Yamaguti, 1963. Antarct Res Ser Wash 11:101–112
Dollfus RP (1965) Acanthocephale d'un Telostéen du genre *Notothenia*i Richardson des Kerguelen. Bull Mus Hist Nat Paris 36:641–646
Dollfus RP, Euzet L (1964) Complément a la description de *Pseudobenedania nototheniae* T.H. Johnston, 1931 (Trematoda, Monogenea) parasite d'un Téléostéen du genre *Notothenia* Richardson des Kerguelen. Bull Mus Hist Nat Paris 36:849–857
Drozdz J (1987) Oocysts of 6 new Coccidiomorpha species from pinnipeds of King George Island (South Shetlands, Antarctic). Acta Protozool 26:263–267
Edmonds SJ (1955) Acanthocephala collected by the Australian National Antarctic Research Expedition on Heard Island and Macquarie Island during 1949–50. Trans R Soc S Aust 78:141–144
Edmonds SJ (1957) Acanthocephala. BANZARE Rep Ser B (Zool) 5:93–98
Eichler W (1949) Notulae mallophagologica. XV. Sturmvogel federlinge. Rev Bras Biol 9:337–347
Epshtein VM, Utevsky AY (1994) A new marine leech *Cryobdella lyadovi* sp. n. from Antarctic seas (Hirudinea). Zoosystematica Rossica 3:23–25
Evans CW (2014) X-cell disease in Antarctic fishes. Polar Biol 3:1261–1269
Fain A, Hyland KE (1963) Un nouveau parasite du Manchot d'Adelie (Acarina: Rhinonyssidae). Bull Soc Roy Zool d'Anvers 32:3–5
Feiler K (1986) Trematodes in *Chionis alba* and *Larus dominicanus* from the South Shetland Islands, Antarctic. Angew Parasitol 27:23–33
Fell HB (1961) Fauna of the Ross Sea. Part 1. Ophiuroidea. Mem N Z Oceanogr Inst 18:1–79
Fredes F, Madariaga C, Raffo E, Valencia J, Herrera M, Godoy C, Alcaino H (2007) Gastrointestinal fauna of gentoo penguins from (*Pygoscelis papua*) from the Peninsula Monita, Bahia Paraiso, Antarctica. Antarct Sci 19:93–94
Fredes F, Raffo E, Munoz P, Herrera M, Godoy C (2008) Gastrointestinal parasite fauna in Adelia penguin (*Pygoscelis adeliae*) in an Antarctic territory protected zone. Parasitologia Latinoamericana 63:64–68

Furuya H, Hochberg FG (2002) New species of *Dicyemennea* (Phylum: Dicyemida) in deep-water *Graneledone* (Mollusca: Cephalopoda: Octopoda) from the Antarctic. J Parasitol 88:330–336
Gaevskaya AV (1982) Discovery of trematode metacercariae in mysids of South Georgia Island. Biol Nauki No 8:27–29 [In Russian]
Gaevskaya AV, Rodyuk GN, Parukhin AM (1990) Peculiarities and formation of parasitofauna of the Patagonian toothfish *Dissostichus eleginoides*. Biol Morya Vladivostok 4:23–28 [In Russian]
Gibson D (1976) Monogenea and Digenea from fishes. Discovery Rep XXXVI:179–266
Golemansky V (2003) *Eimeria pygosceli* n. sp. (Coccidia: Eimeriidae) from the penguins (Pygoscelidae) of the Livingston Island (the Antarctic). Acta Zool Bulg 55:3–8
Golemansky V (2008) Diversity and prevalence of the coccidians (Apicomplex: Eucoccidia) of Gentoo penguins (*Pygoscelis papua* FORSTER, 1781) from King George Island (South Shetland Archipelago, the Antarctic). Acta Zool Bulg 60:71–75
Golemansky V (2011) Coccidian parasites (Apicomplexa) of penguins (*Pygoscelis* spp.) from Livingston Island and King George Island, Antarctica. Pol Polar Res 32:263–268
Graefe G (1968) *Paramonostomum antarcticum* n. sp. (Trematoda: Notocotylidae) und Beobachtungen zur Larvenentwicklung in der Antarktis. Z ParasitKde 30:207–232
Graff L (1884) The voyage of H.M.S. 'Challenger' XXVII. Report on the Myzostomida collected during the voyage of H.M.S. 'Challenger' during the years 1873–1876, 82 pp
Gravier C (1912a) Sur l'histoire de un Crustace parasite annelidicole rapporte par la 2e expedition antarctique francaise. Bull Mus Paris 1912:26–30
Gravier C (1912b) Sur un Crustace parasite d'un Polynoidien de l'antarctique sudamericaine (*Selioides tardus* n. sp.). Bull Mus Paris 1912:63–67
Gravier C (1912c) Sur un type nouveau de Crustace parasite d'un Serpulien de l'antarctique sud-Americaine (*Bactropus* nov. g. *cystopomati* nov. sp.). Bull Mus Paris 1912:67–71
Gravier C (1912d) Sur un nouveau genre de Crustace parasite d'un Syllidien de l'antarctique sud-Americaine (*Thylacoides* nov. g. *sarsi* nov. sp.). Bull Mus Paris 1912:71–74
Gravier C (1912e) Sur une Copepode (*Zanclopus antarcticus* nov. sp.) parasite d'un *Cephalodiscus* recueilli par la seconde Expedition antarctique francaise et sur l'evolution du genre *Zanclopus* Calman. Bull Mus Paris 1912:240–245
Gravier C (1912f) On some parasitic annelidicolic crustaceans from the second French Antarctic expedition. C R hebd Séanc Acad Sci Paris 154:830–832
Gravier C (1912g) On the pterobranches brought back by the French Antarctic expedition and on a parasitic crustacean on one of them. C R hebd Séanc Acad Sci Paris 154:1438–1440
Gravier C (1913) *Crustacea parasites*. 2e expedition antarctique francaise (1908–1910) Paris:27–78.
Gravier C (1914) *Isidicola antarctica* Gravier. Crustace parasite de quelques Isidae de l'Antarctique sud-Americaine. C R hebd Séanc Acad Sci Paris 158:354–356
Grygier MJ (1981) A representative of the genus *Dendrogaster* new record (Cirripedia, Ascothoracica) parasitic in an Antarctic starfish *Acodontaster conspicuous*. Antarct Res Ser 32:1–15
Grygier MJ (1987) Antarctic records of asteroid-infecting Ascothoracida (Crustacea), including a new genus of Ctenosculidae. Proc Biol Soc Wash 100:700–712
Gusev AV (1958) Parasitological investigations in the Antarctic. Inform Bull Soviet Antarct Exped 3:71–72 [In Russian]
Gusev AV (1960) Parasitological investigations. Sov Antarct Exped 7:105–110 [In Russian]
Gusev AV (1967) Two new species of genus *Gyrodactylus* Nordmann (Monogenoidea) from the coasts of Antarctica. Issled Fauny Morei 4:187–189 [In Russian]
Hargis WJ, Dillon WA (1968a) Helminth parasites of Antarctic vertebrates. Part II. Monogenetic trematodes from Antarctic fishes: the superfamily Gyrodactyloidea Johnston and Tiegs, 1922. In: Schmitt WL, Llano GA (eds) Antarctic research series 11. Biology of the Antarctic Seas III. American Geophysical Union, Washington, D.C. pp 91–99
Hargis WJ, Dillon WA (1968b) Helminth parasites of Antarctic vertebrates. Part IV. Monogenetic trematodes from Antarctic fishes: the superfamily Capsaloidea Price, 1936. Proc Biol Soc Wash. American Geophysical Union, Washington, D.C. 81:403–412

Hargis WJ, Zwerner DE (1968) Parasites of Antarctic vertebrates and invertebrates. Trematodes, copepods and isopods on *Notothenia* spp., *Trematomus* spp., *Chaenocephalus aceratus, Harpagifer bispinis*. Antarctic J U S 3:163–164

Heegard P (1951) Antarctic parasitic copepods and an ascothoracid cirriped from brittle -stars. Vidensk Meddr dansk naturh Foren 113:117–190

Hoberg EP (1984) Trematode parasites of birds in Antarctica: the distribution of *Gymnophallus deliciosus* (Olsson, 1893). Antarctic J U S 19:159–160

Hoberg EP (1985) *Reticulotaenia* gen. nov. for *Lateriporus australis* Jones & Williams, 1967, and *L. mawsoni* Prudhoe, 1969 (Cestoda: Dilepididae) from sheathbills, *Chionis* spp. in Antarctica, with a consideration of interspecific variation and speciation. J Parasitol 71:319–326

Hoberg EP (1986) Aspects of ecology and biogeography of Acanthocephala in Antarctic seabirds. Ann Parasit Hum Comp 61:199–214

Hoberg EP (1987) *Reighardia sternae* (Diesing, 1864) (Pentastomida) from seabirds in Antarctica. Can J Zool 65:1289–1291

Holloway HJ Jr (1967) Endoparasites of Antarctic vertebrates. Antarctic J U S 2:199

Holloway HL, Klewer HL, Husain A (1967) Notes on the genus *Ascarophis* Beneden, 1871 in Antarctic fishes. Proc Biol Soc Wash 34:222–227

Holloway HL (1968) Endoparasites of Antarctic vertebrates *Corynosoma hammani* and *Ascarophis* spp. in fish and bird hosts. Annu Rev Psychol 2:199

Holloway HL Jr, Bier JW (1968) *Helicometra antarcticae* sp. nov. from Antarctic coastal fishes. Proc Helminthol Soc Wash 35:30–34

Holloway HL Jr, Bier JW (1967) Notes on the host specificity of *Corynosoma hamanni* (Linstow, 1892). Bull Wildlife Dis Ass 3:76–78

Holloway HL Jr, Klewer HL (1969) *Rhabdochona beatriceinsleyae* n. sp. (Nematoda: Spiruridea: Rhabdochonidae) from the Antarctic zoarcid, *Rhigophila dearborni*. Trans Am Microsc Soc 88:460–471

Hoogesteger JN, White MG (1981) Notes on parasite infestation of inshore fish at Signy Island, South Orkney Islands. Brit Antarct Surv Bull 54:23–31

Horne PA, Rounsevell D (1982) A collection of feather mites (Acari, Astigmata) from Greater (Eastern) Antarctica. Pac Insects 24:196–197

Janusz J, Sosinski J (2000) *Eubrachiella antarctica* (Quidor, 1906) (Copepoda) – levels of infection in selected fish species of the family Notothenidae. Acta Ichthyol Piscat 29:43–52

Jezewski W, Zdzitowiecki K, Laskowski Z (2014) Digenea in notothenioid fish in the Beagle Channel (Magellanic sub-region, sub-Antarctica). Acta Parasitol 59:42–49

Johnston TH (1930) The anatomy of the trematode *Macrophyllida antarctica* (Hughes). Aust J Exp Biol Med Sci 7:101–107

Johnston TH (1931) New trematodes from subantarctic and Antarctic. Aust J Exp Biol Med Sci 8:91–98

Johnston TH (1937a) Report on the Trematoda. Sci Results Australas Antarct Exped (C) 10(1):1–29

Johnston TH (1937b) The Cestoda of the Australasian Antarctic Expedition. Sci Rep Aust Antarct Exped 10(4):1–74

Johnston TH (1938) Report on the parasitic nematodes of the Australasian Antarctic Expedition. Sci Rep Aust Antarct Exped 1911–14 (C) 10(5):1–31

Johnston TH, Best EWD (1937) Report on the Acanthocephala. Aust Antarct Exped (1911–1914). Sci Rep Ser C Zool Bot 10C:1–20

Johnston TH, Mawson PM (1945) Parasitic nematodes. BANZARE 1929–1931. Rep Ser B V, Part 2:73–160

Jones HI (1988) Notes on parasites in penguins (Spheniscidae) and petrels (Procellariidae) in the Antarctic and sub-Antarctic. J Wildl Dis 24:166–167

Jones NV, Williams IC (1969) The nematode and acanthocephalan parasites of the sheathbill, *Chionis alba* (Gmelin), at Signy Island, South Orkney islands and a summary of host-parasite relationships in the sheathbill. J Helminthol 43:59–67

Kagei N, Asano K, Kihata M (1978) On the examination against the parasites of Antarctic krill, *Euphausia superba*. Scient Rep Whales Res Inst 30:311–313

Kagei N, Kureha K (1970) Study of anisakid Nematoda (Anisakinae).1. Survey of Anisakis sp. on marine mammals collected in the Antarctic Ocean. Bull Inst Public Health Tokyo 19:193–196

Kagei N, Watanuki T (1977) On the parasites of fishes from the Antarctic Ocean. Antarct Rec 54:84–93

Kleinertz S, Christman S, Silva LMR, Hirzmann J, Hermosilla C, Taubert A (2014) Gastrointestinal parasite fauna of Emperor penguins (*Aptenodytes forsteri*) at the Arkta Bay, Antarctica. Parasitol Res 113:4133–4139

Klimpel S, Busch MW (2008) Antarctic fish parasite fauna: life cycle patterns and genetic species characterization. Berichte zur Polar-und Meeresforschung 569:106–110

Klöser H, Plötz J (1992) Morphological distinction between adult *Contracaecum radiatum* and *Contracaecum osculatum* (Nematoda, Anisakidae) from the Weddell seal (*Leptonychotes weddelli*). Zool Scr 21:129–132

Klöser H, Plötz J, Palm HW, Bartsch A, Hubold G (1992) Adjustment of anisakid nematode life cycles to the high Antarctic food web as shown by *Contracaecum radiatum* and *C. osculatum* in the Weddell Sea. Antarct Sci 4:171–178

Kovaleva AA, Gaevskaya AV (1974) New representatives of the genus *Plagioporus* (Trematoda, Opecoelidae) from Antarctic fishes. Zool Zhurnal 53:1407–1409 [In Russian]

Kovaleva AA, Gaevskaya AV (1977) Two new species of monogeneans, parasites of Antarctic fishes. Zool Zhurnal 61:783–786 [In Russian]

Kovaleva AA, Gaevskaya AV (1984) New species of the genera *Kudoa* and *Pentacapsula* (Myxosporidia, Multivalvulea) with unusual location. Zool Zhurnal 63:1090–1092

Kovaleva AA, Rodjuk GN, Grudnev MA (2002) Myxosporeans (Cnidospora: Myxosporea) of Antarctic fishes. Parazitologiya 36:502–513

Kurochkin YV, Nikol'skii OR (1972) On the trematode fauna of Antarctic seals. *Izvestiya tikhookean nauchno-issled.* Inst Ryb Khoz Okeanogr 81:266–269

Laird M (1956) Antarctic parasitic protozoa. Special report (Royal Society of New Zealand. Antarctic Research Committee) no. 9

Laird, M. (1961). A lack of avian and mammalian haematozoa in Antarctic and Canadian Arctic. Can J Zool 39:209–213

Laskowski Z, Jezewski W, Zdzitowiecki K (2010) New data on the occurrence of Acanthocephala on Antarctic Amphipoda. Acta Parasitol 55:161–166

Laskowski Z, Jezewski W, Zdzitowiecki K (2014) Changes in digenean infection of the Antarctic fish *Notothenia coriiceps* in Admiralty Bay, King George Island, over three decades. Pol Polar Res 35:513–520

Laskowski Z, Korszak-Abshire M, Zdzitowiecki K (2012) Changes in acanthocephalan infection of the Antarctic fish *Notothenia coriiceps* in Admiralty Bay, King George Island, over 29 years. Pol Polar Res 33:91–108

Laskowski Z, Zdzitowiecki K (2005) The helminth fauna of some notothenioid fishes collected from the shelf of Argentine Islands, Antarctica. Pol Polar Res 26:315–324

Leaderwilliams N (1980) Observations on the internal parasites of reindeer introduced into South Georgia. Vet Rec 107:393–395

Leiper RT, Atkinson EL (1914) Helminths of the British Antarctic expedition 1910–1913. Proc Zool Soc Lond (1):222–226.

Leiper RT, Atkinson EL (1915) Parasitic worms with a note on a free-living nematode. Br Mus (nat Hist), Br Antarct (Terra Nova) Exped, 1910. Nat Hist Reps Zool 2:19–60

Linstow OV (1892) Helminthen von Südgeorgien. Nach der Ausbeute der Deutschen Expedition von 1882–1883. Jb Hamb Anst 9:59–77

Linstow OV (1905) Helminthen der Russischen Polar-Expedition 1900–1903. St. Peterburg Memoires Ac Sc Ser 8(18):1–17

Linstow OV (1907) Nematodes of the Scottish National Antarctic Expedition 1902–1904. Proc R Soc Edinb 26:464–472

Linstow OV (1911) Nematoda. Nat Antarct Exped 1901–1904:1–4

Lopez-Gonzalez PJ, Bresciani J, Conradi M (2000) Two new species of *Herpyllobius* Steenstrup and Lutken, 1841 and a new record of *Herpyllobius antarcticus* Vanhoffen, 1913 (parasitic Copepoda) from the Weddell Sea, Antarctica. Polar Biol 23:265–271

Lopez-Gonzalez PJ, Bresciani J, Conradi M (2006) New genus, three new species and new records of Herpyllobiidae Hansen, 1892 (Crustacea, Copepoda), parasites of polychaetes from Antarctica. Sci Mar 70:243–259

Ludwig H (1903) Seesterne. Resultats du voyage du S.H. Belgica. Zoologie:72

Lutzen J, Jones B (1976) 2 new species of *Herpyllobius* parasitic Copepoda from New Zealand and the Antarctic. N Z Jl Mar Freshw Res 10:371–374

Lyadov VN (1985) Zoogeographical characteristics of the helminths of fishes from the Antarctic zone of the World Ocean. In: Hargis WJ Jr (ed) Parasitology and pathology of marine organisms of the world ocean. NOAA Technical Report NMFS, vol 25. pp 41–43

Markowski S (1971) On some species of parasitic worms in the Discovery Collections obtained in the years 1925–1936. Bull Br Mus nat Hist 21:53–65

Mattiucci S, Cipriani P, Paoletti M, Nardi V, Santoro M, Bellisario B, Nascetti G (2015) Temporal stability of parasite distribution and genetic variability values of *Contracaecum osculatum* sp D and C *osculatum* sp E (Nematoda: Anisakidae) from fish of the Ross Sea (Antarctica). Int J Parasitol Parasit Wildlife 4:356–367

Mawson PM (1953) Parasitic nematodes collected by the Australian National Antarctic Research Expedition, Heard Island and Macquarie Island. 1948–1951. Parasitology 43:291–297

Michajlow W (1978) *Dinema antarcticum* sp. n., *Dinema pseudoboeckellae* sp. n. and other Euglenoidina parasites of *Pseudoboeckella silvestri* (Calanoida) from Antarctica. Bull Acad Pol Sci Cl II Sér Sci Biol 26:51–54

Mironov SV (1991) Two new species of the feather mites superfamily Analgoidea from the Antarctic birds. Inf Byull Sov Antarktich Eksped 116:69–75

Mortensen T (1903) The Danish Ingolf-Expedition. Volume iv. 1. Echinoidea (Part 1). Translated by Torben Lundbeck, 198 pp

Moser M, Cowen RK (1991) The effects of periodic eutrophication on parasitism and stock identification of *Trematomus bernacchii* (Pisces: Nototheniidae) in McMurdo Sound, Antarctica. J Parasitol 77:551–556

Murray MD, Nicholls DG (1965) Studies on the ectoparasites of seals and penguins. i. The ecology of the louse *Lepidophtheirus macrorhini* Enderlein on the southern elephant seal *Mirounga leonine* (L.). Aust J Zool 13:437–454

Murray MD, Palma RL, Pilgrim LRC (1993) Ectoparasites of Australian, New Zealand and Antarctic birds. In: Marchant S, Higgins, PJ (eds) Handbook of Australian, New Zealand and Antarctic birds. Volume 2: raptors to lapwings. Oxford University Press, Melbourne, Australia. pp 959–962.

Murray MD, Palma RL, Pilgrim RLC, Shaw MD (1999) Ectoparasites of Australian, New Zealand and Antarctic birds. In: Higgins PJ (ed) Handbook of Australian, New Zealand and Antarctic birds. Volume 4, parrots to dollarbird. Oxford University Press, Melbourne, Australia. pp 1240–1242

Nascetti G, Mattiucci S, Paggi L, Bullini L (1997) Genetic and ecological studies on *Contracaecum* larvae (Nematoda: Anisakidae), parasitic on fish from the Ross Sea, Antarctic. Biol Mar Med 4:664–665 (In Italian)

Nikol'skii OR (1974) Acanthocephalan fauna of the pinnipeds of Pacific Antarctic. Izvest nauch-issl Inst rybn Khoz Okeanogr 88:101–106

Noble ER (1984) Ecologic aspects of protozoan infections in Antarctic fishes. J Protozool 31:575–577

Odening K (1982) Cestodes from birds of the South Shetlands, Antarctic, and the Falkland Islands. Angew Parasitol 23:202–223

Odening K (1986) Tissue cyst-forming Coccidia in Antarctic vertebrates. Symp Biol Hung 33:351–355

Oguz MC, Tepe Y, Belk MC, Heckman RA, Aslan B, Gurgen M, Bray RA, Akgul U (2015) Metazoan parasites of Antarctic fishes. Turk Parazitol Derg 39:174–178

Orecchia P, Mattiucci S, D'Amelio S, Paggi L, Plötz J, Cianchi R, Nascetti G, Arduino P, Bullini L (1994) Two new members in the *Contracaecum osculatum* complex (Nematoda, Ascaridoidea) from the Antarctic. Int J Parasitol 24:367–377

Oresland V, Pleijel F (1991) An ectoparasitic typhloscolecid polychaete on the chaetognath *Eukrohnia hamata* from the Antarctic Peninsula. Mar Biol 108:429–432
Paggi L, Mattiucci S, Gibson DI, Berland B, Nascetti G, Cianchi R, Bullini L (2000) *Pseudoterranova decipiens* species A and B (Nematoda: Ascaridoidea): nomenclatural designation, morphological diagnostic characters and genetic markers. Syst Parasitol 45:185–197
Palm H (1997) A contribution to the parasitic fauna of the Weddell Sea. Ber Polarforschung 249:100–101
Palm HW (1999) Ecology of *Pseudoterranova decipiens* (Krabbe, 1878) (Nematoda: Anisakidae) from Antarctic waters. Parasitol Res 85:638–646
Palm HW, Andersen K, Klöser H, Plötz J (1994) Occurrence of *Pseudoterranova decipiens* (Nematoda) in fish from the south-eastern Weddell Sea (Antarctic). Polar Biol 14:539–544
Palm HW, Klimpel S, Walter T (2007) Demersal fish parasite fauna around the South Shetland Islands: high species richness and low host specificity in deep Antarctic waters. Polar Biol 30:1513–1522
Palm HW, Reimann N, Spindler M, Poltz J (1998) The role of the rock cod *Notothenia coriiceps* Richardson, 1844 in the life cycle of Antarctic parasites. Polar Biol 19:399–406
Parukhin AM, Lyadov VN (1979) New genus and species of trematodes – fish parasites in the sub-Antarctic zone of the Indian Ocean. Zool Zhurnal 58:637–642 [In Russian]
Parukhin AM, Lyadov VN (1981) The parasite fauna of notothenoid fish (Notothenoidei) in the waters of the Atlantic and Pacific Oceans. Vest Zool 3:90–94 [In Russian]
Platonova TA, Potin VV (1972) On new genera *Harpagonchoides* (Nematoda, Chromadorida, Harpagonchidae, fam. n.) living on the parapodia and gills of the Antarctic polychaetes *Aglaophamus* Kinberg and *Hemipodus* Quatrifages. Issled Fauny Morei 19:81–87
Pois NV (1975) Materials on the infestation by cestodes on some commercial fish of the Indian and Pacific sectors of the Antarctic and subantarctic. Prob Parazit 2:113–114
Prudhoe S, Bray RA (1973) Digenetic trematodes from fishes. Rep BANZARE 1929–1931 Ser B 8:195–225
Prudhoe S (1969). Cestodes from fish, birds and whales. Reports B.A.N.Z. Antarctic Research Expedition 1929–1931, Ser. B, 8:171–193
Pugh PJA (1993) A sysonomic catalog of the Acari from Antarctica, the sub-Antarctic islands and the southern ocean. J Nat Hist 27:323–421
Quidor A (1913) Copepodes parasites. 2e expedition antarctique francaise (1908–1910). Masson, Paris, pp 197–215
Railliet A, Henry A (1907) Némathelminthes parasites. Expéd Antarct Française (1903–1905). Masse et Cie, Paris, pp 1–15
Reimer LW (1987) Helminthen von Fischen der Antarktis. Fischerei-Forschung Rostok 25:36–40
Rennie J (1907) “Scotia” collections. On *Echinorhynchus antarcticus* n.sp. and its allies. Proc R Soc Edinb 26:437–446
Rocka A (1999) The tetraphyllidean cercoids from teleosts occurring in the Weddell Sea (Antarctic). Acta Parasitol 44:115–118
Rocka A (2002) Nematodes of fishes in the Weddell Sea (Antarctica). Acta Parasitol 47:294–299
Rocka A (2003) Cestodes of the Antarctic fishes. Pol Polar Res 24:261–276
Rocka A (2004) Nematodes of the Antarctic fishes. Pol Polar Res 25:135–152
Rocka A (2006) Helminths of Antarctic fishes: life cycle biology, host specificity and geographical distribution. Acta Parasitol 51:26–35
Rocka A, Zdzitowiecki K (1998) Cestodes in fishes of the Weddell Sea. Acta Parasitol 43:64–70
Rodyuk GN (1985) Parasitic fauna of the fishes of the Atlantic part of the Antarctic (South Georgia Island and South Shetland Islands). In: Hargis WJ Jr (ed) Parasitology and pathology of marine organisms of the World ocean, vol 25. NOAA technical report NMFS. pp 31–32
Rodyuk GN (1986a) A new representative of the genus *Neopavlovskioides* (Monogenea, Tetraonchoididae) from plugars of the west Antarctic area. Vest Zool 2:79–81 [In Russian]
Rodyuk GN (1986b) New species of Acanthocephala of the genus *Echinorhynchus* (Echinorhynchidae) from the south-western Atlantic. Parazitologiya 20:224–227 [In Russian]

Rohde K (2000) Ecology and zoogeography of parasites of Antarctic fish. Australian Antarctic Data Centre. Source URL: http://gcmd.nasa.gov/KeywordSearch/Metadata.do?Portal=amd_au&MetadataView=Full&MetadataType=0&KeywordPath=&OrigMetadataNode=AADC&Entryld=ASAC_67

Rokicka M (2009) Report on species of *Gyrodactylus* Nordmann, 1832, distribution in polar regions. Polar Sci 3:203–206

Rokicka M, Lumme J, Zietara MS (2009) Two new Antarctic *Gyrodactylus* species (Monogenoidea): description and phylogenetic characterization. J Parasitol 95:1112–1119

Rokicki J (2009) Effects of climatic changes on anisakid nematodes in polar regions. Polar Sci 3:197–201

Rokicki J, Skora KE (1987) The dynamics of occurrence of *Eubrachiella antarctica* (Quidor, 1906) on *Notothenia gibberifrons* Lonnberg, 1905. Wiad Parazytol 32:511–515

Rokicki J, Rodjuk G, Zdzitowiecki K, Laskowski Z (2009) Larval ascaridoid nematodes (Anisakidae) in fish from the South Shetland Islands (Southern Ocean). Pol Polar Res 30:49–58

Rubtsov RA (1977) A new genus and species of parasitic nematode, *Ananus asteroideus* (Nematoda, Marimermithidae), from the asteroid *Diplopteraster peregrinator*. Bull Mus Hist Nat Paris (Zoologie) 345:1113–1116

Santoro M, Cipriani P, Pankov P, Lawton SP (1990) *Aporocotyle michaudi* n. sp. (Digenea: Aporocotylidae) from the emerald rock cod, *Trematomus bernacchii* (Teleostei: Perciformes) in Antarctica. Parasitol Int 64:324–329

Santoro M, Mattiucci S, Work T, Cimmaruta R, Nardi V, Cipriani P, Bellisario B, Nascetti G (2013) Parasitic infection by larval helminths in Antarctic fishes: pathological changes and impact on the host body condition index. Dis Aquat Organ 105:139–148

Santoro M, Mattiucci S, Cipriani P, Bellisario B, Romanelli F, Cimmaruta R, Nascetti G (2014) Parasite communities of icefish (*Chionodraco hamatus*) in the Ross Sea (Antarctica): influence of the host sex on the helminth infracommunity structure. PLoS One 9:e88876

Sawyer RT, White MG (1969) A new genus and species of marine leech, *Glyptonobdella antarctica*, from an Antarctic isopod. Bull Br Antarct Surv 22:1–14

Schultz GA (1980) *Arcturocheres gaussicola* n. sp. (Cabiropsidae), parasite in *Antarcturus gaussianus* Vanhoffen (Arcturidae) from Antarctica (Isopoda). Crustaceana 39:153–156

Shipley AE (1901) On the abysmal fauna of the Antarctic region. Being Chapter XVIII (pp 241–275) of the Antarctic Manual…1901. In: Murray G (ed). Royal Geographical Society, London, unpaginated

Short RB, Hochberg FG (1970) A new species of *Dicyemennea* (Mesozoa, Dicyemidae) from near Antarctic Peninsula. J Parasitol 56:517–522

Siegel V (1980a) Quantitative investigations on parasites of Antarctic channichthyid and nototheniid fishes. Meeresforschung – Reps Mar Res 28:146–156

Siegel V (1980b) Parasite tags for some Antarctic channichthyid fish. Arch FischWiss 31:97–103

Silveira T, Bianchini A, Robaldo R, Colares EP, Costa Muelbert MM, Martinez PE, Pereira E, Schifino Valente AL (2014) *Corynosoma* spp. (Acanthocephala, Polymorphidae) in *Mirounga leonina* (Pinnipedia, Phocidae) of South Shetlands Islands: a new host for *Corynosoma cetaceum*. Pan-Am J Aquat Sci 9:66–69

Skrjabin AS (1967) A gigantic diphyllobothriid *Polygonoporus giganticus* n.g. n. sp., sperm whale parasite. Parazitologiya 1:131–136 [In Russian]

Skrjabin AS (1969) A new species of trematodes of the genus *Ogmogaster* (Notocotylidae), parasite of whalebone whales. Zool Zhurnal 48:1882–1885 [In Russian]

Skrjabin AS (1971a) A new *Crassicauda* (*Crassicauda delamureana* n. sp.), a parasite of the sei whale. Problem Parazit 5:100–107 [In Russian]

Skrjabin AS (1971b) The first finding of larvae of the cestode *Monorygma delphini* (Tetraphyllidea, Phyllobothriidae) in Antarctic whales. Zool Zhurnal 49:1721–1722 [In Russian]

Skrjabin AS (1974) The first record of the trematode *Lecithodesmus goliath* (van Beneden, 1859) Braun, 1902 (fam. Campulidae Odhner, 1926) in Antarctic whales. Izv tikh nauchno-issled Inst Ryb Kho Okeanogr 88:139–144 [In Russian]

Skrjabin AS, Muravieva SI (1970) On the study of cestodes from the genus *Priapocephalus* Nybelin. Biol Morya Kiev 20:217–225 [In Russian]

Skrjabin AS, Muravieva SI (1971) The new cestode *Tetrabothrius egregius* n. sp., a parasite of fin whale. Nauch Doklady Vyssh Shkoly, biol Nauki:17–21. [In Russian]

Skrjabin AS, Muravieva SI (1972) First finding of cestode *Tetrabothrius curilensis* (Tetraphyllidea) in *Physeter catodon* L. caught in the Antarctic waters. Vest Zool 3:85–87 [In Russian]

Skryabin AS, Nikol'skii OR (1971) *Corynosoma singularis* sp. n. (family Polymorphidae), a parasite of Antarctic sea mammals. Nauch Doklady Vyssh Shkoly Biol Nauki 11:7–9 [In Russian]

Sokolov SG, Gordeev II (2013) New data on trematodes (Platyhelminthes, Trematoda) of fishes in the Ross Sea (Antarctic). Zoologiya Bespozvonochnykh 10:255–267

Sokolov SG, Gordeev II (2015) New data on trematodes of Antarctic fish. Parazitologiya 49:12–26

Sosinski J, Janusz J (1986) The occurrence of the parasite *Eubrachiella gaini* (Quidor, 1913) in Antarctic fishes of the family Chaenichthyidae. Acta Ichthyol Piscat 16:87–105

Stryukov AA (2002) About the acanthocephal *Corynosoma arctocephali* (Acanthocephala, Polymorphidae): the parasite of sea leopard *Hydrurga leptonyx* from Pacific sector of Antarctic. Vest Zool 36:71–77

Stryukov AA (2004) Invasion of Antarctic phocid seals by acanthocephals. Vest Zool 38:23–29

Suarez-Morales E, Boxshall GA (2012) A new species of *Sabellacheres* M. Sars, 1862 (Copepoda: Gastrodelphiidae) from a deep-water benthic polychaete in Antarctic waters, with a key to the species of the genus. Syst Parasitol 83:65–75

Sures B, Reimann N (2003) Analysis of trace elements in the Antarctic host-parasite system *Notothenia coriiceps* and *Aspersentis megarhynchus* (Acanthocephala) caught at King George Island, South Shetland Islands. Polar Biol 26:680–686

Szelenbaum-Cielecka D, Zdzitowiecki K (1979) The tape-worm *Nototaenia fileri* Jones et Williams, 1967 from the sheathbill (*Chionis alba*) on King George Island (South Shetlands, Antarctic). Bull Acad Pol Sci Cl II Sér Sci Biol 27:281–284

Szidat L (1965) Estudios sobre la fauna de parásitos de peces antárcticos. I. Los parásitos de *Notothenia neglecta* Nybelin. Servicio de Hidrografía Naval de la Secretaría de Marina de la Rep Argentina. Público H 910:1–84

Szidat L, Graefe G (1967) Estudios sobre la fauna de parásitos de peces antárcticos. II. Los parásitos de *Parachaenichthys charcoti*. Servicio de Hidrografía Naval de la Secretaría de Marina de la Rep Argentina. Público H 911:1–27

Szidat L, Graefe G (1969) The parasites of *Parachaenichthys charcoti*, an Antarctic fish, in relation to problems of zoogeography. In: Symposium on Antarctic oceanography, Santiago, Chile, 1966, International Council of Scientific Unions – Special Committee in Antarctic research, pp 169–170.

Takahashi KT, Kawaguchi S, Kobayashi M, Toda T (2004) The variability in abundance of eugregarines living in the Antarctic krill. Polar Biosci 17:16–25

Takahashi KT, Kawaguchi S, Kobayashi M, Toda T, Tanimura A, Fukuchi M, Odate T (2011) Eugregarine infection within the digestive tract of larval Antarctic krill, *Euphausia superba*. Polar Biol 34:1167–1174

Takahashi KT, Kobayashi M, Kawaguchi S, Saigusa J, Tanimura A, Fukuchi M, Naganobu M, Toda T (2008) Circumpolar occurrence of gregarinid protozoan *Cephaloidophora pacifica* associated with Antarctic krill, *Euphausia superba*. Antarct Sci 20:437–440

Timofeeva TA, Gaevskaya AV, Kovaleva AA (1987) Capsalids of the notothenoid fishes from Atlantic region of Antarctica and sub-Antarctica (Monogenea, Capsalidae). Tr Zool Inst 61:78–93

Utevsky AY (1993) A new marine leech *Nototheniobdella sawyer* gen. et sp. n. from Antarctic seas (Hirudinea, Piscicolidae). Zoosystematica Rossica 2:237–240

Utevsky AY (1995) A new Antarctic leech genus (Hirudinea, Piscicilidae [Piscicolidae]). Vest Zool 5(6):3–12

Utevsky AY (1997) A new species of piscicolid leeches (Hirudinea, Piscicolidae) from Antarctic seas. Vest Zool 31:17–24

Utevsky AY (2005) An identification key to Antarctic fish leeches (Hirudinea: Piscicolidae). Ukr Antark Zh 3:135–144

Utevsky AY (2007) Antarctic piscicolid leeches. Bonn Zool Monographien 54:5–80

Vidal V, Ortiz J, Diaz JL, de Ybanez MRR, Amat MT, Palacios MJ, Benzal J, Valera F, de la Cruz C, Motas M, Barbosa A (2012) Gastrointestinal parasites in Chinstrap penguins from Deception Island, South Shetlands, Antarctica. Parasitol Res 111:723–727

Wägele J-W, Brandt A (1988) *Protognathia bathypelagica* new genus new combination Schultz, 1977 rediscovered in the Weddell Sea, Antarctica: a missing link between the Gnathiidae and the Cirolanidae (Crustacea, Isopoda). Polar Biol 8:359–366

Walter T (1998) The nematode fauna in fish around the Antarctic Peninsula. Berichte zur Polarforschung 274:33–34

Williams IC, Jones NV, Payne MJ, Ellis C (1974) The helminth parasites of the sheathbill, *Chionis alba* (Gmelin), and the diving petrels, *Pelecanoides georgicus* (Murphy and Harper) and *P. urinalrix* (Gmelin) at Bird Island, South Georgia. J Helminthol 48:195–197

Wojciechowska A (1991) Some tetraphyllidean and diphyllidean cestodes from Antarctic batoid fishes. Acta Parasit Pol 36:69–74

Wojciechowska A (1993a) The tetraphyllidean and tetrabothriid cercoids from Antarctic bony fishes. I. Morphology, Identification with adult forms. Acta Parasitol 38:15–22

Wojciechowska A (1993b) The tetraphyllidean and tetrabothriid cercoids from Antarctic bony fishes. II. Occurrence of cercoids in various fish species. Acta Parasitol 38:113–118

Wojciechowska A (1993c) The tetraphyllidean and tetrabothriid cercoids from Antarctic bony fishes. III. Infection of Notothenia neglecta and N. rossiii in Admiralty Bay. Acta Parasitol 38:166–169

Wojciechowska A, Zdzitowiecki K (1995) Cestodes of Antarctic seals. Acta Parasitol 40:125–131

Wojciechowska A, Zdzitowiecki K, Pisano E, Vacchi M (1994) The tetraphyllidean cercoids from bony fishes occurring in the Ross Sea (Antarctic). Acta Parasitol 39:13–15

Woodcock HM, Lodge O (1921) Parasitic Protozoa. British Antarctic (Terra Nova) Expedition 1910:1–24

Yang T (1987) Marine leeches from an Antarctic fish of the genus *Notothenia*. Acta Zool Sinica 33:373–377

Yurakhno MV, Maltsev VN (1997) Cestode infection of Antarctic seals. Parazitologia 31:81–89 [In Russian]

Yurakhno MV, Stryukov AA (2004). About geographical variability in acanthocephalans of Antarctic phocids. In: Belkovich BM (ed) Marine Mammals of the Holarctic. Collection of Scientific Papers. Moscow, KMK, pp 599–601

Zdzitowiecki K (1978a) *Corynosoma shackletoni* sp. n. from hosts in South Shetlands and South Georgia (Antarctica). Bull Acad Pol Sci Cl II Sér Sci Biol 26:629–634

Zdzitowiecki K (1978b) On the occurrence of juvenile acanthocephalans of the genus *Corynosoma* Lühe, 1904 in fishes off South Georgia and South Shetland Islands (the Antarctic). Acta Ichthyol Piscat 8:111–126

Zdzitowiecki K (1978c) Preliminary investigations on fish parasites from off the South Shetland Islands and from off South Georgia (Antarctic). Kosmos (Warsaw) 27:651–659

Zdzitowiecki K (1984) Some Antarctic acanthocephalans of the genus *Corynosoma* parasitizing Pinnipedia with descriptions of 3 new species. Acta Parasit Pol 29:359–378

Zdzitowiecki K (1985) Acanthocephalans of birds from South Shetlands (Antarctic). Acta Parasit Pol 30:11–24

Zdzitowiecki K (1986a) *Acanthocephala* of the Antarctic. Pol Polar Res 7:79–117

Zdzitowiecki K (1986b) Prevalence of acanthocephalans in fishes of South Shetlands (Antarctic). I. Juvenile Corynosoma spp. Acta Parasit Pol 30:143–160

Zdzitowiecki K (1986c) Prevalence of acanthocephalans in fishes of South Shetlands (Antarctic). III. *Metacanthocephalus johnstoni* Zdzitowiecki, 1983, *M. dalmori* Zdzitowiecki, 1983 and notes on other species; general conclusions. Acta Parasit Pol 31:125–141

Zdzitowiecki K (1987) Acanthocephalans of marine fishes in the regions of South Georgia and South Orkneys (Antarctic). Acta Parasit Pol 31:211–217

Zdzitowiecki K (1988) Occurrence of digenetic trematodes in fishes off South Shetlands (Antarctic). Acta Parasit Pol 33:155–167
Zdzitowiecki K (1990) Occurrence of acanthocephalans in fishes of the open sea off the South Shetlands and South Georgia (Antarctic). Acta Parasit Pol 35:131–141
Zdzitowiecki K (1991a) Antarctic Acanthocephala. In: Wägele JW, Sieg J (eds) Synopses of the Antarctic Benthos 3. Koeltz Scientific Books, Königstein, p 116
Zdzitowiecki K (1991b) Occurrence of digeneans in open sea fishes off the South Shetland Islands and South Georgia, and a list of fish digeneans in the Antarctic. Pol Polar Res 12:35–72
Zdzitowiecki K (1997a) Antarctic Digenea, parasites of fishes. In: Wägele JW, Sieg J (eds) Synopses of the Antarctic Benthos, vol 8. Koeltz Scientific Books, Königstein, p 156
Zdzitowiecki K (1997b) Digenea of fishes of the Weddell Sea. IV. Three opecoelid species of the genera *Neolebouria, Helicometra* and *Stenakron*. Acta Parasitol 42:138–143
Zdzitowiecki K (1997c) Digenea of fishes of the Weddell Sea. V. Two new species of the genus *Steringophorus* (Fellodistomidae). Acta Parasitol 42:144–148
Zdzitowiecki K (1997d) Digenea of fishes of the Weddell Sea. VI. The superfamily Hemiuroidea. Acta Parasitol 42:219–224
Zdzitowiecki K (2001a) New data on the occurrence of fish endoparasitic worms off Adelie Land, Antarctica. Pol Polar Res 22:159–165
Zdzitowiecki K (2001b) Acanthocephala occurring in intermediate hosts, amphipods, in Admiralty Bay (South Shetland Islands, Antarctica). Acta Parasitol 46:202–207
Zdzitowiecki K (2002a) Occurrence of Digenea in fishes of the family Nototheniidae in the Weddell Sea. Acta Parasitol 47:154–158
Zdzitowiecki K (2002b) Occurrence of Digenea in fishes of the family Channichthyidae in the Weddell Sea and other subcontinental areas of the Antarctica. Acta Parasitol 47:159–162
Zdzitowiecki K (2002c) Occurrence of Digenea in fishes of the family Artedridaconidae in the Weddell Sea and other areas of Antarctica. Acta Parasitol 47:306–309
Zdzitowiecki K (2002d) Occurrence of Digenea in fishes of the family Bathydraconidae in the Weddell Sea and other areas of Antarctica. Acta Parasitol 47:310–313
Zdzitowiecki K (2003) Occurrence of Digenea in fishes other than Notothenioidei in the Weddell Sea and the whole Antarctica. Acta Parasitol 48:195–199
Zdzitowiecki K, Cielecka D (1997a) Digenea of fishes of the Weddell Sea. I. Parasites of *Macrourus whitsoni* (Gadiformes, Macrouridae). Acta Parasitol 42:23–30
Zdzitowiecki K, Cielecka D (1997b) Digenea of fishes of the Weddell Sea. II. The genus *Macvicaria* (Opecoelidae). Acta Parasitol 42:77–83
Zdzitowiecki K, Cielecka D (1997c) Digenea of fishes of the Weddell Sea. III. The Lepocreadiidae (genera *Neolepidapedon* and *Lepidapedon*), parasites of Notothenioidea. Acta Parasitol 42:84–91
Zdzitowiecki K, Drozdz J (1980) Redescription of *Stegophorus macronectes* and description of *Stegophorus acrotowskii* new species (Nematoda, Spirurida) from birds of South Shetland, the Antarctic. Acta Parasit Pol 27:15–28
Zdzitowiecki K, Niewiadomska K, Drozdz J (1989) Trematodes of birds and mammals in the environs of H. Arctowski Station (South Shetlands, Antarctic). Acta Parasit Pol 34:243–257
Zdzitowiecki K, Presler P (2001) Occurrence of Acanthocephala in intermediate hosts, Amphipoda, in Admiralty Bay, South Shetland Islands. Pol Polar Res 22:205–212
Zdzitowiecki K, Rokosz B (1986) Prevalence of acanthocephalans in fishes of South Shetlands (Antarctic). II. *Aspersentis austrinus* Van Cleave, 1929 and remarks on the validity of *Heteracanthocephalus hureaui* Dollfus, 1965. Acta Parasit Pol 30:161–171
Zlotorzyska J, Modrzejewska M (1992) Contribution to the knowledge of lice from the Antarctic. Pol Polar Res 13:59–63

Chapter 4
Biodiversity and Host Specificity of Monogenea in Antarctic Fish Species

Regina Klapper, Julian Münster, Judith Kochmann, Sven Klimpel, and Thomas Kuhn

4.1 Introduction

Antarctica is a unique environment characterized by extreme physical and chemical conditions. Stable below zero temperatures, an increased oxygen solubility, a narrow shelf area due to a great shelf ice cover, strong seasonal fluctuations in light and productivity in combination with a long evolutionary history of isolation fostered unique adaptations, and an extremely high degree of endemism of both the local flora and fauna and their respective parasite species (Kock 1992; Eastman 1993; Bargelloni et al. 1994; Rocka 2006; Klimpel and Palm 2011).

Monogenean flatworms are typically ectoparasites of all groups of freshwater and marine fishes including teleosts and elasmobranchs, with some of them being radiated onto, e.g., the mouth cavity and urinary bladder of (semi-)aquatic tetrapods (e.g., *Polystoma* sp. on Anura, *Polystomoides* sp. on Chelonia) (van Niekerk et al. 1993; Whittington et al. 2000; Cribb et al. 2002; Du Preez and Van Rooyen 2015). Their biodiversity is estimated at 25,000 with only a fraction (3,000–4,000) having been described to date and a vast majority of them being parasitic on bony fish species (Whittington 1998; Cribb et al. 2002). Two major taxonomic groups exist within the Monogenea, Monopisthocotylea and Polyopisthocotylea. They can be easily distinguished by the external structure of their opisthaptor, the hook or clamp bearing attachment organ at the posterior

R. Klapper (✉) • J. Münster • J. Kochmann • S. Klimpel • T. Kuhn
Goethe-University, Institute for Ecology, Evolution and Diversity,
Max-von-Laue-Str. 13, D-60438 Frankfurt/Main, Germany

Senckenberg Gesellschaft für Naturforschung, Senckenberg Biodiversity and Climate Research Centre, Senckenberganlage 25, D-60325 Frankfurt/Main, Germany
e-mail: r.klapper@em.uni-frankfurt.de

S. Klimpel et al. (eds.), *Biodiversity and Evolution of Parasitic Life in the Southern Ocean*, Parasitology Research Monographs, Vol. 9,
DOI 10.1007/978-3-319-46343-8_4

end of the parasite. The opisthaptor of the Monopisthocotylea (=Polyonchoinea) forms a single attachment unit consisting either of a large sucker bearing various types of hooks, or it consists entirely of large and small hooks (Whittington 2005). Polyopisthocotylean attachment organs bear several to numerous grasping units in the form of sclerotized clamps (Whittington 2005). The two monophyletic lineages evolved independently and have a common ancestor with either cestodes or trematodes (Olson and Tkach 2005).

Their monoxenous life cycle lacks intermediate hosts and asexual reproduction, however, species show a high degree of host specificity. Both groups, Monopisthocotylea and Polyopisthocotylea, are hermaphroditic, with most species being oviparous (egg-laying) and cross-fertilizing. The aquatic ciliated larvae (=Oncomiracidium) hatch from eggs, a process initiated by host-stimulated cues, and are either instantaneously infective for their hosts (same as for adult stages of the worm) or have to find a specific host and attach permanently to complete the life cycle (Bychowsky 1961; Whittington 2005). In most cases, each propagule develops into a single adult parasite. An exception can be found in species of the viviparous genus *Gyrodactylus*, in which, comparable to a "Russian doll" mode of reproduction (i.e., hyperviviparity), a fully grown daughter in utero of a parental individual encloses a developing embryo: boxed inside one another, this viviparous mode of reproduction allows an explosive population growth of Gyrodactylidae (Cable and Harris 2002; Whittington 2005).

Studies on biodiversity and prevalence of Antarctic monogeneans are very scarce. Descriptions were primarily conducted by Hargis and Dillon (1968a, b), Lyadov (1985), Parukhin and Lyadov (1979), and Rohde et al. (1995, 1998). The latter deserves additional mentioning, as they intensively dealt with biogeography of monogeneans and their evolution, with a major focus on Gyrodactylidae. The aim of this chapter is to provide an overview of the biodiversity of monogenean parasites in Antarctic marine vertebrate species. Data on the biodiversity were compiled from the literature and discussed in the context of distinct life cycle adaptations in the extreme Antarctic environment. Furthermore, some preliminary hypotheses on host specificity and parasite biogeography are presented.

4.2 Biodiversity of Antarctic Monogenea

A total of 23 monogenean species of 11 genera, 7 families, 4 orders, and 2 subclasses was found from 16 publications (Table 4.1). Fish hosts were representatives of 25 species, 8 genera, 3 families, 2 suborders, and 2 orders. The majority of fish host species belonged to the order Perciformes and two species to Gadiformes. The vast majority of fish were of the suborder Notothenioidei, which is also the most common suborder in the Antarctic region. Infection sites on the host included, in descending order, the gills (17 parasite species), skin and gills (3 species, *Allotetraonchoides rhigophilae*, *Pavlovskioides trematomi*, *P. wilkesensis*), skin (2 species, *Pseudobenedenia dissostichi*, *P. nototheniae*), and pelvic fin

Table 4.1 Reported monogeneans in Antarctic waters and subregions, including parasite and host species, sampling site, and the respective references

Parasite	Host	Location	Reference
Acanthocotylidae			
Acanthocotyle sp.	Unknown	Kerguelen Subregion	Lyadov (1985)
Tetraonchoididae			
Allotetraonchoides rhigophilae	*Lycodichthys dearborni*	McMurdo Sounds	Dillon and Hargis (1968)
	Lycodichthys antarcticus	Windmill Islands	Dillon and Hargis (1968)
Neopavlovskioides dissostichi	*Dissostichus eleginoides*	Bouvet Island, Crozet Island, Heard Island, Kerguelen Subregion, Lena Seamount, Ob Seamount	Parukhin and Lyadov (1979); Rohde et al. (1998)
	Dissostichus mawsoni	McMurdo Sound	Dillon and Hargis (1968)
	Unknown	Kerguelen Subregion	Lyadov (1985)
Neopavlovskioides georgianus	*Dissotichus eleginoides*	Falkland Islands, Heard Island, Macquarie Island, Prince Edward Island, Ross Sea, Shag Rock, South Georgia	Brickle et al. (2005, 2006)
Neopavlovskioides georgianus	Unknown	Glacial Subregion	Lyadov (1985)
Pavlovskioides antarcticus	*Trematomus bernacchii*	Casey Station, McMurdo Sound, Windmill Islands	Dillon and Hargis (1968); Rohde et al. (1995)
	Trematomus pennellii	McMurdo Sound, Windmill Islands	Dillon and Hargis (1968)
	Pagothenia borchgrevinki	McMurdo Sound	Dillon and Hargis (1968)
Pavlovskioides prudhoei	Unknown	Glacial Subregion	Lyadov (1985)
Pavlovskioides trematomi	*Trematomus loennbergii*	Prydz Bay	Rohde et al. (1998)
	Trematomus bernacchii	McMurdo Sound, Windmill Islands	Dillon and Hargis (1968)
	Trematomus pennellii	McMurdo Sound	Dillon and Hargis (1968)
	Trematomus hansoni	McMurdo Sound, Windmill Islands	Dillon and Hargis (1968)
	Trematomus lepidorhinus	McMurdo Sound	Dillon and Hargis (1968)
Pavlovskioides wilkesensis	*Trematomus bernacchii*	McMurdo Sound, Windmill Islands	Dillon and Hargis (1968)
	Trematomus hansoni	McMurdo Sound, Windmill Islands	Dillon and Hargis (1968)

(continued)

Table 4.1 (continued)

Parasite	Host	Location	Reference
Diclidophoridae			
Diclidophora antarctica	*Macrourus holotrachys*	Heard Island	Rohde et al. (1995)
Diclidophoridae gen. sp.	*Macrourus holotrachys*	Heard Island	Rohde et al. (1998)
Gyrodactylidae			
Gyrodactylus antarcticus	*Trematomus newnesi*	Davis Sea	Gusev (1967)
Gyrodactylus australis	*Trematomus scotti*	Princess Elizabeth Land	Gusev (1967)
	Trematomus eulepidotus	Lars Christensen and Princess Ragnhild coasts	Gusev (1967)
Gyrodactylus byrdi	*Trematomus newnesi*	Windmill Islands	Hargis and Dillon (1968a)
Gyrodactylus centronoti	*Trematomus pennelli*	McMurdo Sound	Hargis and Dillon (1968a)
Gyrodactylus coriicepsi	*Notothenia coriiceps*	Admiralty Bay	Rokicka et al. (2009)
Gyrodactylus nudifronsi	*Lepidonotothen nudifrons*	Admiralty Bay	Rokicka et al. (2009)
Gyrodactylus rhigophilae	*Lycodichthys dearborni*	McMurdo Sound	Hargis and Dillon (1968a)
Gyrodactylus trematomi	*Trematomus newnesi*	Windmill Islands	Hargis and Dillon (1968a)
Gyrodactylus wilkesi	*Trematomus bernacchii*	McMurdo Sound	Hargis and Dillon (1968a)
	Trematomus bernacchii	Windmill Islands	Hargis and Dillon (1968a)
	Trematomus hansoni	Windmill Islands	Hargis and Dillon (1968a)
Gyrodactylus sp.	*Lepidonotothen mizops*	Heard Island	Rohde et al. (1998)
	Lepidonotothen squamifrons	Macquarie Island, Heard Island, Prydz Bay	Rohde et al. (1995, 1998)
	Trematomus eulepidotus	Prydz Bay, Davis Station	Rohde et al. (1998)
Gyrodactylidae spp.	*Trematomus eulepidotus*	Prydz Bay, Davis Station	Rohde et al. (1995)
Diclidophoridae			
Macruricotyle claviceps	*Macrourus whitsoni*	South Shetland Islands, Weddell Sea	Walter et al. (2002)
	Unknown	Kerguelen Subregion	Lyadov (1985)
Mazocraeidae			
Neogrubea seriolellae	Unknown	Glacial Subregion	Lyadov (1985)

Table 4.1 (continued)

Parasite	Host	Location	Reference
Capsalidae			
Pseudobenedenia dissostichi	*Dissotichus eleginoides*	Falkland Islands, Heard Island, Shag Rock	Rohde et al. (1998); Brickle et al. (2005, 2006)
Pseudobenedenia nototheniae	*Notothenia coriiceps*	Admiralty Bay, King George Island, Potter Cove, Vernadsky Station	Palm et al. (1998); Zdzitowiecki and Laskowski (2004)
	Dissotichus mawsoni	Paradise Bay	Oguz et al. (2012)
	Lepidonotothen squamifrons	Heard Island	Rohde et al. (1998)
	Unknown	Patagonian Shelf	Lyadov (1985)
	Paranotothenia magellanica	Antipodes Island (Subantarctic Islands New Zealand)	Johnston (1931)
	Notothenia angustata	Antipodes Island (Subantarctic Islands New Zealand)	Johnston (1931)
	Trematomus bernacchii	McMurdo Sound, Windmill Islands	Hargis and Dillon (1968b)
	Notothenia microlepidota	Antipodes Island (Subantarctic Islands New Zealand)	Hargis and Dillon (1968b)
	Paranotothenia magellanica	Antipodes Island (Subantarctic Islands New Zealand)	Hargis and Dillon (1968b)
	Notothenia rossi	Macquarie Island, Kerguelen Subregion	Hargis and Dillon (1968b); Parukhin and Lyadov (1979)
	Dissotichus eleginoides	Kerguelen Subregion	Parukhin and Lyadov (1979)
	Notothenia rossi	Crozet Island, Heard Island	Parukhin and Lyadov (1979)
	Dissotichus eleginoides	Crozet Island	Parukhin and Lyadov (1979)
	Notothenia rossi	Ob Seamount, Skif Seamount	Parukhin and Lyadov (1979)
	Dissotichus eleginoides	Lena Seamount, Ob Seamount	Parukhin and Lyadov (1979)
	Gobionotothen gibberifrons	South Georgia	Parukhin and Lyadov (1979)
	Patagonotothen ramsayi	Falkland Island	Parukhin and Lyadov (1979)
Pseudobenedenia sp.	*Pagothenia borchgrevinki*	McMurdo Sound	Hargis and Dillon (1968b)
Pseudobenedenoides antarctica	Unknown	Glacial Subregion	Lyadov (1985)

(continued)

Table 4.1 (continued)

Parasite	Host	Location	Reference
Pseudobenedenoides shorti	*Trematomus bernacchii*	Casey Station	Rohde et al. (1998)
	Lycodichthys dearborni	McMurdo Sound	Hargis and Dillon (1968b)
	Trematomus hansoni	Casey Station, McMurdo Sound, Windmill Islands	Hargis and Dillon (1968b); Rohde et al. (1995, 1998)
	Trematomus bernacchii	Casey Station, McMurdo Sound, Windmill Islands	Hargis and Dillon (1968b); Moser and Cowen (1991), Rohde et al. (1995)
	Trematomus pennellii	Windmill Islands	Hargis and Dillon (1968b)
	Trematomus sp.	Ramp Cove	Hargis and Dillon (1968b)
	Unknown	Glacial Subregion	Lyadov (1985)
Hexabothriidae			
Rajonchocotyle sp.	Unknown	Kerguelen Subregion	Lyadov (1985)

Table 4.2 Mean prevalence (P) and intensity (I) range of major monogenean parasites in Antarctic fishes and the respective references

Group	References	mP (range)	*I*
Capsalidae	Rohde et al. (1995); Brickle et al. (2005, 2006); Palm et al. (1998); Zdzitowiecki and Laskowski (2004); Moser and Cowen (1991)	18 (2–17)	1–24
Gyrodactylidae	Rohde et al. (1995); Rokicka et al. (2009)	40 (15–65)	1–400
Tetraonchoididae	Rohde et al. (1995); Brickle et al. (2005, 2006)	41 (8–83)	1–100
Polyopisthocotylea	Rohde et al. (1995); Walter et al. (2002)	61 (15–100)	1–15

(1 species, *P. nototheniae*). Detailed information on parasite prevalence and intensities were only given in few publications. A summary of mean prevalence as well as intensity ranges for the four major monogenean groups in the Antarctic is shown in Table 4.2. Since only few values of small sample sizes were summarized, results can only hint rather than reflect real values. In all publications, species identification was almost exclusively based on morphological characteristics; only two gyrodactylid species were identified using molecular markers (Rokicka et al. 2009).

Monopisthocotylea constitute a high biodiversity in Antarctic waters whereas the subclass of Polyopisthocotylea was only represented by a few species. The biodiversity and taxonomy of both subclasses from the Antarctic are briefly introduced in the following sections.

4.2.1 Monopisthocotylea

Estimations assume that over 10,000 species of Monopisthocotylea may exist worldwide, but only approximately 2,500 species are described of which 1,000 species are marine (Whittington 2005). Monopisthocotylens are known to infect hosts on various microhabitats, primarily on external surfaces such as fins, skin, head, gills, and the oral cavity of fishes across many fish groups, feeding on host epithelial cells (Whittington 2005). Twenty species and four families of monogenean Antarctic fish parasites have been documented in the literature so far. Species of the families Gyrodactylidae and their sister group Capsalidae are the most strongly represented monogeneans. These two groups differ extremely in size, ranging from a size of 300 μm in some Gyrodactylidae to more than 3 cm in some Capsalidae (Whittington 2005).

4.2.1.1 Gyrodactylidae

The most diverse monogenean group in the Antarctic habitat is Gyrodactylidae (Table 4.1). Gyrodactylidae possess an exceptive role within the Monogenea: explosive radiation within the genera *Gyrodactylus* and *Dactylogyrus* are thought to be very likely the source of increased divergence (Olson and Tkach 2005).

Descriptions of nine *Gyrodactylus* species on Antarctic fish are available from the literature (Table 4.1). Almost all of them infect hosts of the family Nototheniidae genus *Trematomus*. Only *G. rhigophilae* parasitize *Rhigophila* of the family Zoarcidae. Rokicka et al. (2009) divided the species into two groups: a group with large anterolateral projections (*G. antarcticus*, *G. byrdi*, and *G. wilkesi*), and one without or with small anterolateral projections (*G. australis*, *G. centronoti*, *G. rhigophilae, G. trematomi, G. coriicepsi*, and *G. nudifronsi*). Due to their small size and lack of distinct taxonomic characters, morphological species identifications must be considered as extremely difficult (Hargis and Dillon 1968a).

4.2.1.2 Tetraonchoididae

Species belonging to the Tetraonchoididae are characterized by a single, blind caecum. The female reproductive organs are compact and comprise a single ovary, whereas male testes are elongated (Bychowsky et al. 1967; Whittington 2005). Seven species from Antarctic waters have been recorded. Most of them infected the genus *Dissostichus* and *Trematomus* within the family Nototheniidae, and Zoarcidae. One example is *Neopavlovskioides dissostichi* which was found to infect the skin of *D. elegenoides* at depths of 200–500 m (see Klimpel et al. (2009)). In a study by Brickle et al. (2005), a significant positive correlation between host length and abundance and an increase in prevalence with increasing length up to 40 cm was shown. The authors suggested that the increase might be attributed to the available surface area of larger fish (Brickle et al. 2005). *Neopavlovskioides georgianus* also parasitizes *D. elegenoides*, but abundance is not correlated with increasing host length (Brickle et al. 2005). This monogenean probably causes cysts of unknown etiology (CUEs) which are called “tumor of attachment” (Brickle et al. 2005, 2006).

4.2.1.3 Capsalidae

Capsalidae are characterized by highly branched intestines (Whittington 2005). About 200 capsalid species from nine subfamilies are described with characteristic camouflage, large size, and large host range (Whittington 2004). They are usually located on skin, fins, and gills of marine fish (Whittington 2004). Similar to the Gyrodactylidae, infection of Capsalidae also leads to high aggregations on infections on a particular host (Rohde et al. 1995). Four species of this family were found in Antarctic fish. *Pseudobenedenia nototheniae* is a common, large monogenean of Antarctic fishes which can attach to either skin or gills (Oguz et al. 2012). The monogenean was found in 11 host species of the family Nototheniidae and thus, had the highest number of different hosts. An exceptional case in larval development is *Pseudobenedenia shorti*: contrasting to most species of Capsalidae, *P. shorti* has nonciliated, fully developed larvae which hatch and remain on the same host and can only switch hosts by direct contact transfer (Rohde 1985).

4.2.2 Polyopisthocotylea

The group of Polyopisthocotylea consists of approximately 800 described, mainly marine living species (Hayward 2005). Polyopisthocotylea use fish, and rarely invertebrates such as copepods and isopods, as hosts. Despite their cosmopolitan distribution, most species of the group show high host specificity (Hayward 2005). The infestation sites of Polyopisthocotylea are gills and oral cavity where they feed on the host’s blood. Although Polyopisthocotylea are not as diverse as Monopisthocotylea, they can have a high prevalence in Antarctic fish (e.g., *Macruricotyle claviceps* in *Macrourus whitsoni*) (Walter et al. 2002). There are only a few documentations on Polyopisthocotylea from Antarctic fish. Today, four genera (*Diclidophora, Macruricotyle, Neogrubea, Rajonchocotyle)* of three families (Diclidophoridae, Mazocraeidae, Hexabothriidae) are known to occur in Antarctic and Sub-Antarctic waters. An example is *Neogrubea seriolellae,* a typical parasite of the gills. Its opisthaptor consist of clamps, suitable to stick to gill filaments, but not smooth body parts (Rohde 1984).

Macruricotyle claviceps is reported to parasitize different macrourid species. It was found on *Macrourus whitsoni* in Antarctic waters and on *M. carinatus* and *M. holotrachys* in Sub-Antarctic regions (Gaevskaya and Rodjuk 1988; Walter et al. 2002; Klimpel et al. 2009).

Macruricotyle whitsoni was originally described as *Diclidophora*, and later, after a revision by Rubec and Dronen (1994), transferred to the genus *Macruricotyle*. *Macrourus whitsoni*, a species endemic to the Antarctic Convergence, is the only known host of this species. The only record of *Macruricotyle whitsoni* on *Macrourus whitsoni* stems from the south-west coast of Argentina and Falkland Islands (Suriano and Martorelli 1984). However, Rubec and Dronen (1994) mentioned that the parasite should be compared with specimen of the closely related species *M. clavipes* also reported from these waters to avoid misidentification.

4.3 Host Specificity of Antarctic Monogenea

In general, monogeneans are considered to be among the most host specific parasites. Ectoparasites are usually more host- and site-specific which is related to the attachment organs and direct life cycle (Rohde and Heap 1998). In different studies on Antarctic monogeneans, some species were exclusively found on one host species while others infected a variety of different host fish species (Fig. 4.1). *Pseudobenedenia dissostichi* infected 11 different fish host species. Most Antarctic monogeneans parasitize fish of the suborder Nototheniidae, with *Trematomus bernacchii* being host to the highest number of different monogenean species. Members of the Capsalidae were found on Nototheniidae, Tetraonchoididae occurred mainly on the genus *Trematomus*. Macrouridae were only parasitized by Polyopisthocotylea. All species belonging to the subclass Polyopisthocotylea were reported from one host species (Lyadov 1985; Rohde et al. 1995; Walter et al. 2002) and thus, seem to be host-specific. Nototheniidae were generally infected, but are at the same time also best studied fishes, most likely due to their wide distribution and commercial importance in this area. Despite the limited data availability and resulting difficulty for assessment, it appears that Antarctic monogeneans tend to infect hosts that are closely related to each other. Such "phylogenetic specificity" has also been described by Rohde (1979), who suggested differing degrees of host specificity for marine monogeneans from all world oceans with 78 % restricted to single host species, 89 % to one genus, 96 % to one family, 98 % to one order (Rohde 1979).

Environmental conditions in Antarctic waters require special adaptations for reproduction and localization of a suitable host (Rohde and Heap 1998). Whittington et al. (2000) reviewed important factors that contribute to host specificity. Both reproduction and attachment have a decisive role to successful completion of the life cycle.

In comparison to cestodes and digeneans, monogeneans produce only few eggs, which requires larvae to possess successful strategies to find suitable hosts (Whittington et al. 2000). Under conditions of slowed metabolism such as in Antarctic waters, rhythmical hatching and hatching by host-generated cues are examples for such biological life cycle strategies. Monogenean larvae are attracted by the species-specific chemical composition of the host epidermis and mucus (Buchmann and Uldal 1997). Reports exist on monogenean eggs which only hatch in the presence of host mucus (Whittington et al. 1999). Another example is the

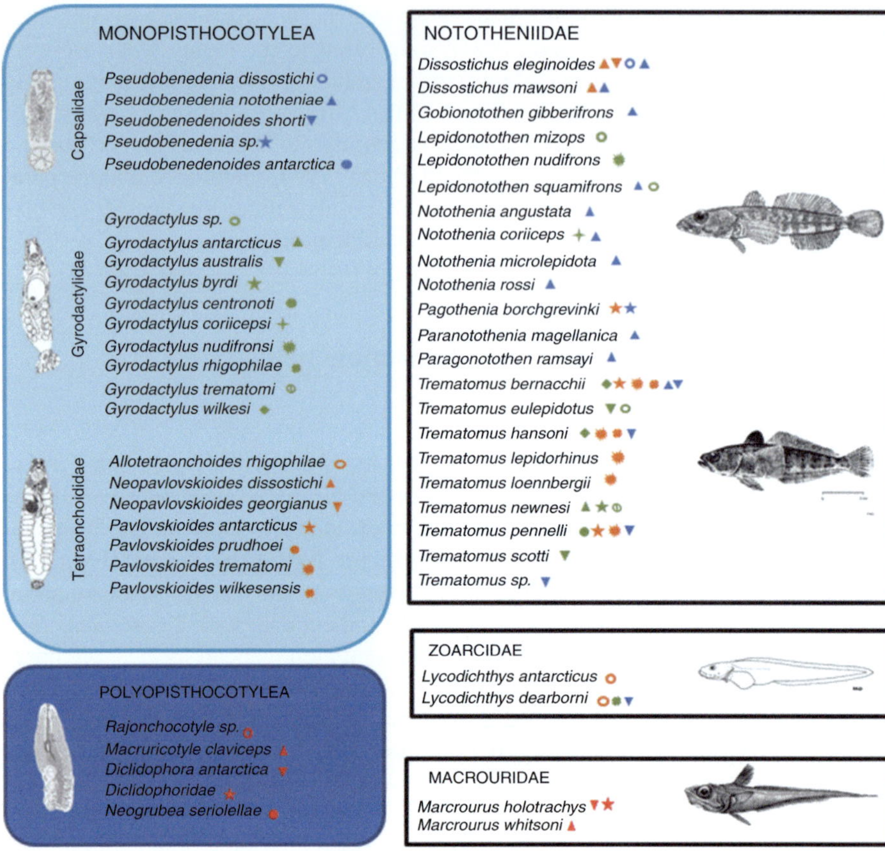

Fig. 4.1 Parasite-host documentation from Antarctic waters. *Left*: Monogenea classified by suborder and family. *Right*: Fish hosts classified by families. Symbols behind host names indicate a record of the parasite species as given by the symbol on the left side. Drawings of monogeneans (*top-down*): *P. shorti*, *G. wilkesi*, *P. antarctica*, *N. seriolellae* (Dillon and Hargis 1968; Hargis and Dillon 1968a, b; Hernández-Orts et al. 2014); *right*: *D. elegenoides*, *T. newnesi*, *L. antarcticus*, *M. whitsoni* (FAO species catalogue)

viviparity of Gyrodactylidae. This reproductive mode leads to aggregation of the parasite on its host with rapid population growth due to a reduced generation time (Rohde et al. 1995). The success of viviparity in cold-water habitats may be explained by the difficulty of small larvae to infect a suitable host in the vast areas of the ocean under reduced metabolism, and the difficulty of producing a sufficient number of eggs to "guarantee" survival of the next generation (Rohde 1985).

Besides larval host recognition, attachment plays an important role for the completion of the life cycle. Chemical recognition by the tegument of the adhesive area and chemical adhesives play an important role and are characteristic features of host-specificity (Buchmann and Uldal 1997). Epidermis and host products such as mucus can either be attractant to monogeneans or constitute an inhospitable habitat due to immunological activities (Whittington et al. 2000). Morphological specialization of the haptor results from adaptations to attachment and lead to high monogenean family-level diversity (Cribb et al. 2002).

Furthermore, environmental factors led to speciation, e.g., the Antarctic circumpolar current forms a physical barrier and functions as a driver for local populations of fish species and endemic parasitic helminths (Lyadov 1985). However, it has been noticed that sample sizes might have biased these conclusions as parasites previously considered as species-specific were less specific when more host species were examined (Whittington et al. 2000). This may certainly be the case for the sparsely studied Antarctic Monogenea (Table 4.1, Fig. 4.1).

4.4 Biogeography of Antarctic Monogenea

The distribution of sampled monogenean parasites in Antarctic and Subantarctic waters is shown in Fig. 4.2. Highest species number was recorded for the Ross Sea (McMurdo Sound), with representatives of three families. Most polyopisthocotyleans were documented around the Kerguelen Islands. Capsalidae were distributed

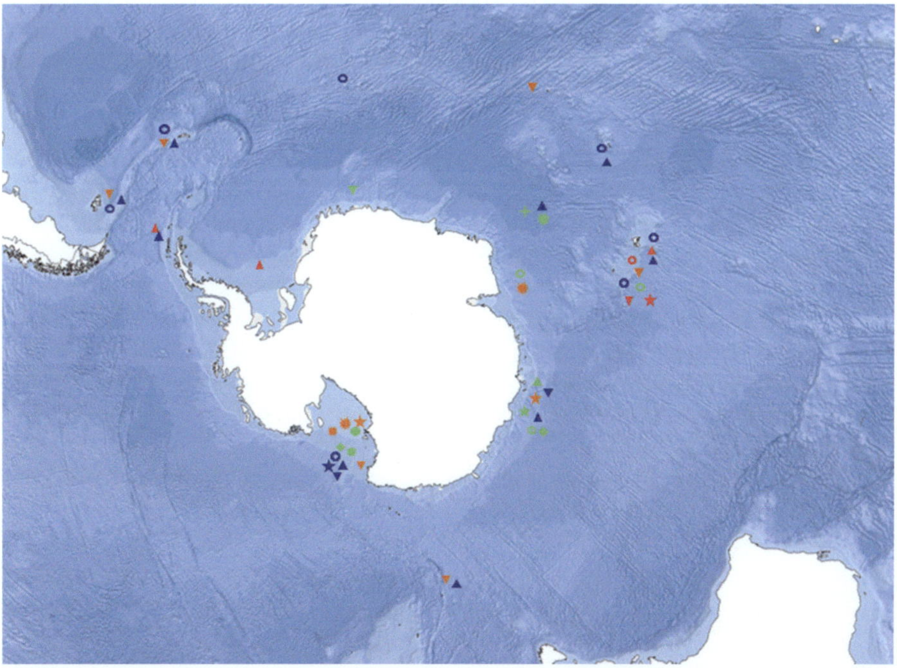

Fig. 4.2 Map of monogeneans from Antarctic waters. Each symbol represents a sampling site of a documented monogenean species. Color and symbol code in Fig. 4.1. Map source: ArcGIS

circumpolar, while Gyrodactylidae were mainly found on the Indo-Pacific site of the Southern Ocean and less on the Atlantic site. However, based on the limited sampling in only few locations, it is difficult to draw any conclusions on the general biogeography of monogenean species in this region. Possible patterns may as well be related to different sampling efforts. Hypotheses in the literature about parasitic traits of monogenean biogeography and diversity can be classified into two categories: differences in diversity of latitudinal gradients and oceans.

In this context, Rohde et al. (1995) should be mentioned due to his extensive work on biogeography and diversity of monogeneans. They conducted a metaanalysis on metazoan ectoparasites of marine fishes and found lowest ectoparasite richness and prevalences for Antarctic and deep-sea fishes. Of 102 fish species they examined, 86 are parasitized by at least one monogenean species. Only prevalences in Antarctic and New South Wales deep-water fishes are low, with approximately one third of uninfected fish species. In contrast to the ectoparasite species richness, relative and absolute numbers of Gyrodactylidae increase from a latitude of 64–65°N to more northern cold waters by 90 % (Rohde 1985). In Antarctic waters, this proportion is not as high with an increase of only about one third (Rohde 1985; Rohde et al. 1998). However, whether these differential observations are rather an artifact of different sampling efforts, with Arctic fish parasites being well sampled in contrast to Antarctic species, could not be evaluated (Rohde 1985). Based on these observations of Gyrodactylidae in both hemispheres, Rohde (1985) hypothesized that the increase of viviparity with latitude could serve as evidence for Thorson's rule. Thorson's rule states that nonpelagic development increases with latitude which had previously been shown for benthic invertebrates (e.g.,Thorson 1950; Mileikovsky 1971; Arnaud 1977). As temperature is correlated with latitude and has a major influence on metabolic and chemical processes, temperature was assumed to be one major influencing factor for monogenean distribution (Rohde et al. 1995).

In the Pacific and Atlantic Ocean, relative species diversity of gill monogenean of coastal marine fishes was greater in the northern and southwestern Pacific than in the northeastern and central and southwestern Atlantic (Rohde 1986). The described pattern was illusively recognizable in our map (Fig. 4.2), with a slight trend of more gyrodactylid species occurring in the Indo-Pacific part of the Southern Ocean than in the Atlantic part. Two hypotheses were suggested to explain this pattern. The first states that Gyrodactylidae accumulated in the older Pacific Ocean while the Atlantic Ocean is much younger. Another possibility could be that during the last glaciation the ice sheet cover of the Atlantic was higher than in the Pacific which led to higher abundances followed by more speciation events of Gyrodactylidae in the Pacific (Rohde 1986). No hypotheses were suggested for other monogenean families.

4.5 Concluding Remarks

Despite the limited number of studies, Antarctic monogeneans have shown a great diversity with most of them being host-specific. Adaptations to the extreme environment are reflected by their characteristic specializations in reproduction and attachment. However, due to the limited studies on Antarctic monogeneans, it is difficult to draw conclusions and hypotheses on host specificity and biogeography remain solely theoretical. Biodiversity may be underestimated: 197 marine fish species are currently known to exist in the Antarctic Convergence (Froese and Pauly 2016), and only 25 fish species have been described as hosts to monogeneans. Although morphological identification is often difficult due to their small sizes and poor morphological conservation, little effort has been made on the genetic validation of monogenean records.

References

Arnaud PM (1977) Adaptations within the Antarctic marine benthic ecosystem. Adapt Antarct Ecosyst Wash DC Smithson Inst 135:57

Bargelloni L, Ritchie PA, Patarnello T et al (1994) Molecular evolution at subzero temperatures: mitochondrial and nuclear phylogenies of fishes from Antarctica (suborder Notothenioidei), and the evolution of antifreeze glycopeptides. Mol Biol Evol 11:854–863

Brickle P, MacKenzie K, Pike A (2005) Parasites of the Patagonian toothfish, *Dissostichus eleginoides* Smitt 1898, in different parts of the Subantarctic. Polar Biol 28:663–671

Brickle P, MacKenzie K, Pike A (2006) Variations in the parasite fauna of the patagonian toothfish (*Dissostichus elegenoides* Smitt, 1898), with length, season, and depth of habitat around the Falkland Islands. J Parasitol 92:282–291

Buchmann K, Uldal A (1997) *Gyrodactylus derjavini* infections in four salmonids: comparative host susceptibility and site selection of parasites. Dis Aquat Organ 28:201–209

Bychowsky B (1961) Monogenetic trematodes: their systematics and phylogeny. American Institute of Biological Sciences, Washington

Bychowsky BE, Gusev AV, Nagibina LF (1967) Monogenetic trematodes of the family Tetraonchoididae Bychowsky, 1951. Trudy Zool Inst 35:140–166

Cable J, Harris PD (2002) Gyrodactylid developmental biology: historical review, current status and future trends. Int J Parasitol 32:255–280

Cribb TH, Chisholm LA, Bray RA (2002) Diversity in the Monogenea and Digenea: does lifestyle matter? Int J Parasitol 32:321–328

Dillon WA, Hargis WJ (1968) Helminth parasites of Antarctic vertebrates: part III. Monogenetic trematodes from Antarctic fishes: the superfamily Tetraonchoidea Yamaguti, 19631. Biol Antarct Seas III:101–112

Du Preez LH, Van Rooyen M (2015) A new polystomatid (Monogenea, Polystomatidae) from the mouth of the North American freshwater turtle *Pseudemys nelsoni* (539):1–9

Eastman JT (1993) Antarctic fish biology: evolution in a unique environment. Academic, San Diego

Froese R, Pauly D (2016) Fishbase. http://fishbase.org. Accessed Jan 2016

Gaevskaya AV, Rodjuk GN (1988) Ecological characteristics of the parasitofauna of the macrourid *Macrourus carinatus* Günther in the South Atlantic. Nauchnye Dokl Vysshei Shkoli Biol Nauki 2:21–25

Gusev AV (1967) Two new species of genus Gyrodactylus Nordmann (Monogenoidea) near off the coast of the Antarctic continent. Zoologicheskiy Institut Akademii Nauk SSSR. Explorations of the fauna of the seas IV (XII). Biol Rep Sov Antarct Exped 1955–1958 3:187–189

Hargis WJ, Dillon WA (1968a) Helminth parasites of Antarctic vertebrates: part II. Monogenetic trematodes from Antarctic fishes: the superfamily Gyrodactyloidea Johnston and Tiegs, 19221. Biol Antarct Seas III:91–99

Hargis WJ, Dillon WA (1968b) Helminth parasites of Antarctic vertebrates. Part IV. Monogenetic trematodes from Antarctic fishes: the superfamily Capsaloidea Price, 1936. Proc Biol Soc Wash 81:403–412

Hayward C (2005) Monogenea Polyopisthocotylea (ectoparasitic flukes). Mar Parasitol CSIRO Collingwood Aust CABI Oxon UK 55–63
Hernández-Orts JS, Alama-Bermejo G, Crespo EA et al (2014) A revision of the genus Neogrubea Dillon & Hargis, 1968 (Monogenea: Mazocraeidae): new morphological and molecular data from off the Patagonian coast of Argentina. Systematic parasitology 89: 59–72
Johnston TH (1931) New trematodes from the Subantarctic and Antarctic. Aust J Exp Biol Med Sci 8:91–98
Klimpel S, Busch MW, Kellermanns E et al (2009) Metazoan Deep Sea Fish Parasites, vol 11. Verlag Natur & Wissenschaft, Solingen
Klimpel S, Palm HW (2011) Anisakid nematode (Ascaridoidea) life cycles and distribution: increasing zoonotic potential in the time of climate change? In: Progress in parasitology. Springer, Berlin, pp 201–222
Kock K-H (1992) Antarctic fish and fisheries. Cambridge University Press, Cambridge
Lyadov VN (1985) Zoogeographical characteristics of the helminths of fishes from the Antarctic zone of the World Ocean. Parasitol Pathol Mar Org World Ocean NOAA Tech Rep NMFS 25:41–43
Mileikovsky SA (1971) Types of larval development in marine bottom invertebrates, their distribution and ecological significance: a re-evaluation. Mar Biol 10:193–213
Moser M, Cowen RK (1991) The effects of periodic eutrophication on parasitism and stock identification of *Trematomus bernacchii* (Pisces: Nototheniidae) in McMurdo Sound, Antarctica. J Parasitol 77:551–556
Oguz MC, Heckmann RA, Cheng CC et al (2012) Ecto and endoparasites of some fishes from the Antarctic Region. Sci Parasitol 13:119–128
Olson PD, Tkach VV (2005) Advances and trends in the molecular systematics of the parasitic Platyhelminthes. Adv Parasitol 60:165–243
Palm HW, Reimann N, Spindler M, Plötz J (1998) The role of the rock cod *Notothenia coriiceps* Richardson, 1844 in the life-cycle of Antarctic parasites. Polar Biol 19:399–406
Parukhin AM, Lyadov VN (1981) Parasitofauna of Notothenioidei from waters of the Atlantic and Indian Oceans. Vestn Zool 3: 90–94 (In Russian)
Rocka A (2006) Helminths of Antarctic fishes: life cycle biology, specificity and geographical distribution. Acta Parasitol 51:26–35
Rohde K (1985) Increased viviparity of marine parasites at high latitudes. Hydrobiologia 127:197–201
Rohde K (1984) Zoogeography of marine parasites. Helgol Mar Res 37:35–52
Rohde K (1986) Differences in species diversity of Monogenea between the Pacific and Atlantic Oceans. Hydrobiologia 137:21–28
Rohde K (1979) A critical evaluation of intrinsic and extrinsic factors responsible for niche restriction in parasites. Am Nat 114:648–671
Rohde K, Hayward C, Heap M (1995) Aspects of the ecology of metazoan ectoparasites of marine fishes. Int J Parasitol 25:945–970
Rohde K, Heap M (1998) Latitudinal differences in species and community richness and in community structure of metazoan endo-and ectoparasites of marine teleost fish. Int J Parasitol 28:461–474
Rohde K, Ho J-S, Smales L, Williams R (1998) Parasites of Antarctic fishes: Monogenea, Copepoda and Acanthocephala. Mar Freshw Res 49:121–125
Rokicka M, Lumme J, Zietara MS (2009) Two new Antarctic *Gyrodactylus* species (Monogenoidea): description and phylogenetic characterization. J Parasitol 95:1112–1119
Rubec LA, Dronen NO (1994) Revision of the genus *Diclidophora* Krøyer, 1838 (Monogenea: Diclidophoridae), with the proposal of *Macrouridophora* n.g. Syst Parasitol 28:159–185
Suriano DM, Martorelli SR (1984) Monogeneos parásitos de peces pertenecientes al Orden Gadiformes de la plataforma del mar Argentino. Rev Mus La Plata (Nueva Ser) 13 Secc Zool 140:195–210

Thorson G (1950) Reproductive and larval ecology of marine bottom invertebrates. Biol Rev 25:1–45

van Niekerk S, Kok DJ, Seaman MT (1993) A new species of Polystoma (Monogenea: Polystomatidae) parasitic in *Hyperolius marmoratus* (Anura: Hyperoliidae) in South Africa. Syst Parasitol 25:73–80

Walter T, Palm H, Piepiorka S, Rückert S (2002) Parasites of the Antarctic rattail *Macrourus whitsoni* (Regan, 1913)(Macrouridae, Gadiformes). Polar Biol 25:633–640

Whittington I (2005) Monogenea Monopisthocotylea (ectoparasitic flukes). In: Rohde K (ed) Marine Parasitology. CSIRO Publishing, Canberra

Whittington ID (1998) Diversity “down under”: monogeneans in the Antipodes (Australia) with a prediction of monogenean biodiversity worldwide. Int J Parasitol 28:1481–1493

Whittington ID (2004) The Capsalidae (Monogenea: Monopisthocotylea): a review of diversity, classification and phylogeny with a note about species complexes. Folia Parasitol (Praha) 51:109–122

Whittington ID, Chisholm LA, Rohde K (1999) The larvae of Monogenea (Platyhelminthes). Adv Parasitol 44:139–232

Whittington ID, Cribb BW, Hamwood TE, Halliday JA (2000) Host-specificity of monogenean (platyhelminth) parasites: a role for anterior adhesive areas? Int J Parasitol 30:305–320

Zdzitowiecki K, Laskowski Z (2004) Helminths of an Antarctic fish, Notothenia coriiceps, from the Vernadsky Station (Western Antarctica) in comparison with Admiralty Bay (South Shetland Islands). Helminthologia 41:201–207

Chapter 5
Biodiversity and Evolution of Digeneans of Fishes in the Southern Ocean

Anna Faltýnková, Simona Georgieva, Aneta Kostadinova, and Rodney A. Bray

5.1 Introduction

5.1.1 Historical Notes

Edward L. Atkinson (1881–1929), a Royal Navy surgeon and Antarctic explorer, was the doctor attached to Captain Robert Falcon Scott's tragic Antarctic Expedition (1910–1913) that sailed south aboard the ship '*Terra Nova*'. He accompanied the group of explorers who set off for the South Pole on the 1st November 1911. As planned, he went as far as the Beardmore Glacier and returned to base, leaving the Polar Party to go on to the South Pole, where they were forestalled, by 34 days, by a Norwegian party led by Roald Amundsen. Atkinson led two attempts to rescue Scotts's party, in February and March 1912, but was beaten back by the bad weather of the encroaching austral winter. Eventually, in October 1912, the winter over, he set out again, only to find the frozen bodies of the Polar Party on the 12th November (Cherry-Garrard 1922). While waiting at base camp at Cape Evans on the coast of the Ross Sea in the harsh 'exceedingly unfavourable' winter of 1911, he collected parasites from specimens of the nototheniid fish *Trematomus bernacchii* Boulenger, 1902, which he caught by 'digging a hole through the ice, and lowering a trap baited with seal-meat' (Leiper and Atkinson 1915). Thus was the study of digeneans of Antarctic marine fishes started. Five digeneans were found and returned to England,

A. Faltýnková • S. Georgieva • A. Kostadinova
Institute of Parasitology, Biology Centre of the Czech Academy of Sciences,
Branišovská 31, 370 05 České Budějovice, Czech Republic
e-mail: faltyn@paru.cas.cz; simona.georgieva@gmail.com; aneta.kostadinova@uv.es

R.A. Bray (✉)
Department of Life Sciences, Natural History Museum,
Cromwell Road, London SW7 5BD, UK
e-mail: r.bray@nhm.ac.uk

S. Klimpel et al. (eds.), *Biodiversity and Evolution of Parasitic Life in the Southern Ocean*, Parasitology Research Monographs, Vol. 9,
DOI 10.1007/978-3-319-46343-8_5

where they were described by R.T. Leiper and Atkinson, briefly in 1914 (Leiper and Atkinson 1914) and then in greater detail in 1915 (Leiper and Atkinson 1915). These worms are now known as *Elytrophalloides oatesi* (Leiper & Atkinson, 1914), *Genolinea bowersi* (Leiper & Atkinson, 1914), *Lepidapedon garrardi* (Leiper & Atkinson, 1914) and *Macvicaria pennelli* (Leiper & Atkinson, 1914) (syn. *Allocreadium fowleri* Leiper & Atkinson, 1914). The first two species were named for members of the fatal Polar Party, Lawrence ('Titus') Oates and Henry ('Birdie') Bowers.

No further significant work on Antarctic digeneans was published until after the Second World War, when Byrd (1963) described six (four new) worms from McMurdo Sound and Szidat (1965) and Szidat and Graefe (1967) described worms from the South Orkney and South Shetland Islands. The British Australian (and) New Zealand Antarctic Research Expedition (BANZARE) led by Sir Douglas Mawson (whose daughter, Patricia Mawson Thomas, became a parasitologist) collected fish parasites in the Antarctic between 1929 and 1931 and the collection eventually made its way to the Natural History Museum in London where it was written up by Prudhoe and Bray (1973). The most important worker on Antarctic fish digeneans, however, is, without doubt, Krzysztof Zdzitowiecki, who has published many papers, and in doing so has sorted out many of the taxonomic problems associated with the earlier descriptions of worms, often based on poorly prepared specimens. Much of his work is summarised in the important volume (Zdzitowiecki 1997), although he has published many significant studies subsequently.

5.1.2 *The Southern Ocean*

In terms of this study, the 'Southern Ocean' refers to those coastal and shelf waters listed under this heading by Spalding et al. (2007), which follow the molluscan zones suggested by Linse et al. (2006). For the pelagic zones we include the Antarctic and Antarctic Polar Front as delimited by Spalding et al. (2012). Our study covers three 'Provinces' of Spalding et al. (2007). Province number 59: 'Subantarctic Islands' covers the islands south of the Indian Ocean, such as Kerguelen, Crozet Island, Heard and MacDonald Islands and Prince Edward Island and some of those south of the Atlantic Ocean, such as Bovet Island and Peter the First Island. Province number 60: 'Scotia Sea' includes other island groups south of the Atlantic Ocean, such as the South Sandwich, South Georgia, South Orkney and South Shetland Islands, along with the Antarctic Peninsula. Province number 61: 'Continental High Antarctic' covers, as its name suggests the entire coast of continental Antarctica, except the Antarctic Peninsula, including the large inlets of the Ross and Weddell Seas. The final 'Southern Ocean' Province, number 62, 'Subantarctic New Zealand', has, as far as we are aware, no records of marine fish digeneans. In fact, most of the collecting has been done near the Antarctic mainland and the sub-Antarctic Islands south of the Indian and Atlantic Islands.

Each of the Provinces are divided into 'ecoregions', and we have used these to refine the distribution data.

5.1.3 Developing Knowledge of Digenean Diversity in the Southern Ocean

5.1.3.1 Exploring the Taxonomy of the Digeneans in the Southern Ocean

After the descriptions of the first five (later recognised as four valid) digenean species from the Ross Sea by Leiper and Atkinson (1914, 1915) their generic status was refined by Byrd (1963). Szidat (1965) and Szidat and Graefe (1967) reported eight species from fishes off the South Orkney and South Shetland Islands two of which, *Lecithaster macrocotyle* Szidat & Graefe, 1967 and *Neolebouria antarctica* (Szidat & Graefe, 1967), are now recognised under their original specific names. Prudhoe and Bray (1973) re-described a number of species and described seven species as new from off the Antarctic continent and Kerguelen Island. Gibson (1976) described two new species, *Discoverytrema markowskii* Gibson, 1976 and *Neolebouria georgiensis* Gibson, 1976, from South Georgia and erected the genus *Neolebouria* Gibson, 1976 to differentiate it from the similar genera *Podocotyle* Dujardin, 1845 and *Plagioporus* Stafford, 1904 splitting this complex into three. Kovaljova and Gaevskaya described *Macvicaria antarctica* (Kovaljova & Gaevskaya, 1974), *M. georgiana* (Kovaljova & Gaevskaya, 1974) and *Neolepidapedon magnatestis* (Gaevskaya & Kovaljova, 1976) from the southern part of the Atlantic Ocean (Gaevskaya and Kovaljova 1976; Kovaljova and Gaevskaya 1974).

In the 1980s a further three species of the families Derogenidae Nicoll, 1910 and Lepidapedidae Yamaguti, 1958 from the Sub-Antarctic Islands were described by Gaevskaya and Rodyuk (1983, 1988). Up to that time a total of 23 species had been described as new from the Antarctic or sub-Antarctic and seven species, *Aporocotyle nototheniae* Parukhin, 1985, *Derogenes varicus* (Müller, 1784), *Gonocerca haedrichi* Campbell & Munroe, 1977, *Gonocerca phycidis* Manter, 1925, *Genolinea nototheniae* (Kurochkin, 1975), *Glomericirrus macrouri* (Gaevskaya, 1973) and *Otodistomum cestoides* (van Beneden, 1871), which were originally described from elsewhere (North Atlantic Ocean, Indian Ocean), were recorded in the Antarctic or sub-Antarctic region.

Starting in 1977, Krzysztof Zdzitowiecki took part in a number of Polish Antarctic expeditions and participated in founding the Polish Antarctic Station of Henryk Arctowski in the Admiralty Bay of the King George Island (South Shetland Islands). He devoted his research to the Antarctic helminth fauna for nearly 40 years. His contribution is tremendous, within 11 years he described, alone or with co-authors, a total of 27 digenean species from fishes in South Georgia, South Shetland Islands, the Antarctic Peninsula, Weddell Sea, East Antarctic Wilkes Land and the Ross Sea. He was the first who collected the fish material personally, and thus obtained exact data on the prevalence and abundance, the previous studies being based on material collected mainly by non-specialists due to the harsh conditions (Zdzitowiecki 1978).

Zdzitowiecki (1987) transferred *Podocotyle pennelli* Leiper & Atkinson, 1914, one of the most common digeneans of the family Opecoelidae Ozaki, 1925, to *Macvicaria* Gibson & Bray, 1982. Zdzitowiecki (1990a) proposed the new combination *Macvicaria antarctica* (Kovaljova & Gaevskaya, 1974) and described two new species, *Macvicaria muraenolepidis* Zdzitowiecki, 1990 and *M. ophthalmolyci* Zdzitowiecki, 1990 from off the South Shetland Islands and South Georgia. Later, as he examined more material of *Macvicaria* spp., including the material of Leiper and Atkinson (1914), Zdzitowiecki et al. (1992) suggested that *M. pennelli* is restricted to the eastern Antarctica and *M. georgiana*, which he proposed as a new combination, to the western Antarctica. He considered as distinguishing features the larger sucker ratio in *M. pennelli* and 'probably differences in the arrangement of the vitelline follicles'. However, judging from the published figures, the differences in sucker ratios are not unambiguous. Furthermore, the specimens of *M. georgiana* in the original description of Kovaljova and Gaevskaya (1974) were found together with *M. pennelli* in the same host, *Notothenia rossii*, and in the 'Antarctic sector of the Atlantic Ocean', which apparently belongs to western Antarctica; these facts shed doubt on this geographical delimitation of the distribution of the two species.

Zdzitowiecki (1990a) constructed a key to the five known species based mainly on the size of the eggs, sucker ratio and the relative length of the forebody. Zdzitowiecki and co-authors (Zdzitowiecki 1990a, 1999; Zdzitowiecki & Cielecka 1997b; Laskowski et al. 2013) described in total six species of *Macvicaria* (including from off the Falkland Islands) and provided a key to the nine known Antarctic species (Laskowski et al. 2013; Zdzitowiecki 1997).

Neolebouria georgiensis Gibson, 1976 is another species which is very common in the Scotia Sea and Zdzitowiecki et al. (1993) believed that this species was restricted to the western Antarctic, whereas the smaller *N. terranovaensis* Zdzitowiecki, Pisano & Vacchi, 1993 was thought to occur only in the eastern Antarctic. However, later Zdzitowiecki (1997, 2003) and Sokolov and Gordeev (2015a) recorded *N. terranovaensis* in the Weddell Sea (western Antarctic). We should note that our new material of *N. georgiensis* sequenced was also collected from the western Antarctic (see below).

Zdzitowiecki also significantly contributed to the knowledge of the diversity of *Lepidapedon* Stafford, 1904 in the Antarctica. Zdzitowiecki (1990b) and Zdzitowiecki and Cielecka (1997a) described six of the seven known Antarctic species of the genus: *Lepidapedon notogeorgianum* Zdzitowiecki, 1990, *L. paralebouri* Zdzitowiecki 1990 and *L. tertium* Zdzitowiecki, 1990 from off South Shetland Islands (Zdzitowiecki 1990) and *L. balgueriasi* Zdzitowiecki & Cielecka, 1997, *L. brayi* Zdzitowiecki & Cielecka, 1997 and *L. ninae* Zdzitowiecki & Cielecka, 1997 from the Weddell Sea. These authors also provided a key to the Antarctic species of the 'Beveridgei subgroup' of *Lepidapedon*.

5.1.3.2 Faunistic and Ecological Studies

Zdzitowiecki has not only described more than half of the digeneans in fishes off Antarctica, but also contributed significantly to the knowledge of their ecology and distribution within the Antarctic and sub-Antarctic. As he was able to collect fishes in areas which were poorly studied (e.g. Adélie Land and the Ross Sea, which was neglected after the pioneer studies), the relatively evenly distributed records from all Antarctic and sub-Antarctic regions are greatly to his credit. By examining fishes during the whole year in South Shetlands and South Georgia, Zdzitowiecki (1988) obtained data on the prevalence and intensity of infection of digeneans, and also the size and habitat (depth) of the fish hosts. This allowed him to speculate on the most probable intermediate hosts of the digeneans, which are usually copepods, amphipods, benthic gastropods or bivalves and annelids. Zdzitowiecki (1991) found that demersal fish species were most heavily infected whereas pelagic fishes were usually not infected with digeneans.

Zdzitowiecki (1991) summarised the digenean species occurring in the Antarctic and Zdzitowiecki and White (1992) provided a list of species with a comparison of South Georgia with South Shetland Islands and found out that the species composition was the same in both areas, but there were substantial differences in prevalence and intensity. Zdzitowiecki et al. (1997) reported that the digenean fauna found off South Orkney Islands was similar to that off South Shetland Islands. Zdzitowiecki and Pisano (1996) examined fishes from Heard Island, Kerguelen sub-region (sub-Antarctica) and found new species records for this region.

Zdzitowiecki et al. (1998) examined the parasite fauna of fishes collected off Adélie Land (East Antarctic Wilkes Land), an area poorly studied before, and reported ten digenean species (seven new records); later Zdzitowiecki (2001) added more data on the prevalence and intensity of digeneans in fishes. Laskowski et al. (2007) continued collecting data from off Adélie Land and provided a checklist of helminths from *Trematomus newnesi* Boulenger, 1902 including seven digenean species. Zdzitowiecki and Ozouf-Costaz (2013) added more records from off Adélie Land providing a parasite-host list with *Genolinea bowersi* and *Neolebouria terranovaensis* being the most common species.

Zdzitowiecki et al. (1999) investigated the parasites of fishes in the Ross Sea, which by that time was much less studied than the Weddell Sea, and found 11 digenean species. Laskowski et al. (2005) added more records, including three digenean species from Bathydraconidae from the Ross Sea.

A series of papers in 2002–2003 by Zdzitowiecki (2002a, b, c, d, 2003) was devoted to the digeneans in the Weddell Sea with digeneans being most abundant in fishes of the families Nototheniidae and Channichthyidae.

Laskowski and Zdzitowiecki (2005) examined fishes from off the coastal area of the Antarctic Peninsula and found six species of digeneans with *Macvicaria georgiana* Kovaljova & Gaevskaya, 1974 being most abundant and indicated that the species spectrum was the same as off the South Shetland Islands, only the prevalences were lower.

In their most recent paper, Laskowski et al. (2014) evaluated changes in the digenean species spectrum in *Notothenia coriiceps* Richardson, 1844 from the Admiralty Bay, South Shetland Islands, after 30 years (samples from 1978 and 1979 compared with 2007 and 2008) and found that the species were more numerous and more diverse in the past. These authors concluded that pollution caused by human presence may have affected the invertebrates which serve as intermediate hosts in the life-cycles of the Antarctic digeneans.

Recently only two new species, *Aporocotyle michauda* Santoro, Cipriani, Pankov & Lawton, 2015 and *Paralepidapedon variabile* Sokolov & Gordeev, 2015 have been described, from the Ross Sea and the Amundsen Sea, respectively (Santoro et al. 2015; Sokolov and Gordeev 2015b). Sokolov and Gordeev (2013) described, but did not give a specific name to, the first zoogonid reported from the Antarctic, namely *Proctophantastes* sp. from the Ross Sea.

Few studies have been focused primarily on the ecology of Antarctic digeneans as an important element of the helminth faunas and communities. Holloway and Spence (1980) analysed the helminth fauna in three fish species in the Ross Sea and considered the differences in infection related to phylogenetic and ecological parasite-host relationships as the native *Trematomus pennellii* Regan, 1914 hosted the widest spectrum of digeneans in comparison with *Trematomus borchgrevinki* (Boulenger, 1902) which switched from benthic to pelagic life and *Licodichthys dearborni* (DeWitt, 1962), an immigrant to the Antarctic area. Differences in infection levels in *Trematomus bernacchii* from different localities in the Ross Sea were explained as related to eutrophication by Moser and Cowen (1991). The importance of *Notothenia coriiceps*, one of the most common nototheniid fishes with a circumpolar distribution, in the transmission of Antarctic parasites was examined by Palm et al. (1998). Santoro et al. (2014) studied the helminth infracommunities in *Chionodraco hamatus* (Lönnberg, 1905) in the Ross Sea and found that differences in behaviour between sexes during spawning most probably influence parasite infracommunities.

5.2 Digeneans in Fishes of the Southern Ocean: Taxonomic Diversity and Patterns of Host-Specificity

Our review of the diversity of digeneans in fishes of the Southern Ocean is based on a newly-developed database comprising 1204 host-parasite-locality records based on an extensive search of 80 publications in the literature, including the monograph of Zdzitowiecki (1997). Most of the papers deal with taxonomy, including descriptions of 53 new species, and the remaining provide faunistic and ecological data, including prevalence and intensity of infection with digeneans. We have updated the taxonomy and nomenclature of both the fish and digenean species, while excluding the records of innominate species (these were kept in a few cases where no other species of the genus have been recorded). Here we use the currently accepted names for fish and parasite species, following Froese and Pauly (2015) and in the World Register of Marine Species (WoRMS Editorial Board 2015), respectively. For the analysis of faunal richness distribution and similarity we used the classification (and coding) of provinces and marine ecoregions by Spalding et al. (2007) (see Fig. 5.2); this limited the data available for analyses to the records with precise locality descriptions. Nevertheless, the present data are characterised by a sufficiently ample coverage with respect to both taxonomic diversity and regional distribution to allow the endemic faunas in the endemic Antarctic fish hosts to be assessed with the application of quantitative approaches.

A total of 60 digenean species allocated to 28 genera in ten families have been described or reported from fishes in 15 marine ecoregions of the three provinces of the Southern Ocean, i.e. Sub-Antarctic Islands (Province 59), Scotia Sea (Province 60) and Continental High Antarctic (Province 61). No records exist for five ecoregions: Bouvet Island (217); Peter the First Island (218); South Sandwich Islands (219); East Antarctic Enderby Land (225); and East Antarctic Dronning Maud Land (226). Two digenean families, the Opecoelidae and the Lepidapedidae Yamaguti, 1958 with more than 60 % of the records, are distinctly the most diverse in the database (Table 5.1). The Hemiuridae Looss, 1899 appears well represented in the region (second according to the number of records) but with a much lower richness (only seven species). The most frequently recorded species are *Elytrophalloides oatesi* (Leiper & Atkinson, 1914) (13.0 % of all records), *Gonocerca phycidis* Manter, 1925 (10.8 %), *Lecithaster macrocotyle* Szidat & Graefe, 1967 (9.0 %) and *Genolinea bowersi* (Leiper & Atkinson, 1914) (8.9 %).

The 60 digenean species known from the Southern Ocean have been reported from 76 fish species of 41 genera in 13 families (Table 5.2), resulting in a mean richness of 0.80 digenean species per fish species. Fishes act as definitive hosts for all species; the only records of metacercariae are for the azygiid *Otodistomum cestoides*, a parasite of *Bathyraja* spp. using *Artedidraco skottsbergi* Lönnberg, 1905 and *Racovitzia glacialis* Dollo, 1900 as intermediate hosts in the region. By far the most species-rich fish family in the database is the Nototheniidae Günther, 1861 comprising nearly 40 % of the host species and represented in 60 % of the records. Three other families, the Channichthyidae Gill, 1861, Bathydraconidae Regan, 1913 and Artedidraconidae Eakin, 1988 appear relatively species-rich (9–10 species, collectively representing 31 % of the records), whereas the remaining 9 fish families comprise 1–4 species (Table 5.2).

The Nototheniidae is also the family harbouring the highest digenean diversity (38 species of 18 genera in 7 families), followed by the Channichthyidae (18 species of 12 genera in 5 families) and the Bathydraconidae (16 species of 13 genera in 7 families). The relative composition and digenean species richness for the three most species-rich host families are provided in Fig. 5.1. Five digenean families, are represented in all three fish families with opecoelids clearly prevailing with respect to species richness: 15 species in hosts of the Nototheniidae and 6 species in hosts of the Channichthyidae and Bathydraconidae each. Species of three digenean families are recorded in a single host family, the Aporocotylidae Odhner, 1912 and Monorchiidae Odhner, 1911 in hosts of the Nototheniidae and the Fellodistomidae

Table 5.1 Trematode families reported from fishes of the Southern Ocean, their generic and species richness and representation in the database

Family	No. of genera	No. of species	No. of records
Aporocotylidae Odhner, 1912	1	2	2
Azygiidae Lühe, 1909	1	1	8
Derogenidae Nicoll, 1910	2	5	182
Fellodistomidae Nicoll, 1909	2	3	8
Hemiuridae Looss, 1899	6	7	305
Lecithasteridae Odhner, 1905	2	3	134
Lepidapedidae Yamaguti, 1958	6	18	207
Monorchiidae Odhner, 1911	1	1	12
Opecoelidae Ozaki, 1925	6	19	345
Zoogonidae Odhner, 1902	1	1	1

Table 5.2 Fish families reported as hosts in Southern Ocean, their generic and species richness and representation in the database

Family	No. of genera	No. of species	No. of records
Arhynchobatidae	1	3	4
Artedidraconidae	3	9	47
Bathydraconidae	7	9	104
Bathylagidae	1	1	2
Channichthyidae	9	10	223
Congiopodidae	1	1	4
Harpagiferidae	1	2	12
Liparidae	1	3	9
Macrouridae	1	2	28
Muraenolepididae	1	2	40
Myctophidae	1	1	1
Nototheniidae	11	29	718
Zoarcidae	3	4	12

Nicoll, 1909 in hosts of the Bathydraconidae. Although the digenean fauna of nototheniids is generally more diverse in terms of means of number of species per family, the overall compositional pattern is similar (Fig. 5.1) indicating low effects of host-specificity at the host-family level.

The host-parasite list compiled from the present data was used to examine the patterns of host-specificity of the digeneans in the region. The database comprises a total of 416 host-parasite associations with an uneven distribution, i.e. 17 fish species (22 %) reported to harbour 10–17 digenean species and another 16 species (21 %) reported as hosts of single species. Given the limited overall digenean diversity, these data also indicate low levels of host-specificity. Indeed, 14 digeneans were reported in 10–38 fish hosts with eight species exhibiting an extremely low host-specificity: *Elytrophalloides oatesi* (with 38 hosts reported), *Genolinea bowersi* (33 hosts), *Lepidapedon garrardi* (29 hosts), *Gonocerca phycidis* (28 hosts), *Lecithaster macrocotyle* and *Macvicaria georgiana* (26 hosts each), *Neolebouria*

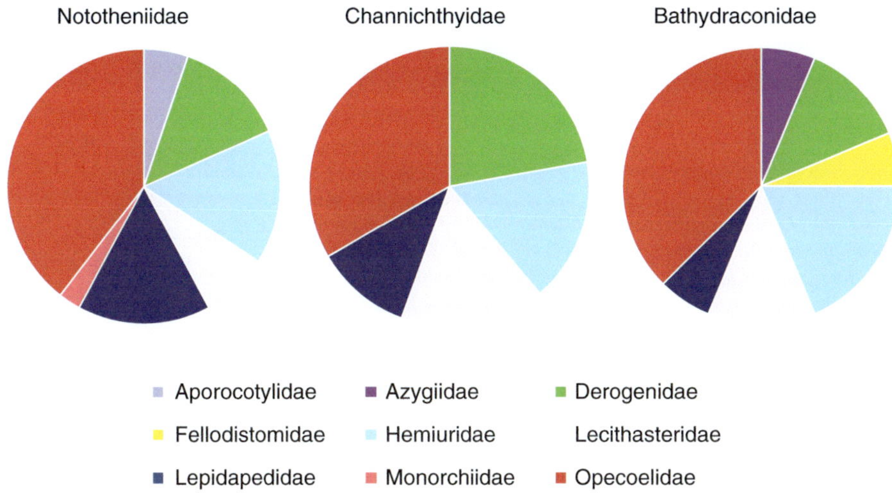

Fig. 5.1 Relative composition of the digenean faunas in fishes of the three most species-rich host families

georgiensis (21 hosts) and *Neolebouria antarctica* (20 hosts). Another group of 25 digenean species has been recorded in 2–9 fish hosts (Table 5.3). Although 22 digenean species have been recorded in a single host species, this figure reflects poor sampling rather than a pattern of strict host-specificity since more than a half of the species in this group (13) have been recorded only once. These data suggest that digeneans in fishes of the Southern Ocean typically exhibit low levels of host-specificity.

5.3 Contemporary Diversity Assessment: Current Baselines

5.3.1 Importance of Baselines for Prediction of Global Changes

The completion of the complex digenean life-cycle relies on the health and availability of all the hosts in the life-cycle. Any reduction in the viability of a host in the sequence will result in the reduction, or elimination, of the digeneans in the other hosts. Any perturbation of the environment, such as the effects of global warming on the Antarctic Seas and ice-sheets, will inevitably be reflected in the occurrence, prevalence and intensity of the digeneans found in the definitive hosts, fishes in the case of the present study. For this to be a useful indicator of environmental health a baseline is needed, but the picture of diversity, distribution and host associations of digeneans in the Southern Ocean is still disturbingly incomplete. Due to logistic challenges for field assessments a rather small number of regionally limited inventories have been carried out and then predominantly focused on taxonomy,

Table 5.3 Host-specificity of the digeneans reported in the Southern Ocean based on the records in the present database

Digenean species	No. of host species
Aporocotyle michauda; *Aporocotyle nototheniae*; *Boreascotia megavesicula*; *Genolinea nototheniae*; *Gibsonia hastata*; *Gonocerca haedrichi*; *Helicometra pisanoae*; *Lecithochirium whitei*; *Lepidapedon brayi*; *Lepidapedon ninae*; *Lepidapedon paralebouri*; *Lepidapedon tertium*; *Macvicaria skorai*; *Neolepidapedon macquariensis*; *Paralepidapedon antarcticum*; *Paralepidapedon awii*; *Paralepidapedon variabile*; *Steringophorus arntzi*; *Aphanurus* sp.; *Fellodistomum* sp.; *Proctophantastes* sp.; *Stenakron* sp.	1
Genolinea bowersi; *Gonocerca muraenolepisi*; *Helicometra rakusai*; *Macvicaria longibursata*; *Macvicaria muraenolepidis*; *Macvicaria ophthalmolyci*; *Muraenolepitrema magnatestis*; *Paralepidapedon dubium*; *Paralepidapedon lepidum*; *Steringophorus liparidis*	2
Discoverytrema gibsoni; *Lepidapedon notogeorgianum*; *Postlepidapedon opisthobifurcatum*	3
Caudotestis glacialis; *Caudotestis kerguelensis*	4
Discoverytrema markowskii; *Lecithophyllum champsocephali*; *Macvicaria antarctica*; *Otodistomum cestoides*	5
Helicometra antarcticae	6
Lepidapedon balgueriasi; *Neolepidapedon magnatestis*	7
Lecithaster micropsi; *Postmonorchis variabilis*	8
Derogenes varicus	9
Macvicaria microtestis; *Macvicaria pennelli*	10
Derogenes johnstoni; *Neolebouria terranovaensis*	12
Glomericirrus macrouri; *Neolepidapedon trematomi*	13
Neolebouria antarctica	20
Neolebouria georgiensis	21
Lecithaster macrocotyle; *Macvicaria georgiana*	26
Gonocerca phycidis	28
Lepidapedon garrardi	29
Genolinea bowersi	33
Elytrophalloides oatesi	38

identification and morphological characterisation of the digeneans. As a result, practically no historical baseline data exist that could be used for assessing trends of change in digenean diversity and distribution and/or host associations and prevalence of individual species. Nevertheless, the number of species recorded in the present database shows an increase of 36 % in the nearly 20 years since the last review of the digenean fauna of Antarctic fishes (Zdzitowiecki 1998; see also detailed baseline in Zdzitowiecki 1997). Therefore, we provide here a list of species in addition to the species richness mapped by ecoregion to serve as potential contemporary baselines for digenean diversity in the ecoregions of the Southern Ocean (Table 5.4). Almost all of the species (93.3 %) are endemic to the Southern Ocean (see Zdzitowiecki 1997, 1998).

Table 5.4 Potential contemporary baselines for digenean diversity in the ecoregions of the Southern Ocean

Digenean species	Ecoregion
Caudotestis kerguelensis; *Derogenes varicus*; *Glomericirrus macrouri*; *Gonocerca phycidis*; *Lecithaster macrocotyle*; *Lecithophyllum champsocephali*	212 Macquarie Island
Caudotestis kerguelensis; *Derogenes varicus*; *Elytrophalloides oatesi*; *Glomericirrus macrouri*; *Gonocerca phycidis*; *Lecithaster macrocotyle*; *L. micropsi*; *Lecithophyllum champsocephali*; *Macvicaria antarctica*; *Otodistomum cestoides*	213 Heard and Macdonald Islands
Caudotestis kerguelensis; *Derogenes varicus*; *Elytrophalloides oatesi*; *Genolinea bowersi*; *Gonocerca muraenolepisi*; *G. phycidis*; *Lecithaster macrocotyle*; *Lecithophyllum champsocephali*; *Macvicaria antarctica*; *Neolebouria antarctica*; *Neolepidapedon macquariensis*; *N. magnatestis*; *Postmonorchis variabilis*	214 Kerguelen Islands
Aporocotyle nototheniae; *Derogenes varicus*; *Elytrophalloides oatesi*; *Genolinea nototheniae*; *Gonocerca phycidis*; *Lecithaster macrocotyle*; *Lecithophyllum champsocephali*; *Macvicaria antarctica*; *Neolepidapedon macquariensis*; *N. magnatestis*; *Postmonorchis variabilis*	215 Crozet Islands
Derogenes varicus; *Elytrophalloides oatesi*; *Glomericirrus macrouri*; *Gonocerca phycidis*; *Lecithophyllum champsocephali*; *Neolepidapedon magnatestis*; *Stenakron* sp.	216 Prince Edward Islands
Boreascotia megavesicula; *Discoverytrema gibsoni*; *D. markowskii*; *Elytrophalloides oatesi*; *Fellodistomum* sp.; *Genolinea bowersi*; *Gibsonia hastata*; *Gonocerca phycidis*; *Lecithaster macrocotyle*; *L. micropsi*; *Lecithochirium whitei*; *Lepidapedon garrardi*; *L. notogeorgianum*; *L. paralebouri*; *Macvicaria antarctica*; *M. georgiana*; *M. muraenolepidis*; *M. skorai*; *Muraenolepitrema magnatestis*; *Neolebouria antarctica*; *N. georgiensis*; *Neolepidapedon magnatestis*; *N. trematomi*; *Paralepidapedon lepidum*; *Postlepidapedon opisthobifurcatum*; *Postmonorchis variabilis*	220 South Georgia
Genolinea bowersi; *Lecithaster macrocotyle*; *Lepidapedon garrardi*; *Macvicaria georgiana*; *Neolebouria antarctica*	221 South Orkney Islands
Caudotestis glacialis; *Derogenes varicus*; *Discoverytrema gibsoni*; *D. markowskii*; *Elytrophalloides oatesi*; *Genolinea bowersi*; *Glomericirrus macrouri*; *Gonocerca haedrichi*; *G. phycidis*; *Lecithaster macrocotyle*; *Lecithophyllum champsocephali*; *Lepidapedon garrardi*; *L. notogeorgianum*; *L. tertium*; *Macvicaria georgiana*; *M. ophthalmolyci*; *Neolebouria antarctica*; *N. georgiensis*; *Neolepidapedon trematomi*; *Paralepidapedon awii*	222 South Shetland Islands

(continued)

Table 5.4 (continued)

Digenean species	Ecoregion
Caudotestis glacialis; *Elytrophalloides oatesi*; *Genolinea bowersi*; *Gonocerca phycidis*; *Lecithaster macrocotyle*; *Lepidapedon garrardi*; *Macvicaria georgiana*; *Neolebouria antarctica*; *Neolepidapedon trematomi*	223 Antarctic Peninsula
Caudotestis glacialis; *Derogenes johnstoni*; *Discoverytrema gibsoni*; *Elytrophalloides oatesi*; *Genolinea bowersi*; *Gonocerca phycidis*; *Helicometra antarcticae*; *H. pisanoae*; *Lecithochirium whitei*; *Lepidapedon balgueriasi*; *L. garrardi*; *Macvicaria microtestis*; *M. muraenolepidis*; *M. pennelli*; *Muraenolepitrema magnatestis*; *Neolebouria terranovaensis*; *Neolepidapedon trematomi*; *Paralepidapedon antarcticum*; *P. dubium*; *Postlepidapedon opisthobifurcatum*	224 East Antarctic Wilkes Land
Aphanurus sp.; *Caudotestis glacialis*; *Derogenes johnstoni*; *Elytrophalloides oatesi*; *Genolinea bowersi*; *Glomericirrus macrouri*; *Gonocerca phycidis*; *Helicometra rakusai*; *Lepidapedon balgueriasi*; *L. brayi*; *L. garrardi*; *L. ninae*; *Macvicaria georgiana*; *M. longibursata*; *M. microtestis*; *Muraenolepitrema magnatestis*; *Neolebouria terranovaensis*; *Neolepidapedon trematomi*; *Otodistomum cestoides*; *Paralepidapedon awii*; *Postlepidapedon opisthobifurcatum*; *Steringophorus arntzi; S. liparidis*	227 Weddell Sea
Discoverytrema markowskii; *Glomericirrus macrouri*; *Gonocerca phycidis*; *Lecithaster micropsi*; *Paralepidapedon variabile*	228 Amundsen/Bellingshausen Sea
Aporocotyle michauda; *Derogenes johnstoni*; *D. varicus*; *Discoverytrema gibsoni*; *D. markowskii*; *Elytrophalloides oatesi*; *Genolinea bowersi*; *Glomericirrus macrouri*; *Gonocerca phycidis*; *Helicometra antarcticae*; *H. pisanoae*; *H. rakusai*; *Lecithaster macrocotyle*; *L. micropsi*; *Lepidapedon balgueriasi*; *L. garrardi*; *Macvicaria georgiana*; *M. muraenolepidis*; *M. pennelli*; *Muraenolepitrema magnatestis*; *Neolebouria antarctica*; *N. georgiensis*; *N. terranovaensis*; *Neolepidapedon trematomi*; *Otodistomum cestoides*; *Paralepidapedon dubium*; *P. lepidum*; *Postlepidapedon opisthobifurcatum*; *Proctophantastes* sp.; *Steringophorus liparidis*	229 Ross Sea

The impact of climate change on digeneans can be so far assessed only indirectly from the impact on their putative hosts, as their life-cycles in Antarctic waters are not known. It is reasonable to assume that trematodes are bound to gastropods, bivalves, amphipods and annelids, which are used as first and second intermediate hosts (Zdzitowiecki 1988). Only in the case of *Neolebouria georgiensis* has it been shown that the intermediate hosts are crustaceans of the family Mysidae (see Gaevskaya 1982). The only study on the influence of temperature on cercariae of

Antarctic digeneans was carried out by Graefe (1971) who examined unidentified opecoelid cercariae from the gastropod *Margarella* Thiele, 1893 from the coast of the Antarctic Peninsula and found a striking adaptation to cold temperatures (–4 to –1.7 °C in water during the austral summer) when emergence of cercariae from snails took place at 0–1 °C under experimental conditions. Usually, in temperate climatic conditions trematode development within snail hosts stops at temperatures below 12–14 °C (Dönges 1964; Olson 1966). Further, Graefe (1971) recorded a pronounced mortality of the opecoelid cercariae at artificially elevated temperatures (30–32 °C). At slightly lower temperatures (up to 27 °C) they were able to survive for a short time indicating, perhaps, a potential resilience to elevated temperatures.

5.3.2 Spatial Distribution of Host and Parasite Diversity

The present database provides useful information for an assessment of the regional distribution of host and digenean species. The digenean and host richness mapped in Fig. 5.2, relative to the 18 ecoregions considered, reveal a generally concordant pattern with high richness of both faunas in three ecoregions of the Continental High Antarctic Province (224, 227 and 229) and two ecoregions of the Scotia Sea Province (220 and 222). In an analysis of the geographical patterns of species richness of gastropods and bivalves (intermediate hosts for the digeneans) in the Southern Ocean, Linse et al. (2006) identified as hotspots of taxonomic richness South Georgia (ecoregion 220), Weddell Sea (ecoregion 227) and Ross Sea (ecoregion 229), all characterised by high pelagic productivity (Atkinson et al. 2001; Brierley and Thomas 2002). These findings are concordant with the digenean richness pattern inferred from the present database, identifying these three ecoregions as host fish and digenean richness hotspots.

A breakdown of fish and digenean diversity by province (Table 5.5) indicates an overall higher taxonomic diversity in the Continental High Antarctic Province and the Scotia Sea Province compared with the third (Sub-Antarctic Islands). There was a strong significant correlation between species richness and sampling effort (measured as the number of studies) across ecoregions (Spearman's rho 0.873 and 0.916 for digenean and host species, respectively; both $P<0.05$), the means for both measures of richness being the highest for the Scotia Sea Province. Nevertheless, no significant differences between provinces were found for all measures listed in Table 5.5.

The multidimensional scaling plot of ecoregional faunas based on Jaccard similarities (presence-absence data) in Fig. 5.3 provides a graphical illustration of the patterns of geographical variation of host fish and digenean faunas. The ecoregions of the three provinces formed three clusters except for the fish and digenean faunas of the Amundsen/Bellingshausen Sea (ecoregion 228) which appeared dissimilar from those in the remaining ecoregions. Surprisingly, faunas in distant ecoregions

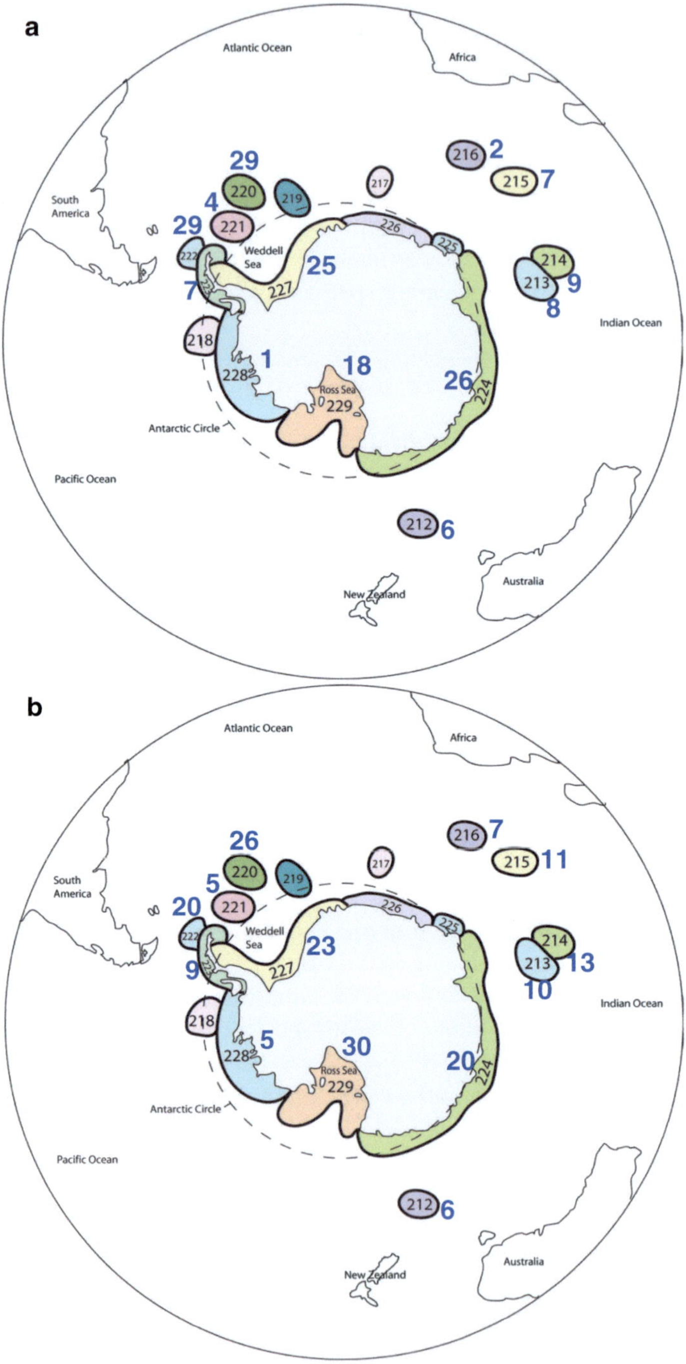

Fig. 5.2 Fish (**a**) and digenean (**b**) species richness in 13 ecoregions of the Southern Ocean based on the present database

Table 5.5 Distribution of host fish and digenean diversity in the three provinces represented in the database

	Subantarctic Islands		Scotia Sea		Continental high Antarctic	
Variable/Province	Range	Mean	Range	Mean	Range	Mean
No. of fish families	2–4	3.0	2–8	5.0	1–9	6.8
No. of fish genera	2–8	5.4	3–21	12.3	1–16	10.8
No. of fish species	2–9	6.4	4–29	17.3	1–26	17.5
No. of digenean families	4–7	5.4	4–7	5.3	4–9	6.0
No. of digenean genera	6–11	8.6	5–17	11.3	5–19	14.0
No. of digenean species	6–13	9.4	5–26	15.0	5–30	19.5
No. of studies	1–7	4.0	3–21	12.3	2–17	10.3
No. of records	7–104	43.6	15–308	145.5	5–154	101.0

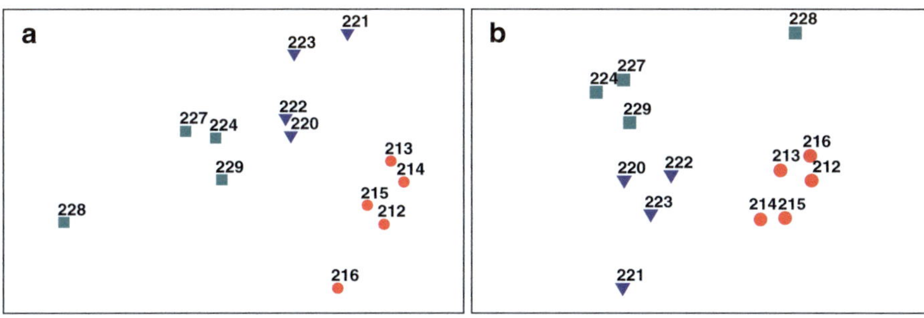

Fig. 5.3 Multidimensional scaling plot of the 13 ecoregional host fish (**a**) and digenean (**b**) faunas based on Jaccard similarities (presence-absence data)

were closely linked (227 Weddell Sea vs. 229 Ross Sea and 224 East Antarctica Wilkes Land for both hosts (Fig. 5.3a) and digeneans (Fig. 5.3b) whereas faunas in neighbouring ecoregions appeared more dissimilar than those in more distant ecoregions (fish faunas: 220–221 and 222–223 vs. 220–222 and 215–216 vs. remaining ecoregions within the Sub-Antarctic Islands Province; digenean faunas: 220–221 vs. remaining ecoregions within the Scotia Sea Province).

Even considering in the further analyses the clear outlier (ecoregion 228) where a single host species has been sampled in just two studies, there was a significant differentiation of both fish and digenean faunas of the three provinces (ANOSIM test, Global $R=0.800$ and 0.722, respectively; both $P=0.001$) (Fig. 5.3). The poorer faunas of the Sub-Antarctic Islands Province formed a tight cluster based on significant dissimilarity in the contrasts with the faunas in the ecoregions of the Scotia Sea Province ($R=0.894$ and 0.925, respectively, both $P=0.008$) and the Continental High Antarctic Province ($R=0.950$ and 0.763, respectively; both $P=0.008$). Finally, although there was a significant differentiation between fish faunas of the Scotia Sea and the Continental High Antarctic provinces ($R=0.536$, $P=0.029$), their digenean faunas were similar ($R=0.359$, $P=0.057$). Given the distinctly low levels of host-specificity of the most widely distributed digenean species, the regional distribution of the digenean faunas appears to support the major divisions within the Southern Ocean in the classification of Spalding et al. (2007). Furthermore, the concordant similarity patterns of host and digenean faunas indicate that these may reflect, in spite of the sampling bias, underlying general biogeographical patterns.

It is worth noting that ecoregion 228 (Amundsen/Bellingshausen Sea) located in the Pacific sector of the Southern Ocean is difficult to assess because of its remote location and pack ice coverage throughout most of the year (Barnes and Hillenbrand 2010); it is the least studied, in every aspect, Antarctic continental shelf region (apart from those regions, 217, 218, 219, 225 and 226, not studied at all). One of the few expeditions dealing with examination of the macro- and megafaunal benthic assemblages in the Amundsen Sea revealed a low species richness of gastropods and bivalves (acting as first intermediate hosts for digeneans) as opposed to the high richness of bryozoans, echinoids and ophiuroids and a numerical dominance by echinoderms. Furthermore, these authors indicated a clear difference in species composition and structure of the benthic assemblages in the Amundsen Sea from those in the Weddell, Ross and Scotia Seas as well as a much lower species richness (Linse et al. 2016). One possible explanation of the lower benthic species diversity could be that the Amundsen/Bellingshausen Sea shelf was probably completely covered by the grounded West Antarctic Ice Shield (see Barnes and Hillenbrand 2010; Lowe and Anderson 2002; Dowdeswell et al. 2006; Evans et al. 2006; Larter et al. 2009) and there were no glacial refugia, thus the area had to be re-colonised. Whatever the cause, the low mollusc richness and abundance can substantially influence digenean diversity and transmission efficiency, respectively, and it is plausible to suggest that the low richness and dissimilarity of the digenean fauna of fishes in the Amundsen Sea reflect the structure and specific composition of the benthic assemblages in this ecoregion. Future sampling is required, however, to test this hypothesis.

Whereas the finding that the digenean faunas of the Sub-Antarctic Islands Province are both species-poor and dissimilar from those of the mainland apparently reflects general island-mainland connectivity patterns, the high similarity between the faunas of Weddell Sea (ecoregion 227) and those of the non-connected distant Ross Sea (229) and East Antarctic Wilkes Land (224), is striking. Barnes and Hillenbrand (2010) compared the composition of modern bryozoan assemblages around Antarctica and revealed a similarly striking similarity between the bryozoan faunas of the shelves of the Weddell and Ross Seas which was, as in the present study (see Fig. 5.3), greater than between those from regions in close proximity, the Antarctic Peninsula shelf (ecoregion 223) and the South Shetland Islands (ecoregion 222). These authors suggested that this key finding implies a past direct connectivity between bryozoans from the two regions ('Weddell-Ross link') favouring the hypothesis for a late Quaternary trans-Antarctic seaway connecting the Weddell and Ross Seas which opened in response to a collapse of the West Antarctic Ice Sheet (WAIS) and allowing dispersal and faunal exchange between the two regions.

The finding that the bryozoan faunas of Weddell and Ross Sea are more similar than any other area in close proximity and areas assumed to be connected by refuges during the Last Glacial Maximum (Barnes and Hillenbrand 2010) agrees with the assumption of at least one WAIS collapse during the last 1.1 million years (Hillenbrand et al. 2009; Pollard and DeConto 2009). Another study corroborating the hypothesis of Barnes and Hillenbrand (2010) is the report of strong similarities between the shelf gastropod and bivalve faunas of the Ross Sea and Weddell Sea (see Linse et al. 2006). Past connectivity of the gastropod and bivalve faunas agrees well with the similarity patterns in our dataset thus providing support for the trans-Antarctic seaway hypothesis from yet another taxonomically and phylogenetically unrelated fauna.

5.4 Phylogenetic Framework: First Molecular Data for Digeneans in the Southern Ocean

The predominantly endemic character of the digenean fauna in fishes of the Southern Ocean raises the question of whether the phylogenetic relationships within the corresponding higher taxa (or the Digenea in general) would change as a result of increased taxon sampling in this region. In the phylogenetic hypotheses presented here, we used the first sequence data for the 28S rRNA gene from our recent study on fish digeneans from off James Ross Island (ecoregion 223) and attempted an assessment of the interrelationships of the digeneans within the three major digenean families characterised by a high taxonomic diversity in the Southern Ocean. Published 28S rDNA sequences for a range of species within the three higher-level taxa available from the GenBank database (sequence IDs incorporated in the figures) are also included in the phylogenetic analyses.

The phylogenetic hypothesis on the relationships within the Hemiuridae provided in Fig. 5.4 is based on an analysis of 15 species of seven subfamilies of the Hemiuridae and eight species of the Lecithasteridae Odhner, 1905 (4 spp.), Derogenidae (3 spp.) and Sclerodistomidae Odhner, 1927 (1 sp.). This dataset represents an expanded subset of the hemiuroidean taxa analysed by Pankov et al. (2006) (however omitting the Didymozoidae Monticelli, 1888, one accacoeliid and one syncoeliid) to which we have added newly-generated sequences for three digeneans: the elytrophalline *Elytrophalloides oatesi* (type-species) ex *Notothenia coriiceps*, the glomericirrine *Glomericirrus macrouri* ex *Trematomus newnesi* and the opisthadenine *Genolinea bowersi* ex *T. newnesi*. To improve sampling within the Dinurinae, we also added a sequence for *Ectenurus lepidus* Looss, 1907 (type-species) ex *Spicara maena* (Linnaeus, 1758) from the western Mediterranean. The species from the four hemiuroid families formed a strongly supported clade.

Hemipera manteri (Crowcroft, 1947) is sister to the remaining hemiuroids which form two major divisions separating the remaining Derogenidae + Sclerodistomidae and the Hemiuridae + Lecithasteridae (as in Pankov et al. 2006). Within the latter, representatives of four hemiurid subfamilies, the Lecithochiriinae, Dinurinae, Elytrophallinae, Glomericirrinae and the plerurine *Plerurus digitatus* (Looss, 1899) and most of the lecithasterids formed a strongly supported major grouping. Perhaps

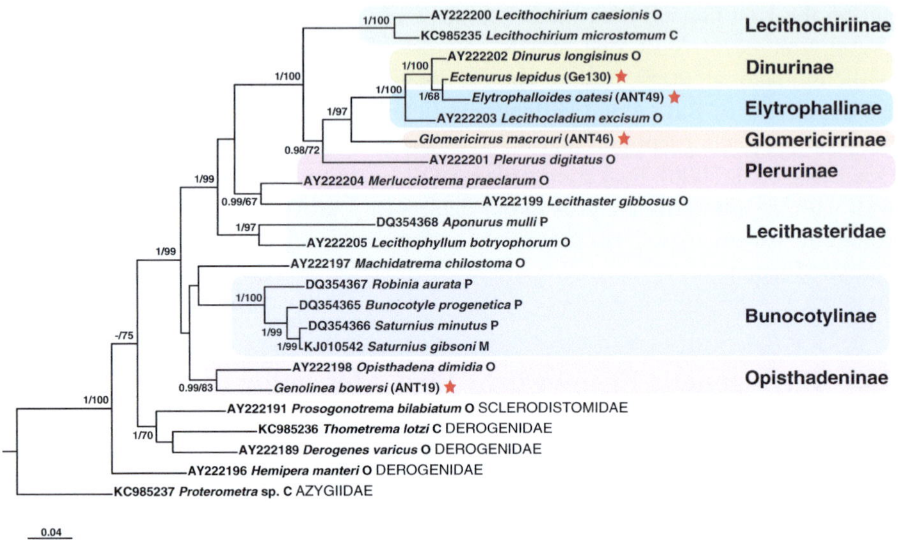

Fig. 5.4 Bayesian tree for the Hemiuroidea inferred from partial sequences of the nuclear 28S rRNA gene (1182 nt positions) constructed with MrBayes, v3.2.6 (Ronquist et al. 2012) under the GTR+I+Γ model. Bayesian inference (BI) analysis was run for 10,000,000 generations, with 25 % of the sampled trees discarded as 'burn-in'. Nodal support is given as posterior probabilities (BI) followed by non-parametric bootstrap validation values based on 1000 replicates in a maximum likelihood (ML) analysis performed with PhyML 3.0 (Guindon et al. 2010); only values>0.95 (BI) and>70 (ML) are shown. The tree is rooted against *Proterometra* sp. (Azygiidae Lühe, 1909). The scale-bar indicates the expected number of substitutions per site. Sequence identification is as in GenBank, followed by a letter: C, Calhoun et al. (2013); M, Marzoug et al. (2014); O, Olson et al. (2003); P, Pankov et al. (2006). *Stars* indicate the new sequences provided in the present study

the most important finding resulting from the addition of novel sequences is the non-monophyly of the Dinurinae and the Elytrophallinae which jointly form a clade with a maximum support (Fig. 5.4). The key difference between the two subfamilies appears to be the thickness of the wall of the seminal vesicle (thick and muscular in the elytrophallines vs. thin in the dinurines). However, the elytrophallines *Lecithocladium excisum* (Rudolphi, 1918) and *Elytrophalloides oatesi*, hitherto united by the possession of a thick-walled seminal vesicle, are not sister taxa, with *L. excisum* basal in this clade and *E. oatesi* as sister to *E. lepidus*. Obtaining sequences for more species within these two subfamilies will help resolve the relationships between the Dinurinae and Elytrophallinae, which may well ultimately be considered synonymous.

Further, adding the first sequence of a glomericirrine species revealed that the Glomericirrinae is the closest sister taxon to the Dinurinae+Elytrophallinae with strong support and the new sequence for the opisthadenine *Genolinea bowersi* helped clarify the unresolved position of *Opisthadena dimidia* in the analysis of Pankov et al. (2006): the two opisthadenines formed a strongly supported clade within the second major clade of the Hemiuridae+Lecithasteridae, sister to the

Bunocotylinae albeit with a poor support. The position of *Machidatrema chilostoma* (Machida, 1980) was also unresolved in the analysis by Pankov et al. (2006). This species exhibited an association with the Bunocotylinae in the present analysis but with poor support. *Machidatrema* León-Règagnon, 1998 was erected as a bunocotyline by León-Règagnon (1998) and transferred to the Hysterolecithinae (Lecithasteridae) by Gibson (2002) following Bray and Cribb (2000). Further sampling within the Hysterolecithinae may help resolve the relationships within the second major clade of the Hemiuridae + Lecithasteridae. The results of this study along with earlier phylogenetic studies (e.g. Blair et al. 1998; Pankov et al. 2006) cast doubt on the validity of the Lecithasteridae.

The phylogenetic hypothesis on the relationships within the family Opecoelidae presented in Fig. 5.5 includes the available sequences for 28 species of 16 genera; these include sequences for three species sampled by us: *Macvicaria* sp. (a new species) ex *Trematomus newnesi*, *M. pennelli* ex *T. bernacchii* and *Neolebouria georgiensis* ex *Trematomus pennellii*. Both BI and ML analyses (alignment comprising 1214 nt positions; 42 ambiguously aligned positions excluded from the analyses) resulted in generally well resolved and congruent trees with minor topological differences (Fig. 5.5). The Opecoelidae was resolved as monophyletic with *Biospeedotrema* spp. as earliest divergent (as in Bray et al. 2016, 2014) and the remaining species grouped into two strongly supported clades with *Helicometra manteri* Ozaki, 1925 as the closest sister taxon. The phylogenetic study of Shedko et al. (2015) indicated that *Biospeedotrema* Bray, Waeschenbach, Dyal, Littlewood & Morand, 2014 may not be closely related to the remaining opecoelids. The first major clade includes all species of *Macvicaria*, two species of *Cainocreadium* Nicoll, 1909, plus single species of the genera *Peracreadium* Nicoll, 1909, *Gaevskajatrema* Gibson & Bray, 1982, *Hamacreadium* Linton, 1910, *Bentholebouria* Andres, Pulis & Overstreet, 2014, and *Pseudopycnadena* Saad-Fares & Maillard, 1986. The Mediterranean species of *Macvicaria* form a strongly supported clade with *Peracreadium idoneum* (Nicoll, 1909) + *Gaevskajatrema perezi* (Mathias, 1926) nested within it, whereas the two newly-sampled Antarctic *Macvicaria* spp. appear in a separate clade with maximum support and *Macvicaria macassarensis* (Yamaguti, 1952) exhibits a strong association with *Hamacreadium mutabile* Linton, 1910 within a strongly supported subclade also containing *Cainocreadium* spp. and *Bentholebouria colubrosa* Andres, Pulis & Overstreet, 2014 (maximum support from BI analysis).

The second strongly supported opecoelid clade comprises three subclades receiving maximum support in BI and ML analyses: (i) the newly-sampled *Neolebouria georgiensis* (type-species) + (*Gaevskajatrema halosauropsi* Bray & Campbell, 1996 + *Buticulotrema thermichthysi* Bray, Waeschenbach, Dyal, Littlewood & Morand, 2014); (ii) *Neolebouria lanceolata* (Price, 1934) + *Podocotyloides brevis* Andres & Overstreet, 2013; and *Opecoeloides* spp. + *Dimerosaccus oncorhynchi* Eguchi, 1931.

The most important point in the present phylogenetic hypothesis for the Opecoelidae is the polyphyly of *Macvicaria*, *Gaevskajatrema* and *Neolebouria*. This pattern for the first two genera has been observed in previous phylogenies

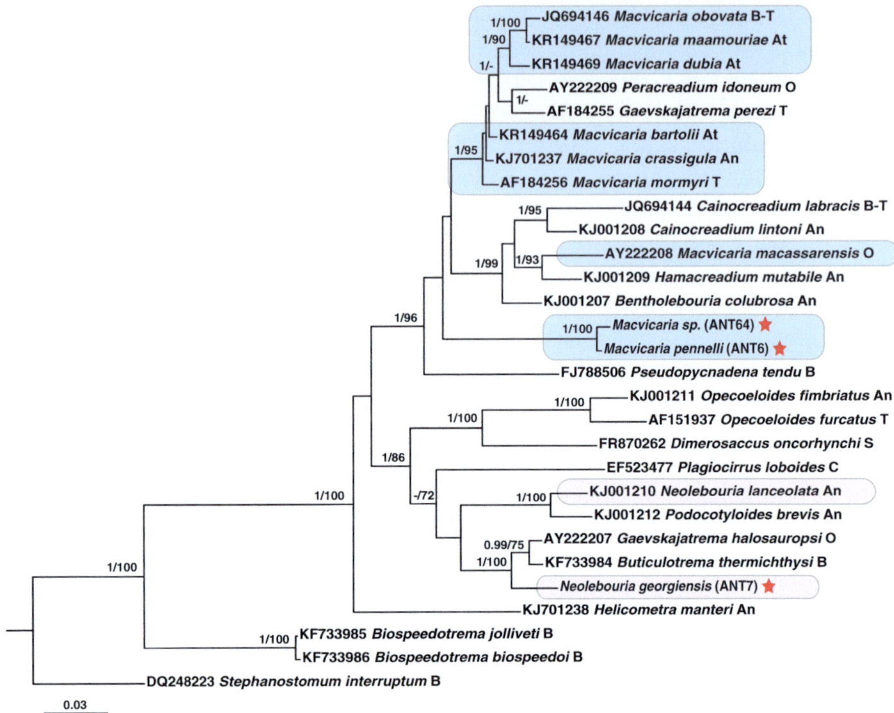

Fig. 5.5 Bayesian tree for the Opecoelidae inferred from partial sequences of the nuclear 28S rRNA gene (1177 nt positions) constructed with MrBayes, v3.2.6 (Ronquist et al. 2012) under the GTR+I+Γ model. Bayesian inference (BI) analysis was run for 10,000,000 generations, with 25 % of the sampled trees discarded as 'burn-in'. Nodal support is given as posterior probabilities (BI) followed by non-parametric bootstrap validation values based on 1000 replicates in a maximum likelihood (ML) analysis performed with PhyML 3.0 (Guindon et al. 2010); only values > 0.95 (BI) and >70 (ML) are shown. The tree is rooted against *Stephanostomum interruptum* Sparks & Thatcher, 1958 (Acanthocolpidae Lühe, 1906). The scale-bar indicates the expected number of substitutions per site. Sequence identification is as in GenBank, followed by a letter: An, Andres et al. (2014a, b); At, Antar et al. (2015); B-T, Born-Torrijos et al. (2012); B, Bray et al. (2009, 2014, 2005); C, Curran et al. (2007); O, Olson et al. (2003); S, Shedko et al. (2015); T, Tkach et al. (2000, 2001). *Stars* indicate the new sequences provided in the present study.

(Andres et al. 2014a; Bray et al. 2016), the latter authors also proving the polyphyly of a third opecoelid genus, *Allopodocotyle* Pritchard, 1966. However, adding just three novel sequences for digeneans from Antarctic fishes helped elucidate yet another polyphyletic genus within the family, *Neolebouria*. It is worth noting that the association of *M. macassarensis* with *H. mutabile* has already been advanced by Andres et al. (2014a) and more recently by Bray et al. (2016). The latter authors developed a phylogenetic hypothesis for the Opecoelidae based on concatenated data for the 18S and 28S rRNA gene for a much larger set of taxa including three species of *Hamacreadium*. Although the relationship between the Antarctic and Mediterranean species of *Macvicaria* was not resolved in the present phylogenetic hypothesis, the former appear to have evolved much faster than the latter thus indicating a probable distinction at the generic level (Fig. 5.5). Further sampling of the endemic Antarctic species of *Macvicaria* would be required to test this hypothesis. Although increased sampling of *Neolebouria* spp. is clearly required, the fact that we have sequenced the type-species, *N. georgiensis*, will help define the boundaries of *Neolebouria* in future phylogenetic studies. The present hypothesis suggests that *N. lanceolata* sequenced by Andres et al. (2014a, b) (KJ001210) does not belong to *Neolebouria*.

The consensus trees from BI and ML analyses of sequences for 18 species of the family Lepidapedidae (alignment comprising 910 nt positions; six ambiguously aligned positions excluded from the analyses) provided little resolution of the relationships at the generic level (Fig. 5.6). The Lepidapedidae was resolved as mono-

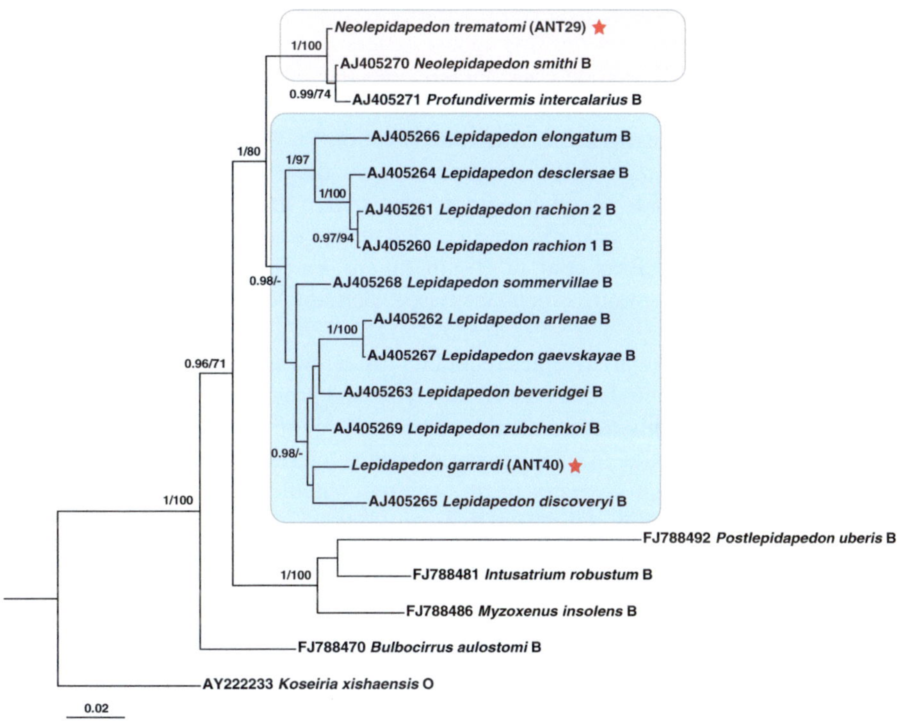

Fig. 5.6 Bayesian tree for the Lepidapedidae inferred from partial sequences of the nuclear 28S rRNA gene (904 nt positions) constructed with MrBayes, v3.2.6 (Ronquist et al. 2012) under the GTR+I+Γ model. Bayesian inference (BI) analysis was run for 10,000,000 generations, with 25 % of the sampled trees discarded as 'burn-in'. Nodal support is given as posterior probabilities (BI) followed by non-parametric bootstrap validation values based on 1000 replicates in a maximum likelihood (ML) analysis performed with PhyML 3.0 (Guindon et al. 2010); only values > 0.95 (BI) and > 70 (ML) are shown. The tree is rooted against *Koseiria xishaensis* Gu & Shen, 1983 (Enenteridae Yamaguti, 1958). The scale-bar indicates the expected number of substitutions per site. Sequence identification is as in GenBank, followed by a letter: B, Bray et al. (1999, 2009); O, Olson et al. (2003). *Stars* indicate the new sequences provided in the present study

phyletic with *Bulbocirrus aulostomi* Yamaguti, 1965 as earliest divergent and a clade formed by *Myzoxenus insolens* (Crowcroft, 1945), *Intusatrium robustum* Durio & Manter, 1968 and *Postlepidapedon uberis* Bray, Cribb & Barker, 1997 as sister to a strongly supported clade formed by: (i) a clade comprising two *Neolepidapedon* spp. (one newly-sampled, *N. trematomi* ex *Notothenia coriiceps*)+*Profundivermis intercalarius* Bray & Gibson, 1991; and (ii) a clade comprising *Lepidapedon* spp., the latter containing two main clades (*L. elongatum* (Lebour, 1908)+*L. desclersae* Bray & Gibson, 1995+*L. rachion* (Cobbold, 1858) 1+*L. rachion* 2 and *L. arlenae* Bray & Gibson, 1995+*L. gaevskayae* Campbell & Bray, 1993+*L. beveridgei* Campbell & Bray, 1993+*L. zubchenkoi* Campbell & Bray, 1993+*L. garrardi*+*L. discoveryi* Bray & Gibson, 1995) (BI support only). *Lepidapedon somervillae* Bray & Gibson, 1995 was the earliest divergent to the later clade, albeit with poor support. The newly-sampled *L. garrardi* ex *Trematomus bernacchii* clustered with *L. discoveryi* within *Lepidapedon* spp. but with poor support. The few resolved relationships appear generally congruent with the analyses of Bray et al. (2009) based on concatenated datasets of the partial 28S DNA and the partial mitochondrial *nad*1 gene sequences for a much larger set of taxa. The main difference represents the different position of *Neolepidapedon* spp. + *Profundivermis intercalarius*, which appeared embedded within *Lepidapedon* in a strongly supported clade together with *L. zubchenkoi* and *L. beveridgei* in their hypothesis. *Lepidapedon garrardi* is embedded within a group of deep-sea species, supporting the view of Bray et al. (1999) that there is a close relationship between some deep-sea and Antarctic digeneans based, perhaps, on their tolerance of low temperatures.

In summary, the first phylogenies including just a few sequences for the endemic digeneans of the Southern Ocean are encouraging and provide a promise that further molecular data from this region would contribute to a better understanding of the digenean relationships at the suprageneric level. Our results indicate that future exploration of digenean diversity in the Antarctic should be based on well-fixed specimens in combination with 28S rDNA sequences.

5.5 Future Directions

This study is the first to include molecular phylogenetic evidence on Antarctic digeneans. Clearly, as molecular techniques become standard, easier and relatively less expensive, this is an area that should be pursued in conjunction with morphological, faunistic and ecological investigations. The inclusion of this comparatively objective evidence will lend weight to the findings. Nevertheless, the investigations already made possible by intrepid collectors has enabled us to develop a rudimentary understanding of the digenean fauna of this remote and demanding environment.

Acknowledgement AF, SG and AK acknowledge partial support by the Czech Science Foundation (Grant ECIP P505/12/G112).

References

Andres MJ, Pulis EE, Overstreet RM (2014a) New genus of opecoelid trematode from *Pristipomoides aquilonaris* (Perciformes: Lutjanidae) and its phylogenetic affinity within the family Opecoelidae. Folia Parasitol 61:223–230

Andres MJ, Ray CL, Pulis EE, Curran SS, Overstreet RM (2014b) Molecular characterization of two opecoelid trematodes from fishes in the Gulf of Mexico, with a description of a new species of *Helicometra*. Acta Parasitol 59:405–412

Antar R, Georgieva S, Gargouri L, Kostadinova A (2015) Molecular evidence for the existence of species complexes within *Macvicaria* Gibson & Bray, 1982 (Digenea: Opecoelidae) in the western Mediterranean, with descriptions of two new species. Syst Parasitol 91:211–229

Atkinson A, Whitehouse MJ, Priddle J, Cripps GC, Ward P, Brandon MA (2001) South Georgia, Antarctica: a productive, cold water, pelagic ecosystem. Mar Ecol Prog Ser 216:279–308

Barnes DKA, Hillenbrand C-D (2010) Faunal evidence for late quarternary trans-Antarctic seaway. Global Change Biol 16:3297–3303

Blair D, Bray RA, Barker SC (1998) Molecules and morphology in phylogenetic studies of the Hemiuroidea (Digenea: Trematoda: Platyhelminthes). Mol Phylogenet Evol 9:15–25

Born-Torrijos A, Kostadinova A, Raga JA, Holzer AS (2012) Molecular and morphological identification of larval opecoelids (Digenea: Opecoelidae) parasitising prosobranch snails in a Western Mediterranean lagoon. Parasitol Int 61:450–460

Bray RA, Cribb TH (2000) The status of the genera *Hysterolecithoides* Yamaguti, 1934, *Neotheletrum* Gibson & Bray, 1979 and *Machidatrema* León-Règagnon, 1998 (Digenea: Hemiuroidea). Syst Parasitol 46:1–22

Bray RA, Cribb TH, Littlewood DTJ, Waeschenbach A (2016) The phylogeny of the digenean family Opecoelidae Ozaki, 1925: molecular phylogenetic inference and the value of morphological characters. Folia Parasitol 63:013

Bray RA, Littlewood DTJ, Herniou EA, Williams B, Henderson RE (1999) Digenean parasites of deep-sea teleosts: a review and case studies of intrageneric phylogenies. Parasitology 119:S125–S144

Bray RA, Waeschenbach A, Cribb TH, Weedall GD, Dyal P, Littlewood DTJ (2009) The phylogeny of the Lepocreadiidae (Platyhelminthes: Digenea) inferred from nuclear and mitochondrial genes: implications for their systematics and evolution. Acta Parasitol 54:310–329

Bray RA, Waeschenbach A, Dyal P, Littlewood DT, Morand S (2014) New digeneans (Opecoelidae) from hydrothermal vent fishes in the south eastern Pacific Ocean, including one new genus and five new species. Zootaxa 3768:73–87

Bray RA, Webster BL, Bartoli P, Littlewood DTJ (2005) A molecular phylogenetic study of the Acanthocolpidae (Digenea). Acta Parasitol 50:281–291

Brierley AS, Thomas DN (2002) Ecology of Southern Ocean pack ice. Adv Mar Biol 43:171–276

Byrd MA (1963) Helminth parasites of Antarctic vertebrates. Part 1. Digenetic trematodes of marine fishes. Proc Helminthol Soc Wash 30:129–148

Calhoun DM, Curran SS, Pulis EE, Provaznik JM, Franks JS (2013) *Hirudinella ventricosa* (Pallas, 1774) Baird, 1853 represents a species complex based on ribosomal DNA. Syst Parasitol 86:197–208

Cherry-Garrard A (1922) The worst journey in the world. Constable & Company Ltd., London

Curran SS, Overstreet RM, Tkach VV (2007) Phylogenetic affinities of *Plagiocirrus* Van Cleave and Mueller, 1932 with the description of a new species from the Pascagoula River, Mississippi. J Parasitol 93:1452–1458

Dönges J (1964) Der Lebenszyklus von *Posthodiplostomum cuticola* (v. Nordmann 1832) Dubois 1936 (Trematoda, Diplostomatidae). Z Parasitenk 24:169–248

Dowdeswell JA, Evans J, Cofaigh CÓ, Anderson JB (2006) Morphology and sedimentary processes on the continental slope off Pine Island Bay, Amundsen Sea, West Antarctica. Geolo Soc Am Bull 118:606–619

Evans J, Dowdeswell JA, Cofaigh CÓ, Benham TJ, Anderson JB (2006) Extent and dynamics of the West Antarctic Ice Sheet on the outer continental shelf of Pine Island Bay during the last glaciation. Mar Geol 230:53–72

Froese R, Pauly D (eds) (2015) FishBase. World wide web electronic publication. Retrieved November, 2015, from www.fishbase.org.

Gaevskaya AV (1982) [The discovering of the trematode metacercariae in mysids of the South Georgia Island.] Nauch Dokl Vysh Shk Biol Nauki 8:27–29 (In Russian)

Gaevskaya AV, Kovaljova AA (1976) [The fauna of trematodes of some common fish species in the South West Atlantic.] Trudy Atlant Nauch Issled Inst Ryb Khoz Okeanogr 60:3–14 (In Russian)

Gaevskaya AV, Rodyuk GN (1983) [New material of fauna of trematodes of fish in South West Atlantic.] Nauch Dokl Vysh Shk. Biol Nauki 3:28–32 (In Russian)

Gaevskaya AV, Rodyuk GN (1988) [New and rare Trematoda species from deep-sea fishes of the South-West Atlantic.] Vestn Zool 5:11–15 (In Russian)

Gibson DI (1976) Monogenea and Digenea from fishes. Discov Rep 36:179–266

Gibson DI (2002) Family Hemiuridae Looss, 1899. In: Gibson DI, Jones A, Bray RA (eds) Keys to the Trematoda, vol 1. CAB International, Wallingford, pp 305–340

Graefe G (1971) Die Temperatur des Lebensraumes und ihre Wirkung auf Cercarien. Überlegungen und Versuche im Anschluß an Beobachtungen in der Antarktis. Parasitologische Schriftenreihe 21:151–156

Guindon S, Dufayard JF, Lefort V, Anisimova M, Hordijk W, Gascuel O (2010) New algorithms and methods to estimate maximum-likelihood phylogenies: assessing the performance of PhyML 3.0. Syst Biol 59:307–321

Hillenbrand C-D, Kuhn G, Frederichs T (2009) Record of a Mid-Pleistocene depositional anomaly inWest Antarctic continental margin sediments: an indicator for icesheet collapse? Q Sci Rev 28:1147–1159

Holloway HR, Spence JA (1980) Ecology of animal parasites in McMurdo Sound, Antarctica. Comp Physiol Ecol 5:262–284

Kovaljova AA, Gaevskaya AB (1974) [New representatives of the genus *Plagioporus* (Trematoda, Opecoelidae) from the Antarctic fishes.] Zool Zh 53:1407–1409 (In Russian)

Larter RD, Graham AGC, Gohl K, Kuhn G, Hillenbrand CD, Smith JA, Deen TJ, Livermore RA, Schenke H-W (2009) Subglacial bedforms reveal complex basal regime in a zone of paleo-ice stream convergence, Amundsen Sea Embayment, West Antarctica. Geology 37:411–414

Laskowski Z, Zdzitowiecki K (2005) The helminth fauna of some notothenioid fishes collected from the shelf of Argentine Islands, West Antarctica. Pol Polar Res 26:315–324

Laskowski Z, Jezewski W, Zdzitowiecki K (2013) Description of a new opecoelid trematode species from nototheniid fish in the Beagle Channel (Sub-Antarctica). J Parasitol 99:487–489

Laskowski Z, Jezewski W, Zdzitowiecki K (2014) Changes in digenean infection of the Antarctic fish *Notothenia coriiceps* in Admiralty Bay, King George Island, over three decades. Pol Polar Res 35:513–520

Laskowski Z, Rocka A, Zdzitowiecki K, Ozouf-Costaz C (2007) Occurrence of endoparasitic worms in dusky notothen, *Trematomus newnesi* (Actinopterygii Nototheniidae), at Adélie Land, Antarctica. Pol Polar Res 28:37–42

Laskowski Z, Rocka A, Zdzitowiecki K, Ghigliotti L, Pisano E (2005) New data on the occurrence of internal parasitic worms in the *Gymnodraco acuticeps* and *Cygnodraco mawsoni* (Bathydraconidae) fish in the Ross Sea, Antarctica. Pol Polar Res 26:37–40

Leiper RT, Atkinson EL (1914) Helminthes of the British Antarctic expedition, 1910-1913. Proc Zool Soc Lond 1:222–226

Leiper RT, Atkinson EL (1915) Parasitic worms with a note on a free-living nematode. British Museum Antartic ("Terra Nova") Expedition, 1910. Nat Hist Rep Zool 2:19–60

León-Règagnon V (1998) *Machidatrema* n. gen. (Digenea: Hemiuridae: Bunocotylinae) and phylogenetic analysis of its species. J Parasitol 84:140–146

Linse K, Griffiths HJ, Barnes DKA, Clarke A (2006) Biodiversity and biogeography of Antarctic and Sub-Antarctic mollusca. Deep-Sea Res II 53:985–1008

Linse K, Griffiths HJ, Barnes DKA, Brandt A, Davey N, David B, De Grave S, d'Udekem d'Acoz C, Eléaume M, Glover AG, Hemery LG, Mah C, Martín-Ledo R, Munilla T, O'Loughlin M, Pierrat B, Saucède T, Sands CJ, Strugnell JM, Enderlein P (2016) The macro- and megabenthic fauna on the continental shelf of the eastern Amundsen Sea, Antarctica. Cont Shelf Res 68:80–90

Lowe AL, Anderson JB (2002) Reconstruction of the West Antarctic ice sheet in Pine Island Bay during the Last Glacial maximum and its subsequent retreat history. Q Sci Rev 21:1879–1897

Marzoug D, Rima M, Boutiba Z, Georgieva S, Kostadinova A, Pérez-del-Olmo A (2014) A new species of *Saturnius* Manter, 1969 (Digenea: Hemiuridae) from Mediterranean mullet (Teleostei: Mugilidae). Syst Parasitol 87:127–134

Moser M, Cowen RK (1991) The effects of periodic eutrophication on parasitism and stock identification of *Trematomus bernacchii* (Pisces: Nototheniidae) in McMurdo Sound, Antarctica. J Parasitol 77:551–556

Olson PD, Cribb TH, Tkach VV, Bray RA, Littlewood DTJ (2003) Phylogeny and classification of the Digenea (Platyhelminthes: Trematoda). Int J Parasitol 33:733–755

Olson RE (1966) Some experimental fish hosts of the strigeid trematode *Bolbophorus confusus* and effects of temperature on the cercaria and metacercariae. J Parasitol 52:327–334

Palm HW, Reimann N, Spindler M, Plötz J (1998) The role of the rock cod *Notothenia coriiceps* Richardson, 1844 in the life-cycle of Antarctic parasites. Polar Biol 19:399–406

Pankov P, Webster BL, Blasco-Costa I, Gibson DI, Littlewood DT, Balbuena JA, Kostadinova A (2006) *Robinia aurata* n. g., n. sp. (Digenea: Hemiuridae) from the mugilid *Liza aurata* with a molecular confirmation of its position within the Hemiuroidea. Parasitology 133:217–227

Pollard D, DeConto RM (2009) Modelling West Antarctic ice sheet growth and collapse through the past five million years. Nature 458:329–333

Prudhoe S, Bray RA (1973) Digenetic trematodes from fishes. B.A.N.Z. Antarct Res Exped Rep Ser B (Zool Bot) 8:199–225

Ronquist F, Teslenko M, van der Mark P, Ayres DL, Darling A, Hohna S, Larget B, Liu L, Suchard MA, Huelsenbeck JP (2012) MrBayes 3.2: efficient Bayesian phylogenetic inference and model choice across a large model space. Syst Biol 61:539–542

Santoro M, Cipriani P, Pankov P, Lawton SP (2015) *Aporocotyle michauda* n. sp. (Digenea: Aporocotylidae) from the emerald rock cod, *Trematomus bernacchii* (Teleostei: Perciformes) in Antarctica. Parasitol Int 64:324–329

Santoro M, Mattiucci S, Cipriani P, Bellisario B, Romanelli F, Cimmaruta R, Nascetti G (2014) Parasite communities of icefish (*Chionodraco hamatus*) in the Ross Sea (Antarctica): influence of the host sex on the helminth infracommunity structure. PLoS One 9:1–7

Shedko B, Sokolov SG, Atopkin DM (2015) The first record of *Dimerosaccus onchorynchi* (Trematoda: Opecoelidae) in fishes from rivers of Primorsky territory Russia, with a discussion on its taxonomic position using morphological data. Parazitologiia 49:171–189

Sokolov SG, Gordeev II (2013) New data on trematodes (Plathelminthes, Trematoda) of fishes in the Ross Sea (Antarctic). Invert Zool 10:255–267

Sokolov SG, Gordeev II (2015) [New data on trematodes of Antarctic fish.] Parazitologiya 49:12–27 (In Russian)

Sokolov SG, Gordeev II (2015) [*Paralepidapedon variabile* sp. n. (Trematoda, Lepocreadioidea, Lepidapedidae) and other representatives of the genus *Paralepidapedon* from Antarctic fish.] Zool Zh 94:756–763 (In Russian)

Spalding MD, Agostini VN, Rice J, Grant SM (2012) Pelagic provinces of the world: A biogeographic classification of the world's surface pelagic waters. Ocean Coast Manage 60:19–30

Spalding MD, Fox HE, Allen GR, Davidson N, Ferdaña ZA, Finlayson M, Halpern BS, Jorge MA, Lombana A, Lourie SA, Martin KD, McManus E, Molnar J, Recchia CA, Robertson J (2007) Marine ecoregions of the world: A bioregionalization of coastal and shelf areas. Bioscience 57:573–583

Szidat L (1965) Estudios sobre la fauna de parasitos de peces antarticos. I. Los parasitos de *Notothenia neglecta*. Serv Hidr Naval Secr Marina Publ H-910:1–84

Szidat L, Graefe G (1967) Estudios sôbre la fauna de parásitos de peces Antarcticos. II. Los parásitos de *Parachaenichthys charcoti*. Serv Hidr Naval Secr Marina Publ H-911:4–27
Tkach V, Pawlowski J, Mariaux J (2000) Phylogenetic analysis of the suborder Plagiorchiata (Platyhelminthes, Digenea) based on partial lsrDNA sequences. Int J Parasitol 30:83–93
Tkach VV, Pawlowski J, Mariaux J, Swiderski Z (2001) Molecular phylogeny of the suborder Plagiorchiata and its position in the system of Digenea. In: Littlewood DTJ, Bray RA (eds) Interrelationships of the Platyhelminthes. Taylor & Francis, London, pp 186–193
WoRMS Editorial Board (2015) World register of marine species. Retrieved February 10, 2016, from http://www.marinespecies.org at VLIZ
Zdzitowiecki K (1978) Introductory observations on the parasites of fish in area of South Shetland Islands and South Georgia (Antarctica). Kosmos 27:651–659 (In Polish)
Zdzitowiecki K (1987) Digenetic trematodes from the alimentary tract of fishes off South Shetlands (Antarctic). Acta Parasitol Pol 32:219–232
Zdzitowiecki K (1988) Occurrence of digenetic trematodes in fishes off South Shetlands (Antarctic). Acta Parasitol Pol 33:155–167
Zdzitowiecki K (1990a) Antarctic representatives of the genus *Macvicaria* Gibson & Bray, 1982 (Digenea Opecoelidae), with descriptions of two new species. Syst Parasitol 16:169–179
Zdzitowiecki K (1990b) Little known and new Antarctic Digenea species of the genera *Neolepidapedon* and *Lepidapedon* (Lepocreadiidae). Acta Parasitol Pol 35:19–30
Zdzitowiecki K (1991) Occurrence of digeneans in open sea fishes off the South Shetland Islands and South Georgia, and a list of fish digeneans in the Antarctic. Pol Polar Res 12:55–72
Zdzitowiecki K (1997) Antarctic Digenea, parasites of fishes. In: Wägele JW, Sieg J (eds) Synopses of the Antarctic Benthos, vol 8. Koeltz Scientific Books, Königstein, pp 1–156
Zdzitowiecki K (1998) Diversity of Digenea, parasites of fishes in various areas of the Antarctic. In: di Prisco G, Pisano E, Clarke A (eds) Fishes of Antarctica: a biological overview. Springer-Verlag Italia Sri, Milano, pp 87–96
Zdzitowiecki K (1999) Digeneans of the families Opecoelidae and Lepocreadiidae, parasites of *Lepidonotothen macrophthalma* from the North Scotia Ridge, and remarks on the discrimination of *Neolepidapedon magnatestis* and *N. trematomi*. Acta Parasitol Pol 44:233–240
Zdzitowiecki K (2001) New data on the occurrence of fish endoparasitic worms off Adélie Land, Antarctica. Pol Polar Res 22:159–165
Zdzitowiecki K (2002a) Occurrence of Digenea in fishes of the family Nototheniidae in the Weddell Sea. Acta Parasitol 47:154–158
Zdzitowiecki K (2002b) Occurrence of Digenea in fishes of the family Channichthyidae in the Weddell Sea and other sub-continental areas of the Antarctica. Acta Parasitol 47:159–162
Zdzitowiecki K (2002c) Occurrence of Digenea in fishes of the family Artedidraconidae in the Weddell Sea and other areas of Antarctica. Acta Parasitol 47:306–309
Zdzitowiecki K (2002d) Occurrence of Digenea in fishes of the family Bathydraconidae in the Weddell Sea and other areas of Antarctica. Acta Parasitol 47:310–313
Zdzitowiecki K (2003) Occurrence of Digenea in fishes other than Notothenioidei in the Weddell Sea and the whole Antarctica. Acta Parasitol 48:195–199
Zdzitowiecki K, Cielecka D (1997a) Digenea of fishes of the Weddell Sea. I. Parasites of *Macrourus whitsoni* (Gadiformes, Macrouridae). Acta Parasitol Pol 42:23–30
Zdzitowiecki K, Cielecka D (1997b) Digenea of fishes of the Weddell Sea. II. The genus *Macvicaria* (Opecoelidae). Acta Parasitol Pol 42:77–83
Zdzitowiecki K, Ozouf-Costaz C (2013) Contribution to the knowledge of the parasitic fauna of fish off Adélie Land, Antarctica. Pol Polar Res 34:429–435
Zdzitowiecki K, Pisano E (1996) New records of Digenea infecting elasmobranch and teleost fish off Heard Island (Kerguelen sub-region, sub-Antarctic). Arch Fish Mar Res 43:265–272
Zdzitowiecki K, White MG (1992) Digenean Trematoda infection of inshore fish at South Georgia. Antarct Sci 4:51–55
Zdzitowiecki K, Palladino S, Vacchi M (1999) Preliminary results on Digenea found in fishes in the coastal waters of Terra Nova Bay, Antarctica. Parassitologia 41:575–578

Zdzitowiecki K, Pisano E, Vacchi M (1992) Additional data to Antarctic representatives of the genus *Macvicaria* Gibson et Bray, 1982 (Digenea, Opecoelidae), with a key to species occurring in the Antarctic. Acta Parasitol Pol 37:131–134

Zdzitowiecki K, Pisano E, Vacchi M (1993) Antarctic representatives of the genus *Neolebouria* Gibson, 1976 (Digenea, Opecoelidae), with description of one new species. Acta Parasitol 38:11–14

Zdzitowiecki K, Rocka A, Pisano E, Ozouf-Costaz C (1998) A list of fish collected off Adélie Land (Antarctic). Acta Parasitol 43:71–74

Zdzitowiecki K, White MG, Rocka A (1997) Digenean, monogenean and cestode infection of inshore fish at the South Orkney Islands. Acta Parasitol Pol 42:18–22

Chapter 6
Cestodes and Nematodes of Antarctic Fishes and Birds

Anna Rocka

6.1 Cestodes of Antarctic Fishes and Birds

6.1.1 Introduction

According to Duhamel et al. (2014) fish comprise the most species-rich group of Antarctic vertebrates with 47 families and 374 species; four families of which, including 14 species, being cartilagineous (Chondrichthyes). The most dominant is the endemic family Nototheniidae with 115 species (eight species require validation). Chondrichthyes are represented by sharks (three families, five species) and rays (one family, nine species). Sharks are mainly recorded in the northern part of the Southern Ocean, with only one southerly record in the Ross Sea. All Antarctic skates belong to the family Rajidae and two genera: *Amblyraja* Malm, 1877 (two species) and *Bathyraja* Ishiyama, 1958 (seven species).

The first report on adult cestodes from fish was provided by Linstow in 1907 who described a new species, *Phyllobothrium dentatum*, taken from an unidentified shark founded on the coast of South Georgia, but the description of this species was very poor and *P. dentatum* is considered as *species inquirenda* (Southwell 1925). Until the beginning of the 1990s, Antarctic skates were not examined for parasites. So far, only skates from the region of the South Shetland Islands and South Georgia have been examined, as well as several specimens from the Weddell Sea. These represented four species: *Bathyraja eatonii* (Günther, 1876), *B. maccaini* Springer, 1971, *Amblyraja georgiana* (Norman, 1938) and *Bathyraja* sp.2 Stehmann, 1985 (Rocka 2003).

A. Rocka
W. Stefański Institute of Parasitology, Polish Academy of Sciences,
ul.Twarda 51/55, 00-818 Warsaw, Poland
e-mail: abroccy@poczta.onet.pl

S. Klimpel et al. (eds.), *Biodiversity and Evolution of Parasitic Life in the Southern Ocean*, Parasitology Research Monographs, Vol. 9,
DOI 10.1007/978-3-319-46343-8_6

Skates are infected with ten species of the Tetraphyllidea and one of the Diphyllidea (Rocka 2003, 2006). Teleosts are infected with three cestode species, *Parabothriocephalus johnstoni* Prudhoe, 1969 and two Sub-Antarctic species, *Bothriocephalus kerguelensis* Prudhoe, 1969 and *B. antarcticus* Wojciechowska et al. 1995 (Prudhoe 1969; Wojciechowska et al. 1995; Kuchta et al. 2008).

Recently, a new classification of the order Tetraphyllidea has been proposed (Healy et al. 2009; Ruhnke 2011; Caira et al. 2014; Ruhnke et al. 2015). Three new orders were created, Rhinebothriidea Healy et al. 2009; Onchoproteocephalidea Caira et al. 2014 and Phyllobothriidea Caira et al. 2014, and some of the species remain within the Tetraphyllidea. In the present study, all tetraphyllidean species parasitizing skates were transferred to new orders.

Antarctic avifauna is represented by penguins (nine species of three genera) and over 130 species of flying seabird from nine families in three orders (Ropert-Coudert et al. 2014). Bird-parasitizing cestodes belong to 24 species, three families and two orders (Temirova and Skrjabin 1978; Barbosa and Palacios 2009).

Penguins from different localities of Antarctica were found to be infected with other helminths such *Contracaecum heardi*, *Stomachus* sp., *Streptocara* sp., *Ascaridia* sp., *Capillaria* sp. egg, *Tetrabothrius* spp. eggs, *Diphyllobothrium* spp. (and their eggs). However, almost all these records referred to immature specimens and could be occasional and/or accidental findings (Barbosa and Palacios 2009; Gonzáles-Acuña et al. 2013; Kleinertz et al. 2014).

Also, larval forms of cestodes are very common among the Antarctic teleosts and mammals (see: Rocka 2003) They include larvae of cestodes parasitizing skates, plerocercoids of the Diphyllobothriidae and larvae of the Tetrabothriidae, in adult stage parasites of marine birds and mammals.

6.1.2 Systematic Review of Cestodes Parasitizing Fishes

6.1.2.1 Cestodes of Skates

Order Onchoproteocephalidea Caira et al., 2014
Family Onchobothriidae Braun, 1900
Genus *Onchobothrium* de Blainville, 1828
Onchobothrium antarcticum Wojciechowska, 1990
(description Wojciechowska 1990a)

Host: *Bathyraja eatonii* (Eaton's skate), *Bathyraja maccaini* (McCain's skate)

Locality: South Shetlands (Wojciechowska 1990a), eastern part of the Weddell Sea (Rocka and Zdzitowiecki 1998).

Strobila 100–150 mm long. Scolex, 0.93–1.15 × 1.06–1.34 mm, with four sessile bothridia. Bothridia 0.75–1.03 × 0.60–0.63 mm. Each bothridium divided into three loculi which varied in size, with a pair of hooks and cushion-like lobe situated on anterior margin of anterior loculus. Hooks equal length, 0.10–0.11 mm, not branched. Testes 125–135 in number. Cirrus pouch 0.34–0.55 × 0.19–0.26 mm. The male genital opening situated on the lobe. Cirrus armed. Ovary 0.25–0.30×0.50–0.65 mm. Vagina with a sphincter in distal part. Vitelline follicles forms two lateral bands. Eggs 0.11–0.16 mm in diameter.

Order Phyllobothriidea Caira et al., 2014
Family Phyllobothriidae Braun, 1900
Genus *Rajicestus* Rocka and Laskowski (2017)
Rajicestus georgiense (Wojciechowska 1991)
(description Wojciechowska 1991a)

syn. *Phyllobothrium georgiense* Wojciechowska, 1991; *Anthocephalum georgiense* Rocka and Zdzitowiecki, 1998

Host: *Amblyraja georgiana* (starry skate)

Locality: shelf around South Georgia (Wojciechowska 1991a)

Strobila 60–170 mm long. Scolex, 1.13–1.50×1.25–1.87 mm, with four sessile bothridia. Each bothridium strongly folded with weak marginal loculi and apical sucker, 0.18–0.23 mm in diameter. Testes 140–190 in number. Cirrus pouch 0.66–0.88×0.22–0.34 mm, situated diagonally. Ovary 0.20–0.80×0.62–1.13 mm. Vitellaria follicular, form two lateral wide bands. Vagina with feeble distal sphincter.

Rajicestus siedleckii (Wojciechowska 1991)
(description Wojciechowska 1991a; Rocka and Zdzitowiecki 1998)

syn. *Phyllobothrium siedleckii* Wojciechowska, 1991; *Anthocephalum siedleckii* Rocka and Zdzitowiecki, 1998

Host: *Bathyraja eatonii*, *B. maccaini*

Locality: South Shetlands (Wojciechowska 1991a), eastern part of the Weddell Sea (Rocka and Zdzitowiecki 1998)

Strobila 36–150 mm long. Scolex, 0.84–1.23×0.86–1.84 mm, with four sessile bothridia. Bothridia, 0.83–1.21×0.46–0.90 mm, with folded margin and an apical sucker, 0.20–0.28 mm in diameter. Marginal loculi weak, 40–45 in number. Testes 85–100 in number. Cirrus pouch 0.40–0.74×0.12–0.22 mm. Cirrus armed. A lobe bearing the male genital opening present. Ovary 0.21–0.62×0.31–0.80 mm. Vitellaria follicular, form a compact layer. Distal vaginal sphincter strong.

Rajicestus rakusai (Wojciechowska 1991a, b)
(description Wojciechowska 1991a)

syn. *Phyllobothrium rakusai* Wojciechowska, 1991; *Anthocephalum rakusai* Rocka and Zdzitowiecki, 1998

Host: *Bathyraja maccaini*

Locality: South Shetlands (Wojciechowska 1991a)

Strobila 50–140 mm long. Scolex, 0.88–1.62×1.37–1.75 mm, with four sessile bothridia. Bothridia with only a folded margin and weak marginal loculi, and an apical sucker, 0.25–0.31 mm in diameter. Testes 120–165 in number. Cirrus pouch, 0.55–0.76×0.18–0.23 mm, situated diagonally. A lobe bearing the male genital opening present. Ovary 0.37–0.92×0.47–0.85 mm. Vitellaria follicular, form a compact layer. Vaginal sphincter feeble.

Rajicestus arctowskii (Wojciechowska 1991)
(description Wojciechowska 1991a)

syn. *Phyllobothrium arctowskii* Wojciechowska, 1991; *Anthocephalum arctowskii* Rocka and Zdzitowiecki, 1998

Host: *Bathyraja* sp. 2

Locality: South Shetlands (Wojciechowska 1991a), eastern part of the Weddell Sea (Rocka and Zdzitowiecki 1998)

Strobila 15–35 mm long. Scolex, 0.75–1.06 × 0.86–1.90 mm, with four sessile bothridia. Bothridia with only a folded margin and weak marginal loculi (30 in number), and an apical sucker, 0.20–0.25 mm in diameter. Testes 60–80 in number. Cirrus pouch 0.37–0.55 × 0.15–0.26 mm. A lobe with the male genital opening present. Ovary 0.12–0.67 × 0.43–0.70 mm. Vitellaria follicular, form a compact layer. Vaginal sphincter strong.

Genus *Guidus* Ivanov, 2006
Guidus antarcticus (Wojciechowska 1991)
(description Wojciechowska 1991b; Ivanov 2006)

syn. *Marsupiobothrium antarcticum* Wojciechowska, 1991

Host: *Bathyraja eatonii*, *B. maccaini*

Locality: South Shetlands (Wojciechowska 1991b)

Strobila 140 mm long. Scolex, 1.6 × 1.8 mm, forms a cephalic peduncle bearing four globular, sac-like, muscular bothridia. Bothridia, 1.34–1.44 × 0.83–0.91 mm, with a muscular sphincter and one accessory sucker, 0.12–0.13 mm in diameter. Apical margin of the bothridium projects anteriorly as a lappet-shaped outgrowth of tissue, opposite to an accessory sucker. Testes 200–220 in number. Cirrus pouch 0.85–0.94 × 0.29–0.44 mm. Ovary, 0.70–0.75 × 1.13–1.25 mm, multilobed. Vitellaria follicular, gathered into two lateral wide bands. Vagina with sphincter at distal part.

Guidus awii (Rocka and Zdzitowiecki 1998)
(description Rocka and Zdzitowiecki 1998)

syn. *Marsupiobothrium awii* Rocka and Zdzitowiecki, 1998

Host: *Bathyraja maccaini*

Locality: the eastern part of the Weddell Sea (Rocka and Zdzitowiecki 1998)

Strobila 30–90 mm long. Scolex, 0.95–1.63 × 0.34–1.54 mm, forms a cephalic peduncle bearing four globular, sac-like, muscular bothridia. Bothridia, 0.95–1.38 × 0.34–0.63 mm, with muscular sphincter and one accessory sucker, 0.086–0.117 mm in diameter. Lappet-shaped outgrowth on the bothrdium present. Testes 120–150 in number. Cirrus pouch 0.47–0.76 × 0.20–0.55 mm. Ovary, 0.28–0.65 × 0.62–0.95 mm, multilobed. Vitelline follicles gathered into two lateral wide bands. Vagina with sphincter at distal part. Eggs 0.016–0.017 mm in diameter.

Order: Rhinebothriidea Healy et al., 2009
Family Echeneibothriidae de Beauchamp, 1905
Genus *Notomegarchynchus* Ivanov and Campbell, 2002

Notomegarchynchus shetlandicum (Wojciechowska 1990)
(description Wojciechowska 1990b; Ivanov and Campbell 2002)

syn. *Pseudanthobothrium shetlandicum* Wojciechowska, 1990
Host: *Bathyraja eatonii*, *B. maccaini*
Locality: South Shetlands (Wojciechowska 1990b)
Strobila 85–125 mm long. Scolex, 0.85–1.30 × 1.24–1.80 mm, composed of four stalked, cylindrical bothridia and a myzorhynchus. Bothridia, 0.7–0.9 mm long, ending in a thick-walled sucker, 0.4–0.5 mm in diameter. Myzorhynchus, 0.6–1.0 × 0.96–1.45 mm, consisting of proscolex and an apical organ. Apical organ muscular, occupying top of myzorhynchus, extending laterally and forming hanging velum. Both, proscolex and apical organ, non-invaginable and non-retractable. Testes 40–60 in number. Cirrus pouch 0.29–0.41 × 0.14–0.23 mm. Cirrus armed. Ovary 0.09–0.32 × 0.12–0.41 mm. Vitellaria follicular, gathered into two narrow lateral bands. Eggs 0.043–0.055 × 0.040–0.047 mm, with filament at one pole.

Genus *Pseudanthobothrium* Baer, 1956
Pseudanthobothrium notogeorgianum Wojciechowska, 1990
(description Wojciechowska 1990b)

Host: *Amblyraja georgiana*
Locality: shelf around South Georgia (Wojciechowska 1990b)
Strobila 18–60 mm long. Scolex, 0.75–1.20 × 1.13–1.43 mm, with four funnel-shaped bothridia and cylindrical myzorhynchus. Bothridia 0.63–0.88 mm long with depression on apex surrounded by thin-walled sucker. Myzorhynchus, 0.16–0.75 mm long with apical disc, 0.13–0.22 mm in diameter. Apical disc invaginable and retractable into myzorhynchus. Testes 25–36 in number. Cirrus pouch 0.20–0.28 × 0.12–0.16 mm. Cirrus armed. Ovary 0.08–0.33 × 0.12–0.30 mm. Vitellaria form two narrow lateral bands. Eggs 0.043–0.046 × 0.038–0.041 mm, passing at one pole into filament.

Pseudanthobothrium minutum Wojciechowska, 1991
(description Wojciechowska 1991b)

Host: *Bathyraja eatonii*
Locality: South Shetlands (Wojciechowska 1991b)
Strobila 22–36 mm long. Scolex, 0.56–0.88 × 0.80–1.33 mm, with four funnel-shaped bothridia and cylindrical myzorhynchus with a disc extruding outside at the tip. Bothridia, 0.50–0.75 × 0.31–0.56 mm, ended with a sucker-like structure. Myzorhynchus 0.38–0.58 × 0.2 mm. Apical disc invaginable and retractable into myzorhynchus. Diameter of the extruded disc 0.28–0.30 mm. Testes 20–26 in number. Cirrus pouch 0.24–0.27 × 0.12–0.15 mm. Ovary 0.08–0.45 × 0.12–0.55 mm. Vitelline follicles gathered into two narrow lateral bands.

Order Diphyllidea Carus, 1863
Family Echinobothriidae Perrier, 1897
Genus *Echinobothrium* Beneden, 1849

Echinobothrium acanthocolle Wojciechowska, 1991
(description Wojciechowska 1991b)

Host: *Amblyraja georgiana*
Locality: shelf near South Georgia (Wojciechowska 1991b)
Strobila 5 mm long. Scolex proper, 0.80×0.59 mm, consists of armed rostellum, 0.16×0.19 mm and two bothria, 0.74×0.60 mm. Hook formula 3(16/15)3. Hook increasing in length toward center of group. Lateral hooklets arranged in two groups. Cephalic peduncle 0.12×0.26 mm, armed with eight longitudinal columns of 2–5 spines. Spines with triradiate bases, 0.028–0.035 mm. Testes 19–25, in 4–5 irregular columns. Cirrus pouch 0.166–0.197×0.113–0.147 mm. Genital opening situated ventrally, on the mid-line of the proglottid, in its posterior part. Ovary 0.185–0.345×0.21–0.34 mm. Vitelline follicles very small, along lateral borders of the proglottid.

6.1.2.2 Cestodes of Bony Fishes

Order Bothriocephalidea Kuchta, Scholz, Brabec and Bray, 2008
Family Bothriocephalidae Blanchard, 1849
Genus *Bothriocephalus* Rudolphi, 1808
Bothriocephalus kerguelensis Prudhoe, 1969
(description Prudhoe 1969)

Host: *Notothenia cyanobrancha* (blue rockcod), *N. rossi* (marbled rockcod)
Locality: Royal Sound, Kerguelen Subregion (Prudhoe 1969)
Strobila 10–82 mm long. Scolex 0.3–0.9×0.4–0.6 mm. Testes 25–30 in number; 4–5 visible in transverse section. Cirrus pouch; its length/width ratio is 4:1. Ovary 0.4×0.3 mm. Eggs 0.057–0.065×0.037–0.042 mm.

Bothricephalus antarcticus Wojciechowska et al., 1995
(description Wojciechowska et al. 1995)

Strobila 110 mm long. Scolex 0.6–1.3 mm long. Testes 80–100 in a proglottid with single genital set; 100–120 in a proglottid with a double set. Six to seven testes visible in transverse section, with 4–5 in sagittal section (and 6–7 in proglottid with double set). Cirrus pouch; its length/width ratio is 2:1. Ovary 0.08–0.17×0.33–0.46 mm. Eggs 0.06–0.07×0.04–0.05 mm, operculate.

(description Wojciechowska et al. 1995)

Host and locality: *Champsocephalus gunnari* (mackerel icefish), *Channichthys rhinoceratus* (crocodile icefish)
Host and locality: Heard Island, Kerguelen Subregion (Wojciechowska et al. 1995)

Family Echinophallidae Schumacher, 1914
Genus *Parabothriocephalus* Yamaguti, 1934
Parabothriocephalus johnstoni Prudhoe, 1969
(description Prudhoe 1969; Rocka and Zdzitowiecki 1998)

Strobila 225 mm long. Scolex 1.75–2.06×0.54–1.00 mm. Testes 100–120 in number. Cirrus pouch 0.65–1.00×0.18–0.32 mm. Cirrus up to 0.62 mm, armed. Ovary 0.19–0.42×0.38–0.74 mm. Eggs 0.072–0.083×0.052–0.062 mm, operculate.

(description Prudhoe 1969; Rocka and Zdzitowiecki 1998)

Host and locality: *Macrourus whitsoni* (Whitson's grenadier), *M. holotrachys* (bigeye grenadier)

Host and locality: Indian sector of the Southern Ocean (Prudhoe 1969), the eastern part of the Weddell Sea (Rocka and Zdzitowiecki 1998) and off Heard and Kerguelen Islands (Kuchta et al. 2008)

6.1.3 Systematic Review of Cestodes Parasitizing Birds

Order Tetrabothriidea Baer, 1954
Family Tetrabothriidae Linton, 1891
Genus *Tetrabothrius* Rudolphi, 1819
Tetrabothrius joubini Railliet and Henry, 1912
(description Cielecka et al. 1992; Georgiev et al. 1996)

Host and locality: *Pygoscelis antarctica* (chinstrap penguin) – South Shetlands (Cielecka et al. 1992; Georgiev et al. 1996)

Strobila 30–40×0.5 mm. Scolex 0.32×0.36 mm, long neck up to 0.5 mm. Auricular appendages of bothridia are remarkable, as they are larger than in other species and wide spreading in relation to small dimensions of scolex as a whole. Genital pores present on one side of margin in the anterior part of proglottid. Testes 4–7 in number, large and irregular in shape. Genital atrium small, 0.042–0.050 mm in diameter. No papilla genitalis. Cirrus pouch, 0.029–0.040×0.035–0.049 mm, with a strong muscular wall. Male duct very short. Vagina lacking distinct widening; forming several loops in the proximal part, attached by a sheath of glandular cells along its whole length. Ovary and vitellarium branched.

Tetrabothrius pauliani (*Uniamniculus*) Joyeux and Baer, 1954
(description Temirova and Skrjabin 1978)

Host and locality: *P. antarctica*, *P. adeliae* (Adélie penguin) – the South Shetlands (Cielecka et al. 1992; Georgiev et al. 1996; Vidal et al. 2012), Bouvet Island (Andersen and Lysfjord 1982); *Pygoscelis papua* (gentoo penguin) and *Aptenodytes patagonicus* (king penguin) – Kerguelen Islands (Prudhoe 1969)

Strobila 10×0.7 mm. Scolex 0.434–0.56×0.45–0.59 mm with four muscular bothridia. Auricular appendages wide. Genital pores present on one side of the margin in the anterior part of the proglottid. Testes 9–14 in number. Genital atrium surrounded by strong muscles, 0.07–0.08 mm in diameter. No papilla genitalis. Male duct 0.026 mm long. Cirrus pouch, 0.045–0.053 mm in diameter, thin-walled. Vagina straight, without widenings and loops. Ovary and vitellarium branched. Oncospheres 0.027×0.023 mm.

Tetrabothrius (*Tetrabothrius*) *diomedea* Fuhrmann, 1900
(description Temirova and Skrjabin 1978)

Host and locality: *Diomedea exulans* (wandering albatross), *Diomedea chlororhynchus* (yellow-nosed albatross) – islands of the Southern Ocean (Temirova and Skrjabin 1978)

Strobila 70–140 mm. Scolex 0.38–0.43 × 0.32–0.35 mm. Testes 16–20 in number. Cirrus pouch 0.045–0.084 × 0.057–0.088 mm. Aperture of male duct on apex of knob, aperture of vagina situated ventrally. Distal part of vagina surrounded by dense layer of muscle fibres. Outside of genital atrium and vagina covered with glands; lumen of vagina with thin spines.

Tetrabothrius (*Tetrabothrius*) *kowalewskii* Szpotanska, 1917
(description Temirova and Skrjabin 1978)

syn. *Porotaenia kowalewskii* Szpotanska, 1917, *P. macrocirrosa* Szpotanska, 1917, *Tetrabothrius kowalewskii* (Szpotanska 1917) Johnston, 1935

Host and locality: *Diomedea chlororhynchus*, *Procellaria aequinoctialis* (white-chinned petrel) – Kerguelen Islands (Temirova and Skrjabin 1978); *Daption capense* (cape petrel) – South Shetlands (Odening 1982)

Strobila (without scolex) 168 × 1.94 mm. Scolex 0.4–0.455 × 0.575–0.59 mm. Testes number 38–40, 0.028 mm in diameter. Genital atrium 0.172 × 0.181 mm. Cirrus pouch 0.053–0.077 × 0.069–0.090 mm. Male duct 0.086 mm. Aperture of male duct in centre of genital atrium. Vaginal aperture ventral to male opening. Ovary, 0.31 × 0.38 mm, multilobed. Vitellarium, 0.043 × 0.069 mm, situated antero-ventrally to ovary. Receptaculum seminis, 0.044 × 0.060 mm, situated behind of ovary. Oncospheres 0.033–0.039 × 0.030–0.033 mm.

Tetrabothrius (*Tetrabothrius*) *umbrella* Fuhrmann, 1899
(description Temirova and Skrjabin 1978)

syn. *Chaetophallus umbrellus* (Fuhrmann, 1899) Nybelin, 1916; *Ch. robustus* Nybelin, 1916; *Ch. musculus* Szpotanska, 1917; *Ch. setigerus* Fuhrmann, 1921; *Ch. setigerus* (Szpotanska 1917) Johnston, 1935; *Ch. fuhrmanni* (Szpotanska 1917) Johnston, 1935; *Ch. siedleckii* (Szpotanska 1917) Johnston, 1935; *Ch. longissimus* (Szpotanska 1917) Johnston, 1935; *Porotaenia setigera* Szpotanska, 1917; *P. fuhrmanni* Szpotanska, 1917; *P. siedleckii* Szpotanska, 1917; *P. longissima* Szpotanska, 1917

Host and locality: *Diomedea exulans*, *D. chlororhynchus*, *Phoebetria palpebrata* (light-mantled albatross), *Macronectes giganteus* (southern giant petrel), *Thalassoica antarctica* (Antarctic petrel), *Procellaria* (*Adamastor*) *cinereas* (grey petrel) – islands of the Southern Ocean (Temirova and Skrjabin 1978)

Strobila 480 × 2.5–4 mm. Scolex 0.34–0.448 × 0.465–0.680 mm. Testes 25–37 in number. Genital atrium, shallow with week musculature, covered with long spines. Cirrus pouch, 0.091–0.114 × 0.114–0.136 mm, spherical. Male duct 0.04 mm long. Vaginal aperture ventral to male opening. covered at the distal part with long spines. Oncospheres 0.032–0.036 mm.

Tetrabothrius (*Biamniculus*) *filiformis* Nybelin, 1916
(description Temirova and Skrjabin 1978)

Host and locality: *Procellaria aequinoctialis* – waters of Antarctica (Temirova and Skrjabin 1978); *Larus dominicanus* – King George Island, South Shetlands.

Strobila very delicate. Scolex 0.3×0.3 mm. Testes 9–11 in number. Genital atrium with week musculature. Cirrus pouch 0.036×0.0288 mm. Male duct 0.028 mm long, opens dorsally to vagina.

Tetrabothrius (*Biamniculus*) *fuhrmanni* Nybelin, 1916
(description Temirova and Skrjabin 1978)

Host and locality: *Diomedea chlororhynchus*, *Procellaria aequinoctialis* – South Georgia, South Shetlands, South Orkneys, Kerguelen Islands (Temirova and Skrjabin 1978)

Strobila 60–200 mm. Scolex 0.33–0.41×0.19–0.27 mm. Testes, 15–16 in number, 0.013 mm in diameter. Genital atrium 0.095×0.142 mm. Cirrus pouch 0.065×0.047 mm. Aperture of male duct located on apex of knob, vaginal aperture at base of knob. Ovary, 0.252×0.211 mm, multilobed. Vitellarium, 0.043×0.047 mm, antero-ventrally to ovary. Oncospheres, 0.016–0.023×0.016–0.021 mm.

Tetrabothrius (*Biamniculus*) *heteroclitus* Diesing, 1850
(description Temirova and Skrjabin 1978)

syn. *Tetrabothrius auriculatus* Linstow, 1888; *T. diomedea* Fuhrmann, 1900; *T. intermedius* Fuhrmann, 1899; *T. valdiviae* Szpotanska, 1917; *T. pseudoporus* Szpotanska 1917; *Porotaenia fragilis* Szpotanska, 1917; *P. fragilis var exulans* Szpotanska, 1917; *P. fragilis var filiginosa* Szpotanska, 1917; *P. fragilis var capensis* Szpotanska, 1917

Host and locality: *Diomedea chlororhynchus*, *D. exulans*, *Phoebetria palpebrata*,*Thalassoica antarctica*, *Pagodroma nivea* (snow petrel), *Daption capense*, *Macronectes giganteus*, *Fulmarus glacialoides* (southern fulmar), *Procellaria* (*Adamastor*) *cinereus*, *Procellaria* sp.

Procellaria sp. -Crozet Islands, Baleny Islands (Temirova and Skrjabin 1978)

Strobila 73–396×3–3.08 mm. Scolex 0.42×0.42 mm. Testes, 26–37 in number, 0.258 mm in diameter. Genital atrium 0.24–0.284 mm in diameter. Apertures of long male duct and vagina on apex of ventrally curved papilla. Cirrus pouch 0.090–0.125×0.12–0.129 mm. Ovary, 2.58×0.16 mm, multilobed. Vitellarium, 0.142–0.18×0.275 mm, ventral to ovary. Oncospheres, 0.043×0.034 mm, with hooks, 0.013–0.017 mm long.

Tetrabothrius (*Biamniculus*) *mawsoni* Johnston, 1937
(description Temirova and Skrjabin 1978)

syn. *Tetrabothrius cylindraceus* Leiper et Atkinson, 1914 nec Rudolphi, 1819

Host and locality: *Catharacta skua* (great skua), Antarctica (Temirova and Skrjabin 1978)

Strobila 128–150×1.7 mm. Scolex 0.35×0.38–0.41 mm. Testes 70 in number. Cirrus pouch 0.11×0.09 mm. Male duct 0.07 mm long with aperture on apex of knob curved ventrally. Vaginal aperture ventral at base of knob.

Tetrabothrius (*Biamniculus*) *nelsoni* Leiper and Atkinson, 1914
(description Temirova and Skrjabin 1978)

syn. *Tetrabothrius glaciloides* Nybelin, 1929
Host and locality: *Diomedea melanophrys* (black-browed albatross), *Phoebetria palpebrata*; Antarctica (Temirova and Skrjabin 1978)
Strobila 39×0.88 mm. Scolex 0.48×0.52 mm. Apical organ 0.56×0.38 mm. Testes 17–18, 0.026 mm in diameter. Genital atrium 0.129×0.107 mm. Cirrus pouch 0.043×0.036 mm. Male duct, 0.056 mm long. Apertures of male duct and vagina located on apex of papilla. Ovary 0.112×0.212 mm with finger-like lobes.

Tetrabothrius (*Culmenamniculus*) *laccocephalus* Spatlich, 1909
(description Temirova and Skrjabin 1978)

Host and locality: *Ardenna griseus* (sooty shearwater), *A. gravis* (great shearwater), *A. creatopus* (pink-footed shearwater), *Calonectris diomedea borealis* (Cory's shearwater), *Pagodroma nivea* (snow petrel), *Fulmarus glacialoides* (slender-billed fulmar), *Procellaria aequinoctialis* – Scott Island (Ross Sea),waters of Antarctica (64°04/156°06) (Temirova and Skrjabin 1978)
Strobila 152×1,62 mm. Scolex 0.52×0.48 mm. Apical organ 0.38×0.46 mm. Testes 26–36, with diameter 0.021 mm. Genital atrium, 0.181–0.233×0.198–0.263 mm, with prominent knob. On apex of this knob located male and female apertures. Male duct 0.172 mm long. Cirrus pouch 0.077×0.060 mm. Ovary, 0.25–0.28×0.56–0.63 mm, multilobed. Vitellarium 0.151–0.168×0.125–0.138 mm. Oncospheres, 0.037–0.045×0.033 mm, with embryonic hooks, 0.013 mm long.

Tetrabothrius (*Culmenamniculus*) *torulosus* Linstow, 1888
(description Temirova and Skrjabin 1978)

syn. *Tetrabothrius polaris* Szpotanska, 1917; *T. intermedius var exulans* Szpotanska, 1917; *T. antarcticus* Fuhrmann, 1921; *T. kowalewskii Szpotanska*, 1925 nec Szpotanska, 1917
Host and locality: *Diomedea exulans*, *Phoebastria nigripes* (black-footed albatross), *P. albatrus* (short-tailed albatross) – Kerguelen Islands (Temirova and Skrjabin 1978)
Strobila 580×2.36 mm. Scolex 0.3×0.54 mm. Apical organ 0.44×0.18 mm. Testes 46–48 in number, 0.034 mm in diameter. Genital atrium 0.095×0.064 mm. Cirrus pouch 0.099–0.12×0.06 mm. Male duct 0.03 mm long. Male and female genital apertures on apex of knob. Ovary 0.62×0.22 mm. Vitellarium 0.15×0.06 mm. Oncospheres 0.02×0.03 mm, embryonic hooks 0.013 mm long.

Tetrabothrius (*Uniamniculus*) *lutzi* Parona, 1901
(description Temirova and Skrjabin 1978)

Host and locality: *Spheniscus magellanicus* (Magellanic penguin), *Pygoscelis papua*; Antarctica (Temirova and Skrjabin 1978)

Strobila 8.5×0.34 mm. Scolex 0.46×0.58 mm. Testes 16–19 in number and 0.017–0.021 mm in diameter. Genital atrium 0.052×0.056 mm. Cirrus pouch 0.034–0.039 mm in diameter. Ovary 0.036×0.072 mm. Vitellarium 0.019×0.022 mm.

Tetrabothrius (*Uniamniculus*) *wrighti* Leiper and Atkinson, 1914
(description Temirova and Skrjabin 1978)

Host and locality: *Pygoscelis adeliae*, *P. papua*, *Aptenodytes forsteri* (emperor penguin); Antarctica (Temirova and Skrjabin 1978)

Strobila 5–8×0.54–0.64 mm. Scolex 0.53–0.56×0.58–0.64 mm. Testes 11–12 in number, 0.017–0.022 mm in diameter. Genital atrium 0.077×0.056 mm. Cirrus pouch 0.056×0.06 mm. Vagina and male duct create a shared canal, opening in the centre of genital atrium. Ovary, 0.125–0.146×0.026–0.039 mm, multilobed.

Tetrabothrius cylindraceus (*Culmenamniculus*) Rudolphi, 1819

Host and locality: *Stercorarius loennbergi* (brown skua) – King George Island, South Shetlands (Odening 1982.)

Strobila 49–72 mm long. Scolex 0.477–0.551×0.323–0.345 mm. Testes, 24–32 in number, 0.035–0.079 mm in diameter. Genital atrium 0.069–0.124×0.072–0.117 mm. Male duct 0.054–0.057 mm long. Cirrus pouch 0.045–0.069×0.031–0.55 mm. Ovary, 0.155–0.32×0.044–0.184 mm, with finger-like projections. Vitellarium 0.066–0.079×0.035–0.55 mm. Oncospheres, 0.038–0.043×0.03–0.04 mm; embryonic hooks 0.013–0.018 mm long.

Tetrabothrius shinni Hoberg, 1987
(description Hoberg 1987)

Host and locality: *Phalacrocorax atriceps bransfieldensis* (blue-eyed shags) – Western Antarctica (Hoberg 1987)

Strobila 274–284 mm long. Scolex 0.284–0.366×0.361–0.376 mm. Testes, 35–61 in number. Muscular sucker-like genital atrium, 0.196–0.254×0.196–0.317 mm. Male duct, 0.086–0.16 mm long, extends through wall of genital atrium, curving ventrally to open anterolaterally near apex of large ventrally directed papilla. Muscular sphincter, 0.029–0.038 mm in diameter, located distally on aperture of male duct. Vagina opens in depression ventral to male papilla. Ovary, 0.96–1.39 mm in width, multilobate, composed of 29–42 lobes; 13–21 porally and 13–25 antiporally. Vitellarium 0216–0.357×0.08–0.157 mm. Oncospheres 0.028–0.044×0.022–0.033 mm; embryonic hooks 0.016–0.0188 mm.

Order Cyclophyllidea van Beneden, 1900
Family Dilepididae Railliet and Henry, 1909
Genus *Parorchites* Fuhrmann, 1932
Parorchites zederi (Baird 1853)
(description Cielecka et al. 1992; Georgiev et al. 1996)

Host and locality: *Pygoscelis antarctica*, *P. papua*, *P. adeliae* – Western Antarctica (Rennie and Reid 1912; Railliet and Henry 1912; Cielecka et al. 1992; Georgiev et al. 1996; Diaz et al. 2013); *Aptenodytes forsteri* (emperor penguin), *P. adeliae* – the eastern coast of Antarctica (Fuhrmann 1921; Johnston 1937; Prudhoe 1969; Holloway 1988, 1989); *Eudyptes schegeli* (royal penguin) – Macquarie Island, Kerguelen subregion (Prudhoe 1969)

Mature specimen up to 15 cm long and 4 mm at maximum width. Scolex 0.36–0.82 mm in width. Diameter of suckers 0.15–0.23 mm. Rostellar hooks, 18–20, arranged in two rows, pseudoscolex present. The genital pores situated irregularly alternating in the anterior part of proglottid. Genital atrium large and well supplied with muscles. Testes, 56–66 in number, in the middle part of proglottid. Cirrus pouch, 0.17–0.23 × 0.44–0.80 mm, with a thick muscular layer. Cirrus unarmed. Female genital system situated on poral side. Ovary in form of irregular, very branched racemose glands. Vitellarium smooth, kindey-shaped and situated at the rear of ovary. Vagina narrow, then widens into receptaculum seminis. Eggs round to oval in shape, 0.115–0.150 × 0.10–0.13 mm. Oncosphere, 0.036–0.039 × 0.031–0.036 mm; embryonal hooks differentiated.

Scolex, neck and juvenile proglottides deeply recessed into a cyst formed by intestinal wall. Juvenile cestodes most often wholly contained in cyst. One cyst contains 1–15 cestodes.

Genus *Anomotaenia* Cohn, 1900
Anomotaenia dominicana (Railliet and Henry 1912)
(description Zdzitowiecki and Szelenbaum-Cielecka 1984)

syn. *Choanotaenia dominicana* Railliet and Henry, 1912; *Anomotaenia antarctica* Fuhrmann, 1921; *A. micracantha dominicana* (Railliet and Henry, 1912); *Paricterotaenia ransomi* (Joyeux and Baer 1954); *P. australis* Szidat, 1964

Host and locality: *Larus dominicanus* – King George Island, South Shetlands (Zdzitowiecki and Szelenbaum-Cielecka 1984)

Strobila 40–60 mm. Scolex, 0.36–0.55 mm in width, with suckers 0.17–0.26 × 0.14–0.22 mm. Rostellar sac reaches beyond the posterior margin of suckers. Rostellum long, bearing 20–26 hooks arranged in two rows at its apex. Hooks length is 0.026–0.036 mm. Genital pore located anteriorly on the lateral margin of proglottid. Genital atrium deep with strongly muscular wall. Testes, 22–27 in number, lie in 2–3 layers dorsally in posterior part of proglottid. Cirrus pouch 0.17–0.25 × 0.30–0.35 in hermaphroditic proglottides. Ovary deeply lobed. Vitellarium lies in median part and has more compact structure than ovary. Oncospheres 0.027–0.034 × 0.023–0.031 mm; median pair of embryonal hooks are 0.013 mm long while two lateral pairs are 0.015 mm long.

Alcataenia dominicana (Railliet and Henry 1912)
(description Georgiev et al. 1996)

Host and locality: *Larus dominicanus* – South Shetlands, Livingston Island (Georgiev et al. 1996)

Strobila 66×0.8 mm. Scolex wider than neck, 0.463×0.456 mm. Suckers situated in middle of scolex, 0.219–0.225 mm in diameter. Rostellum, 0.27 mm long with highly elongate stem and expanded terminal pad. Rostellar pouch 0.20×0.124 mm, extending to posterior half of suckers. Rostellum armed with double crown of 22 hooks with long blade and curved handle. Genital pores irregularly alternating opening into anterior third of lateral proglottid margin. Testes, 24–27 in number, situated in posterior 2/3 of median field dorsally to ovary. Genital atrium large, deep with muscular wall. Cirrus pouch elongate, thin-walled, 0.224–0.237×0.031–0.046 mm. Cirrus armed. Ovary strongly lobed, wide, situated anteriorly and ventrally to testes. Vitellarium reniform, situated postero-ventrally to ovary. Vagina convoluted, surrounded along entire length by intensely stained cellular sleeve, opening posteriorly and slightly ventrally to cirrus pouch.

Genus *Nototaenia* Jones and Williams, 1967
Nototaenia fileri Jones and Williams, 1967
(description Jones and Williams 1967; Cielecka and Zdzitowiecki 1989)

Host and locality: *Chionis alba* (sheathbill) – South Orkneys (Jones and Williams 1967; Howie et al. 1968), South Shetlands (Cielecka and Zdzitowiecki 1989)

Strobila with up to eight proglottides, 2.1–2.4×0.57 mm. Scolex 0.12–0.14×0.19–0.29 mm. Rostellum, 0.21–0.23 mm long, consist of two muscular sacs, one within the other. Rostellar hooks in two rows of ten hooks each, anterior 0.034–0.037 and posterior 0.027–0.031 mm long. Suckers armed on margins with 35–44 hooks, 0.013–0.014 mm long. Testes 37–53, completely encircle female glands. Cirrus pouch 0.20–0.28 mm long, situated diagonally forward across proglottid from genital atrium to the middle of the anterior margin. Cirrus 0.02 mm long with a basal diameter of 0.05 mm, armed with spines. Ovary bilobed with six small, finger-like lobes on each side. Vagina large, armed with spines, 0.007 mm long. Eggs with short filament on each pole.

Reticulotaenia australis (Jones and Williams, 1967)
(description Jones and Williams 1967; Hoberg 1985;
Cielecka and Zdzitowiecki 1989)

Host and locality: *Chionis alba* – South Shetlands (Cielecka and Zdzitowiecki 1989), South Orkneys (Jones and Williams 1967), Palmer Archipelago, South Georgia (Hoberg 1983, 1985)

Strobila 15–60×0.8–2.5 mm. Scolex 0.342–0.460 mm at width. Rostellar hooks, ten in number, arranged in one row, 0.037–0.055 mm long. Genital aperture unilateral. Testes, 32–45 in number, situated in the posterior half of proglottid. Cirrus pouch, 0.08–0.174×0.015–0.040 mm, thin-walled contains the twisted ejaculatory canal. Reticulate ovary strongly branching in the anterior part of proglottid. Vagina thin-walled widens into a transversely elongated receptaculum seminis. Receptaculum seminis, 0.155–0.261×0.052–0.142 mm. Vitellarium multilobular, in the median line of proglottid.

Family Hymenolepididae Ariola, 1899
Genus *Microsomacanthus* Lopez-Neyra, 1942
Microsomacanthus shetlandicus Cielecka and Zdzitowiecki, 1981
(description Cielecka and Zdzitowiecki 1981)

Host: *Larus dominicanus*, *Chionis alba* – King George Island, the South Shetlands (Cielecka and Zdzitowiecki 1981, 1989)

Strobila 2–3 mm (fixative in alcohol) or 10–13 mm (in formalin). Scolex triangular, 0.19–0.25 mm width, with long rostellum and suckers, 0.12–0.14 × 0.09–0.11 mm. Rostellum 0.17–0.19 mm long; rostellar pouch reaches to below the posterior margin of suckers. On rostellum is 10 hooks, 0.046–0.052 mm long. Testes, three in number, situated at a V-type obtuse angle. Cirrus pouch, 0.15–0.205 × 0.03 mm with wall composed of 28–32 bundles. Cirrus small, cylindrical with slightly widening terminal part and slender base covered with thick, small spines. Ovary and vitellarium situated near the mid-line of proglottid, without distinct lobes, ventrally to testes. Vagina thin-walled, wide, without distinct structures in copulatory part. Oncospheres, 0.033–0.04 × 0.03–0.036 mm. Embryonal hooks equal size, 0.015–0.016 mm.

Microsomacanthus secundus Cielecka and Zdzitowiecki, 1989
(description Cielecka and Zdzitowiecki 1989)

Host and locality: *Chionis alba* – King George Island, the South Shetlands (Cielecka and Zdzitowiecki 1989)

Strobila 1.9 × 0.2 mm. Scolex, 0.168–0.18 mm wide with long rostellum, 0.13 mm long. Suckers, 0.067–0.082 mm in diameter. Rostellar hooks, ten in number, 0.037–0.040 mm long. Testes, three in number situated in one transverse line or forming an obtuse angle. Cirrus pouch, 0.085–0.108 × 0.024–0.034 mm. Cirrus thin, short, slightly broader at the basal part and armed with very small spines. On the dorsal side of the cirrus base, in conjunction with the cirrus pouch, an additional duct is present. This duct opens into genital atrium and is similar to sacculus accessorius, but without visible armature. Ovary, 0.03 × 0.045 mm, oval with smooth surface. Vitellarium 0.018 × 0.030 mm, in the central line of proglottid on the ventral side of the ovary.

Branchiopodataenia arctowskii (Jarecka and Ostas, 1984)
(description Jarecka and Ostas 1984)

syn. *Hymenolepis arctowskii* Jarecka et Ostas, 1984; *Wardium arctowskii* (Jarecka and Ostas 1984)

Host and locality: *Larus dominicanus* -King George Island, the South Shetlands (Jarecka and Ostas 1984)

Strobila 70–90 × 0.7 mm. Scolex, 0.24 × 0.18 mm, with four suckers 0.077 × 0.93 mm in diameter. Rostellum, 0.108 × 0.049 mm, armed with a single crown of ten hooks, aploparaxoid-shape, 0.016–0.018 mm long. Genital atrium marginal, unilateral and ventral. Testes, three in number, arranged in a horizontal line in the posterior region of proglottid. Cirrus pouch, 0.18–0.217 × 0.037–

0.045 mm. Cirrus 0.016–0.023 mm long, partially armed. Ovary three-lobed, situated in the postero-ventral space of a proglottid. Vitellarium, oval, 0.023 mm in diameter, ventral to ovary. Receptaculum seminis 0.100–0.13 × 0.04–0.045 mm, ventral to cirrus pouch. Vagina cup-shaped opens into genital atrium ventrally to cirrus. Copulatory portion of vagina is developed into a chitinoid latch-like structure surrounding the vaginal orifice. Vaginal cup, 0.009–0.015 mm long, with chitinoid ornamentation in a form of longitudinal grooves. Oncospheres with filaments. Embryonic hooks asymmetric in length.

6.1.4 Review of Larval Forms Occurring in Teleosts

The larvae of cestodes are very common in the Antarctic teleosts and mammals (see: Rocka 2003), some of which parasitizing skates. The plerocercoids of the Diphyllobothriidae and the Tetrabothriidae are parasites of marine mammals and birds in the adult stage. Plerocercoids are located mainly in the wall of the stomach and the mesentery, but are rare in the liver and the lumen of stomach and small intestine of hosts. Larvae of the Tetrabothriidae occur in the small intestine of fish; they possess an apical sucker and lack bothridia.

Five morphological forms of cestode larvae in adults parasitizing skates have been found in Antarctic and Subantarctic teleosts. These larvae have bothridia divided into one, two and three loculi, bothridia undivided with sucker and hook-like projections, or subcylindrical bothridia. The larvae have typically been found in various parts of the small intestine, but also very rarely in the stomach. Their morphological characteristics only allow identification to genus, family or order, and identification of these larvae is problematic because their scolices are different from those of adult parasites. Fortunately, this problem can be resolved using molecular strategies, but although such studies have been conducted, they have not been used in samples taken from inside Antarctica (see: Jensen and Bullard 2010). Jensen and Bullard (2010) identified eight larval types as belonging to the Tetraphyllidea and the Rhinebothriidea: these being similar to the larvae found in Antarctic teleosts. In Antarctica, only one identification of a larvae with adult form has been confirmed. Laskowski and Rocka (2014) assigned larvae with trilocular bothridia from *Notothenia rossi* to *Onchobothrium antarcticum* from *Bathyraja eatonii* (the South Shetlands).

Larvae with monolocular bothridia (Fig. 6.1)
(description Wojciechowska 1993a)

Body length 1.4–8.3 mm. Scolex with apical sucker and four bothridia. Apical sucker, 0.14–0.24 mm in diameter. Bothridia, 0.33–0.9 × 0.23–0.92 mm, with free posterior ends. Originally, the bothridium was described as monolocular with bothridial sucker but SEM photos suggest that it is not sucker rather small loculi or pad. This structure has 0.13–0.30 mm in diameter.

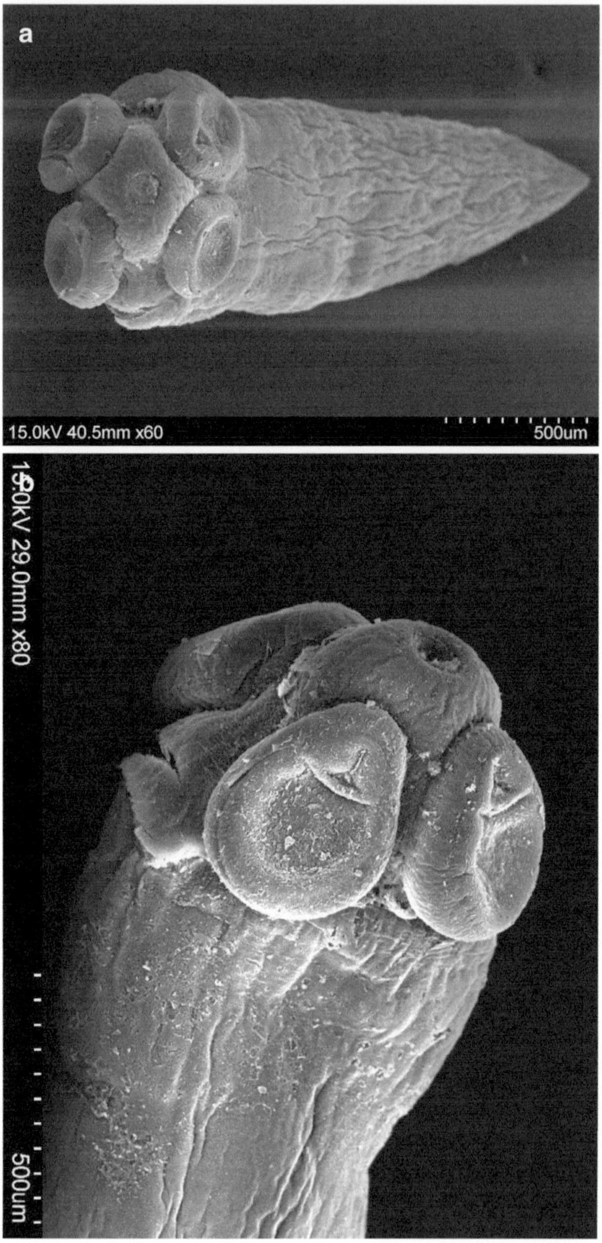

Fig. 6.1 Larva with monolocular bothridia. (**a**) whole larva, (**b**) scolex, apical view scolex (scale bar = 500 μm)

Similar larvae was reported as Type II by Jensen and Bullard (2010).

Host: Notoheniidae, Channichthyidae, Myctophidae, Harpagiferidae and Gemphylidae

Locality: the South Shetlands area – Admiralty Bay, mesopelagic zone at the South Shetlands and Joinville Island, shelves around South Georgia and at Shag Rocks (Wojciechowska 1993a, b; Zdzitowiecki and Zadróżny 1999), off the Heard Island (Wojciechowska et al. 1995), McMurdo Sound (Moser and Cowen 1991).

Larvae with bilocular bothridia (Fig. 6.2)
(description Wojciechowska 1993a; Wojciechowska et al. 1994; Rocka 1999a)

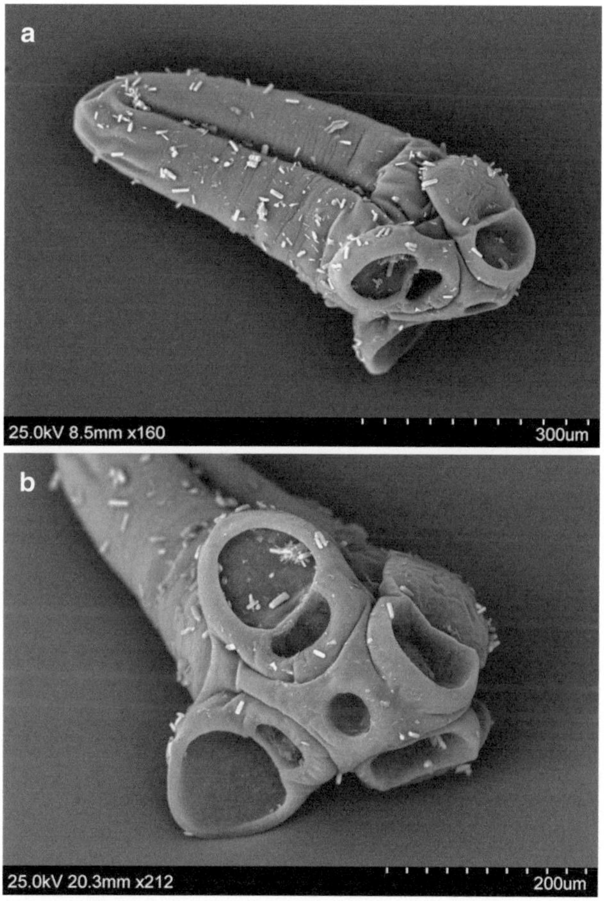

Fig. 6.2 Larva with bilocular bothridia. (**a**) Whole larva (scale bar = 300 μm); (**b**) scolex, apical view (scale bar = 200 μm)

Body length 0.55–1.97 mm. Scolex with apical sucker and four bothridia. Apical sucker, 0.06–0.12 mm in diameter. Bothridia, 0.16–0.35 × 0.09–0.24 mm, sessile or with free posterior ends, divided into two loculi; division between loculi distinct. Anterior loculi 0.07–0.18 × 0.07–0.17 mm; posterior loculi 0.07–0.22 × 0.07–0.19 mm.

Host: Notoheniidae, Artedidraconidae, Channichthyidae, Bathydraconidae, Myctophidae, Macrouridae and Muraenolepididae

Locality: the South Shetlands area – Admiralty Bay, shelves around the South Shetlands and Cumberland Bay (Szidat and Graefe 1968; Parukhin and Sysa 1975; Zdzitowiecki 1978; Rodjuk 1985; Reimer 1987; Wojciechowska 1993a, b; Palm et al. 1998), the Ross Sea (Leiper and Atkinson 1914, 1915; Wojciechowska et al. 1994; Laskowski et al. 2005), off Adeliae Land (Prudhoe 1969; Zdzitowiecki et al. 1998; Zdzitowiecki 2001a; Laskowski et al. 2007), coastal waters at the South Orkneys (Zdzitowiecki et al. 1997), coastal waters at the Davis Sea, the eastern part of the Weddell Sea (Rocka 1999a), off the Heard Island (Wojciechowska et al. 1995) and off Kerguelen Islands (Prudhoe 1969).

Larvae with trilocular bothridia (Fig. 6.3)
(description Wojciechowska 1993a; Wojciechowska et al. 1994; Rocka 1999a)

Body length 1.1–6.3 mm. Apical sucker 0.07–0.16 mm in diameter. Bothridia, 0.23–0.52 × 0.15–0.38 mm, sessile, divided into three loculi. Anterior loculi 0.06–0.15 × 0.06–0.18 mm; middle 0.09–0.30 × 0.12–0.31 mm; posterior loculi 0.07–0.22 × 0.06–0.25 mm.

Host: Nototheniidae, Channichthyidae, Bathydraconidae, Myctophidae, Artedidraconidae, Harpagiferidae

Locality: South Shetlands (Rodjuk 1985; Reimer 1987; Wojciechowska 1993a,b; Palm et al. 1998; Zdzitowiecki and Zadróżny 1999; Zdzitowiecki 2001b); the Ross

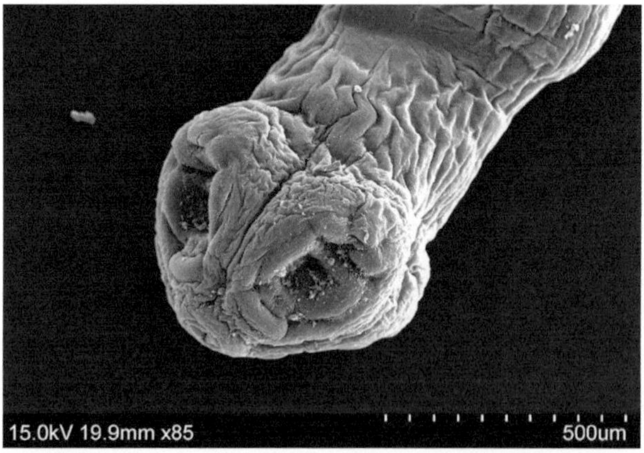

Fig. 6.3 Larva with trilocular bothridia. Scolex (scale bar = 500 μm)

Sea (Wojciechowska et al. 1994); the Weddell Sea (Rocka 1999a); off Adeliae Land (Zdzitowiecki et al. 1998; Zdzitowiecki 2001a; Laskowski et al. 2007), coastal waters at the South Orkneys (Zdzitowiecki et al. 1997)

Larvae with leaf-like bothridia and hook-like projections
(description Wojciechowska 1993a)

Body 0.67–0.92 mm long. Bothridia, 0.22–0.40 × 0.08–0.18 mm, leaf-like, sessile anteriorly. Each bothridium with accessory sucker, 0.06–0.08 mm in diameter, and a pair of hook-like projections, 0.03–0.04 mm long. Apical sucker absent.

Host: *Champsocephalus gunnari* (mackerel icefish) from shelf at Shag Rocks, South Georgian area (Wojciechowska 1993a, b), off the Heard Island (Wojciechowska et al. 1995).

Larvae with subcylindrical bothridia
(description Wojciechowska et al. 1994, Rocka 1999a)

Body length 0.74–1.388 mm. Bothridia, 0.19–0.26 × 0.13–0.19 mm, subcylindrical, free posteriorly. Each bothridium has shallow cavity on its anterior end surrounded by sucker-like structure, 0.11–0.14 mm in diameter.

Host: *Trematomus centronotus* (sharp-spined notothen), *Cryodraco antarcticus* (long-fingered icefish), *Pogonophryne scotti* (saddleback plunderfis)

Locality: the Ross Sea – Terra Nova Bay (Wojciechowska et al. 1994), the eastern part of the Weddell Sea – off Kap Norvegia and north of Halley Bay (Rocka 1999a)

6.1.5 Life Cycle Biology of Antarctic Cestodes

Antarctic teleosts play an important role in the completion of life cycles of many helminth species. They serve as either definitive or intermediate and paratenic hosts. Skates are definitive hosts only for cestodes and one digenean species, *Otodistomum cestoides* (Beneden 1871). As elsewhere, life cycles of the Cyclophyllidea (parasites of birds), and the Bothriocephalidea (parasites of teleost fishes) are probably realized with two hosts. Only one life cycle of the Antarctic cestode, *Branchiopodataenia arctowskii* (Jarecka and Ostas 1984), parasitizing *Larus dominicanus* is known, with *Brachinecta gaini* (Branchiopoda) as its intermediate host (Jarecka 1984). It should be noted that *B. arctowskii* has a bipolar distribution and is not endemic to Antarctica (Bondarenko and Kontrimavichus 2004).

Life cycles of cestodes from skates, and the Diphyllobothriidae and the Tetrabothriidae (parasites of birds and mammals) are probably realized with crustaceans as the first intermediate hosts and with teleosts as the second intermediate and/or paratenic hosts.

6.2 Nematodes of Antarctic Fishes and Birds

6.2.1 Introduction

Only six nematode species have been found as adults in Antarctic teleosts and only five species, belonging to three families and one order, have been found in birds. The life cycles of nematodes from Antarctic teleosts are unknown. Larvae of anisakid genera, *Anisakis* Dujardin, 1845; *Contracaecum* Railliet and Henry, 1912; *Hysterothylacium* Ward and Magath, 1917 and *Pseudoterranova* Mozgovoy, 1950 (adults are parasites of mammals, birds and teleosts) are common in Antarctic bony fishes (Rocka 2006), where they play the role of the second intermediate and paratenic hosts.

6.2.2 Systematic Review of Nematodes Parasitizing Fishes

Order Spirurida
Superfamily Habronematoidea Ivaschin, 1961
Family Cystidicolidae Skrjabin, 1946
Ascarophis Beneden, 1871
Ascarophis nototheniae Johnston and Mawson, 1945
(description Johnston and Mawson 1945; Holloway et al. 1967;
Rocka 1999b, 2002, 2004)

Host: many species of Notothenioidei (Nototheniidae, Harpagiferidae, Channichthyidae, Bathydraconidae), rarely Congiopodidae and Zoarcidae

Locality: West Antarctica: environs of the South Shetland Islands, open sea shelf and fjords at South Georgia, coastal waters at the South Orkney Islands (Szidat and Graefe 1968; Gaevskaya and Rodjuk 1997; Rocka 1999b; Zdzitowiecki and Zadróżny 1999; Zdzitowiecki 2001b), East Antarctica: 64°144′–67°138′S 62°103′–142°136′E (Johnston and Mawson 1945), McMurdo Sound (Holloway et al. 1967; Holloway and Spence 1980), off Adelie Land, Ross Sea (Terra Nova Bay), Davis Sea, the eastern part of the Weddell Sea (Zdzitowiecki et al. 1998; Rocka 1999b, 2002; Zdzitowiecki 2001a)

Subantarctica: the Kerguelen subregion – off the Crozet Islands, off the Heard Island, off the Kerguelen Island, off the Macquarie Island (Johnston and Mawson 1945; Parukhin and Lyadov 1982; Parukhin and Zaitsev 1984; Lyadov 1985; Parukhin 1989; Gaevskaya et al. 1990; Rocka 1999b)

Habitat: mainly stomach, also small intestine and pyloric caeca

Female Body 5.0–24.6 × 0.09–0.22 mm. Buccal capsule 0.11–0.20 mm. Entire oesophagus 2.57–4.98 mm long; muscular and glandular 0.2–0.6 and 2.18–4.48 mm long, respectively. Nerve ring and excretory pore 0.13–0.25 and 0.12–0.36 mm from the anterior end, respectively. Tail 0.03–0.08 mm long with small knob-like terminal projection. Vulva 4.0–7.1 mm from posterior end of the body.

Mature eggs 0.04–0.05 × 0.02–0.03 mm with two filaments arising from prominent knob at each pole.

Male Body 3.7–12.2 × 0.02–0.05 mm. Buccal capsule 0.1–0.2 mm long. Entire oesophagus 1.7–4.4 mm; muscular and glandular 0.26–0.52 and 1.4–3.9 mm long, respectively. Nerve ring and excretory pore 0.12–0.26 and 0.17–0.38 mm from the anterior end, respectively. Tail 0.09–0.15 mm long. Caudal alae narrow. Left spicule 0.4–0.6 mm long with a sharply pointed distal part. Right spicule 0.06–0.12 mm long with blunt distal tip. Area rugosa with nine longitudinal rows of cuticular elevations. Caudal papillae nine pairs; four pairs preanal and five postanal.

Genus *Cystidicola* Fischer, 1798
Cystidicola beatriceinsleyae (Holloway and Klewer, 1969)
(description: Holloway and Klewer 1969; Rocka 2002, 2004)

Syn. *Rhabdochona beatriceinsleyae* Holloway and Klewer, 1969

Host: *Rhigophila* (=*Lycodichthys*) *dearborni*, *Lycodichthys antarcticus* (Zoarcidae)

Locality: McMurdo Sound (Holloway and Klewer 1969), the eastern part of the Weddell Sea (Rocka 2002)

Habitat: mainly various parts of small intestine, stomach

Female Body 15.8–32.2 × 0.1–0.18 mm. Buccal capsule 0.15–0.27 mm long. Entire oesophagus 0.8–1.2 mm long; muscular and glandular 0.16–0.24 and 0.64–0.97 mm long, respectively. Nerve ring and excretory pore 0.26–0.32 and 0.33–0.5 mm from the anterior end, respectively. Tail 0.16–0.24 mm long, curved dorsally, bluntly rounded with a small ventral indentation near tip. Vulva 8.5–11.3 mm from the posterior end of the body. Eggs 0.042–0.049 × 0.021–0.026 mm with two polar caps, each with between two and four filaments.

Male Body 7.9–15.2 × 0.07–0.11 mm. Buccal capsule 0.2–0.26 mm long. Entire oesophagus 0.65–0.98 mm long; muscular and glandular 0.19–0.22 and 0.46–0.76 mm, respectively. Nerve ring and excretory pore 0.26–0.31 and 0.37–0.47 mm from the anterior end, respectively. Tail 0.16–0.29 mm long, bluntly rounded at tip. Posterior end curved ventrally. Caudal alae narrow. Caudal papillae 11 pairs; six preanal and five postanal pairs. Left spicule 0.47–0.77 mm long, anterior half tubular and posterior half concave, ventrally becoming more alate at the level of right spicule; posterior end cupped with irregular sides. Right spicule 0.08–0.12 mm long, broad and cup-shaped to deflect left spicule.

Order Ascaridida
Superfamily Ascaridoidea Baird, 1853
Family Anisakidae Skrjabin and Karokhin, 1945
Genus *Hysterothylacium* Ward and Magath, 1917
Hysterothylacium aduncum (Rudolphi 1802)
(description Mozgovoy 1953; Hartwich 1975; Rocka 2004)

syn. *Ascaris adunca* Rudolphi, 1802

Host: *Dissostichus eleginoides* (Patagonian toothfish), *D. mawsoni* (Antarctic toothfish)

Locality: environs of the South Shetland Islands: open sea shelf at Shag Rocks, off South Georgia (Gaevskaya et al. 1990)

Habitat: stomach and intestine

Female Body 24–48 × 0.72–1.4 mm. Cervical alae 4.5 × 0.046–0.047 mm. Lips up to 0.15 mm, interlabia up to 0.06 mm long. Nerve ring 0.66–0.9 mm from the anterior end. Oesophagus 2.25–4.12 mm long; ventriculus 0.17–0.18 × 0.16–0.18 mm with an appendix, 0.62–0.87 mm long. Intestinal caecum 0.91–1.56 mm long. Tail 0.28–0.42 mm long, conical with a small projection covered with minute spines. Eggs 0.062–0.07 × 0.046–0.047 mm.

Male Body 18–35 × 0.43–0.8 mm. Cervical alae 4.21 × 0.04 mm. Lips up to 0.15 mm, interlabia up to 0.05 mm long. Nerve ring 0.53 mm from the anterior end. Oesophagus 1.94–3.23 mm; ventriculus 0.15–0.17 × 0.12–0.14 mm with an appendix, 0.054–0.81 mm long. Intestinal caecum 0.65–0.93 mm long. Tail 0.11–0.14 mm, curved ventrally, conical with projection covered with minute spines. Caudal papillae 29 pairs; 23 preanal, four postanal and two paraanal pairs. Spicules 2.01–4.65 mm long. Caudal alae absent.

Genus *Paranisakiopsis* Yamaguti, 1941
Paranisakiopsis weddelliensis Rocka, 2002
(description Rocka 2002, 2004)

Host: *Macrourus whitsoni* (Macrouridae)

Locality: the eastern part of the Weddell Sea (Rocka 2002)

Habitat: pyloric caeca

Female Body 55–62 × 0.85–1.55 mm wide near vulva. Head end 0.25–0.27 mm in diameter at base. Lips up to 0.2 mm; interlabia up to 0.11 mm long. Oesophagus 3.64–4.00 mm long and 0.25–0.3 and 0.38–0.41 mm wide at its anterior and posterior part, respectively. Ventriculus 0.32 × 0.34 mm. Vulva preequatorial. Tail conically pointed, 0.25–0.46 mm long. Eggs 0.083–0.085 × 0.063–0.065 mm, thin-walled.

Male Body 40–48 × 0.75–1.0 mm. Head end 0.23–0.27 mm in diameter at base. Lips up to 0.2 mm, interlabia up to 0.14 mm long. Oesophagus 4.2–5.6 mm long, 0.25–0.43 × 0.38–0.46 mm. Tail 0.25–0.38 mm long, pointed, curved ventrally, narrowed immediately posterior to the first pair of postanal papillae. Eight to ten pairs of preanal papillae; four pairs of postanal papillae. First pair of postanals just posterior to the cloacal aperture. Spicules equal or subequal, 0.42–0.62 mm long.

Superfamily Seuratoidea Railliet, 1906
Family Cucullanidae Cobbold, 1864

Genus *Dichelyne* Jagerskiold, 1902
Subgenus *Cucullanellus* (after Petter 1974)
Dichelyne (*Cucullanellus*) *fraseri* (Baylis, 1929)
(description Baylis 1929; Zdzitowiecki and Cielecka 1996)

syn. *Cucullanus fraseri* Baylis, 1929; *Cucullanus fraseri var. nototheniae* Baylis, 1929

Host: Channichthyidae, Nototheniidae, Bathydraconidae, Muraenolepididae, Harpagiferidae

Locality: West Antarctica: off South Georgia, open sea shelf at Shag Rocks, open sea shelves at the South Shetlands and Joinville Island, coastal waters at the South Orkneys (Baylis 1929; Parukhin and Sysa 1975; Zdzitowiecki 1978; Parukhin and Lyadov 1982; Rodjuk 1985; Reimer 1987; Gaevskaya et al. 1990; Zdzitowiecki and Cielecka 1996)

Habitat: intestine

Female Body 3.34–7.34 × 0.22–0.42 mm. Pseudobuccal capsule 0.11–0.18 mm wide. Oesophagus 0.61–1.13 mm long. Intestinal caecum 0.26–0.56 mm long. Tail 0.14–0.22 mm, conical. Anterior ovary almost extends beyond the junction of oesophagus and intestine, posterior ovary almost reaches the anus. Uterus amphidelphic. Vulvar opening on a small protuberance. A pair of phasmids present midway between anus and posterior end.

Male Body 3.06–5.55 × 0.16–0.34 mm. Posterior part of the body ventrally curved. Pseudobuccal capsule 0.1–0.15 mm wide. Oesophagus 0.55–0.97 mm long. intestinal caecum 0.16–0.6 mm long. Tail 0.12–0.21 mm long, conical. Spicules 0.67–1.01 mm with flattened, rounded distal end. Gubernaculum 0.11–0.16 mm long. Caudal papillae 11 pairs; anterior three pairs precloacal, five pairs near the opening of cloaca, two pairs near the posterior end as well as a pair of phasmids. Phasmids midway between paracloacal papillae and those of the two posterior pairs. Testis extending anterially nearly beyond, at the level of, or nearly before the junction of oesophagus and intestine.

Order Enoplida
Superfamily Trichinelloidea Ward, 1907
Family Capillariidae Neveu-Lemaire, 1936
Subfamily Capillariinae Zeder, 1800
Subgenus *Procapillaria* (after Moravec 1987)
Capillaria (*Procapillaria*) sp.
(description Rocka 2002, 2004)

Host: *Macrourus whitsoni*

Locality: the eastern part of the Weddell Sea (Rocka 2002)

Female Body 17.6–26 × 0.08–0.09 mm wide near vulva. Two lateral and one ventral bacillary bands present. Entire oesophagus 8.9–10.4 mm long; its muscular part 0.4–0.53 mm. Stichosome, 8.5–9.9 mm, composed of 46–49 stichocytes. Vulva

0.04–0.11 mm below level of oesophagus and intestine juncture. Bell-shaped vulvar appendage, 0.04–0.06 × 0.018–0.062 mm. Rectum 0.07–0.12 mm long. Anus subterminal, tail 0.009–0.012 mm long. Mature eggs 0.077–0.082 × 0.037–0.042 mm, thin-walled.

Male Not found.

6.2.3 Systematic Review of Nematodes Parasitizing Birds

Order Spirurida
Superfamily Acuarioidea Railliet, Henry and Sisiff, 1912
Family Acuariidae Railliet, Henry and Sisiff, 1912
Genus *Stegophorus* Wehr, 1934
Stegophorus macronectes (Johnston and Mawson 1945)
(description Zdzitowiecki and Dróżdż 1980)

syn. *S. adeliae* (Johnston 1937) sensu Petter, 1959; *S. paradeliae* (Johnston and Mawson 1945)

Host and locality: *Macronectes giganteus*, *Stercorarius skua loennbergi*, *Chionis alba*, *Pygoscelis adeliae*, *P. antarctica*, *P. papua* – the South Shetlands, King George Island (Zdzitowiecki and Dróżdż 1980; Diaz et al. 2013) and Deception Island (Vidal et al. 2012); *Eudyptes chrysocome*, *E. chrysolophus* – Subantarctica, Heard Island and Macquarie Island Mawson (1953)

Habitat: stomach

The collar composed of two lateral lobes each bearing 15–21 denticles (total number from 30 to above 40). Deirids tridentate, with middle denticle shorter than the lateral ones, lie behind the junction of vestibule and oesophagus. Nerve ring posterior to deirids, sometimes at their level. Oesophagus divided into two segments – muscular and glandular one. The glandular is 2.04–3.58 times longer than the muscular.

Female Body 6.4–15.4 × 0.3–13–0.30 mm. Maximum range of collar 0.071–0.109 mm. Deirids located 0.151–0.265 mm from the anterior end, their width being 0.023–0.029 mm. Genital opening on the cuticular protuberance from the midpoint to 2/3 of the body length. Mature eggs 0.042–0.046 × 0.020–0.024 mm.

Male Body 3.7–7.2 × 0.082–0.173 mm. Maximum range of collar 0.051–0.074 mm. Deirids 0.134–0.23 mm from the anterior end, their width is 0.017–0.021 mm. Tail end bears long lateral alae. Cloaca 0.125–0.185 mm from the posterior end. Four pairs of preanal and five pairs of postanal papillae present. The preanal papillae of the 1st and 3rd pairs are thin, the 2nd and 4th preanal and 1st and 3rd postanal pairs more strongly developed. Each papilla dilated in its distal part. Phasmids just behind the last pair of papillae. The longer spicule 0.71–1.23 mm, thin, slightly dilated at the distal end ended with a sharply pointed tip. The shorter spicule 0.074–0.098 mm, slightly bent and ended with cresent-like processus directed anteriad.

Stegophorus arctowskii Zdzitowiecki and Dróżdż, 1980
(description Zdzitowiecki and Dróżdż 1980)

Host and locality: *Macronectes giganteus*, *Stercorarius skua loennbergi* – King George Island, South Shetlands (Zdzitowiecki and Dróżdż 1980)
Habitat: stomach

Female Body 10.1–12.4 mm long and 0.055–0.060 mm wide at the level of the deirids and 0.247–0.333 mm at the level of the oesophagus. Collar composed of two lateral lobes each with 9–12 denticles. Maximum range of collar 0.046–0.039 mm. Deirids tridentate, all denticles equal length, situated before the junction of vestibule and oesophagus. Vulva located 80 % of the total body length, without any protuberance and ornamentation. Mature eggs 0.040–0.043 × 0.020–0.021 mm. Tail 0.083–0.01 mm.

Male Body 5.7 × 0.146 mm. The range of collar 0.033 mm, about ten denticles on each lobe. Tail with two lateral alae. Four pairs of preanal and five pairs of postanal pappilae present, arranged similarly as in males of *S. macronectes*. Cloaca is 0.1 mm from the posterior end. Spicules 0.625 and 0.070 mm long.

Genus *Paracuaria* Rao, 1951
Paracuaria tridentata (Linstow 1877)

syn. *Filaria tridentata* Linstow, 1877; *Spiroptera tridentata* (Linstow 1877) Newman, 1900; *Streptocara tridentata* (Linstow 1877) Skrjabin, 1916; *S. transcaucasica* (Solonitsin 1932); *S. rissae* Kreis, 1958; *Paracuaria macdonaldi* Rao, 1951
Host and locality: *Chionis alba* – South Orkneys (Howie et al. 1968; Jones and Williams 1969)

Female Body 7.3–11.5 × 0.08–0.12 mm. Vulva on two thirds of body length. Eggs 0.035–0.037 × 0.017–0.02 mm. Cephalic structures consist of vestibule, 0.12–0.15 mm long and two trifid spines, 0.017–0.19 mm behind head end. Each trifid spine, 0.016–0.017 × 0.013 mm.

Male Body 7.6 × 0.12 mm. Vestibule 0.13 mm long. Four pairs of preanal papillae present.

Superfamily Habronematoidea Ivaschin, 1961
Family Tetrameridae Travassos, 1914
Genus *Tetrameres* Creplin, 1846
Tetrameres wetzeli Schmidt, 1965
(description Schmidt 1965)

Host and locality: *Eudyptes chrysocome*, *Aptenodytes patagonicus*, *Pygoscelis papua* – South Shetlands, King George Island (Diaz et al. 2013), Kergulen Islands (Schmidt 1965), Crozet Archipelago (Fonteneau et al. 2011)
Habitat: stomach

Female Body 5×6 mm. Nerve ring situated 0.12–0.13 mm from anterior end. Eggs 0.061×0.034 mm.

Male Body 7.34–9×0.48 mm. Nerve ring 0.28 mm from the anterior end. Spicules, 2.73–3.22 mm and 0.17–0.22 mm long, respectively with a length ratio of 1:15. Four pairs of postcloacal papillae and cuticular protuberances wave-like arranged along the dorsal and ventral medium line.

The nearly globular females also possess very poorly developed cuticular protuberances along the dorsal and ventral longitudinal furrow. Male and females in a common cyst.

Superfamily Filarioidea
Family Onchocercidae (Leiper 1911)
Genus *Eulimdana* Founikoff, 1934
Eulimdana rauschorum Hoberg, 1986
(description Hoberg 1986)

Host and locality: *Larus dominicanus*, Palmer Station, Antarctica (Hoberg 1986)
Habitat: subcutaneous connective tissue of the esophageal region
Body robust, bent slightly ventrad with thick cuticle. Anterior and posterior extremities blunt, rounded. Cephalic region slightly expanded, bulbous, divided asymmetrically into small ventral and large dorsal lobe. Mouth simple without oral cuticular inflations, bordered by two large lateral amphids, surrounded by four pairs of small circumoral papillae distributed symmetrically in dorsal and ventral groups. Mouth and papillae surrounded by prominent circumoral groove. Cuticle with fine transverse striations. Irregular longitudinal thickenings of the cuticle sometimes evident.

Female Body 7.6–12.3 mm long. Body 0.140–0.214 mm wide in cephalic region; 0.342–0.490 mm at midbody; 0.151–0.255 mm in caudal region. Nerve ring 0.122–0.184 mm from cephalic extremity. Excretory pore not observed. Esophagus 0.371–0.520×0.029–0.053 mm at base. Vulva postesophageal 0.838–1.620 mm from cephalic extremity. Vagina strongly muscular, 0.72–0.850 mm in length. Ovaries paired, opisthodelphic. Uterine limbs convoluted, maximum extent to nerve ring in anterior and to caudal extremity in posterior. Uterus with developing larvae in delicate thin-shelled eggs 0.058–0.085×0.036–0.055 mm. Anus patent, ventral, subterminal, 0.03–0.07 mm from caudal extremity. Caudal papillae prominent, lateral, seldom median; two-three on the right side; two on left; occasionally single median, papilla present posterior to anus. Phasmids not observed. Caudal extremity usually with cleft, separating tail into dorsal and ventral lobes. Lateral chords with numerous nuclei.

Male Body 3.5–5.8 mm long. Body width 0.127–155 mm in cephalic region; 0.200–0.269 mm at midbody and 0.096–0.120 mm in caudal region. Nerve ring 0.116–0.177 mm from the anterior extremity. Excretory pore not observed. Esophagus 0.354–0.470 × 0.029–0.057 mm at base, clearly demarcated from intestine without appendix. Coelomocytes often prominent in esophageal region. Anus subterminal, 0.028–0.050 mm from caudal extremity; two small unstalked adanal papillae generally present at the postero-lateral margin of anus. Large pedunculate caudal papillae in two rows lateral to anus; four-five papillae present on the right side, and three-five on the left. Spicules equal, asymmetric distally. Right spicule 0.131–0.154 mm long; tip blunt, conical, poorly sclerotized. Left spicule 0.131–0.157 mm; tip sharply pointed with complex sclerotized structure. Phasmids pedunculate, small, indistinct, ventral, subterminal. Alae-like structures on tail absent.

Larvae Ensheathed microfilariae 0.325–0.406 mm long, numerous in vagina; not observed in blood.

References

Andersen KI, Lysfjord S (1982) The functional morphology of the scolex of two *Tetrabothrius* Rudolphi 1819 species (Cestoda: Tetrabothriidae) from penguins. Parasitol Res 67:299–307

Barbosa A, Palacios MJ (2009) Health of Antarctic birds: a review of their parasites, pathogens and diseases. Polar Biol 32:1095–1115

Baylis HA (1929) Parasitic Nematoda and Acanthocephala collected in 1925–7. Discov Rep 1:541–560

Bondarenko S, Kontrimavichus V (2004) On *Branchiopodataenia* n. g., parasitic in gulls, and its type-species, *B. anaticapicirra* n. sp. (Cestoda: Hymenolepididae). Syst Parasitol 57:119–133

Caira JN, Jensen K, Waeschenbach A, Olson PD, Littlewood TJ (2014) Orders out of chaos – molecular phylogenetics reveals the complexity of shark and stingray tapeworm relationships. Int J Parasitol 44:55–73

Cielecka D, Zdzitowiecki K (1981) The tapeworm *Microsomacanthus shetlandicum* sp.n. (Hymenolepididae) from the Dominican Gull of King George Island (South Shetlands, Antarctic). Bull Acad Pol Sci Ser Biol 29:173–180

Cielecka D, Zdzitowiecki K (1989) Tapeworms from the sheathbill (*Chionis alba*) on King George Island (Antarctic). Acta Parasitol Pol 34:15–25

Cielecka D, Wojciechowska A, Zdzitowiecki K (1992) Cestodes from penguins on King George Island (South Shetlands, Antarctic). Acta Parasitol 37:65–72

Diaz JN, Fusaro B, Longarzo L, Coria NR, Vidal V, Jerez S, Ortiz J, Barbosa A (2013) Gastrointestinal helminths of Gentoo penguins (*Pygoscelis papua*) from stranger Point, 25 de Mayo/King George Island, Antarctica. Parasitol Res 112:1877–1881

Duhamel G et al (2014) Chapter 7. Biogeographic patterns of fish. In: De Broyer C, Koubbi P, Griffiths HJ, Raymond B, Udekem d'Acoz C d' et al (eds) Biogeographic Atlas of the Southern Ocean. Scientific Committee on Antarctic Research, Cambridge, pp 328–362

Fonteneau F, Geiger S, Marion L, Le Maho Y, Robin JP, Kinsella JM (2011) Gastrointestinal helminths of King penguins (*Aptenodytes patagonicus*) at Crozet Archipelago. Polar Biol 34:1249–1252

Fuhrmann O (1921) Die Cestoden der Südpolar Expedition 1901–1903. Dt Südpol Exped 1901–1903 16(Zool. 8):467–524

Gaevskaya AV, Rodjuk GN (1997) Ecological-geographical characters of parasite fauna of muraenolepids (Pisces: Muraenolepididae). Ekologiya Morya 46:28–31

Gaevskaya AV, Rodjuk GN, Parukhin AM (1990) Peculiarities and formation of parasitofauna of the Patagonian toothfish *Dissostichus eleginoides*. Biol Morya 4:23–28 (in Russian)

Georgiev BB, Vasileva GP, Chipev NH, Dimitrova ZM (1996) Cestodes of seabirds at Livingston Island, South Shetlands. Bulg Antarct Res Life Sci 1:111–127

Gonzáles-Acuña D, Hernández J, Moreno L, Herrmann B, Palma R, Latorre A, Medina-Vogel G, Kinsella M, Martin N, Araya K, Torres I, Fernandez N, Olsen B (2013) Health evaluation of wild gentoo penguins (*Pygoscelis papua*) in the Antarctic Peninsula. Polar Biol 36:1749–1760

Hartwich G (1975) Parasitic nematodes of vertebrates. I. Rhabditida und Ascaridida. Die Tierwelt Deutschlands 62:256

Healy CJ, Caira JN, Jensen K, Webster BL, Littlewood TJ (2009) Proposal for a new tapeworm order, Rhinebothriidea. Int J Parasitol 39:497–511

Hoberg EP (1983) Preliminary comments on parasitological collections from seabirds at Palmer Station, Antarctica. Antarct J 18:206–208

Hoberg EP (1985) *Reticulotaenia* n. gen. for *Lateriporus australis* Jones et Williams, 1967, and *Lateriporus mawsoni* Prudhoe, 1969 (Cestoda: Dilepididae) from sheathbills, *Chionis* spp., in Antarctica, with a consideration of infraspecific variation and speciation. J Parasitol 71:319–326

Hoberg EP (1986) *Eulimdana rauschorum* n. sp., a filaroid nematode (Lemdaninae) from *Larus dominicanus* in Antarctica, with comments on evolution and biogeography. J Parasitol 72:755–761

Hoberg EP (1987) *Tetrabothrius shinni* sp.nov. (Eucestoda) from *Phalacrocorax atriceps bransfieldensis* (Pelecaniformes) in Antarctica with comments on morphological variation, host–parasite biogeography, and evolution. Can J Zool 65:2969–2975

Holloway HL (1988) Parasites of emperor penguins, *Aptenodytes forsteri*. Trans Am Microsc Soc 107:105

Holloway HL (1989) *Parorchites zederi* cysts in emperor penguins. Trans Am Microsc Soc 108:109

Holloway HL, Klewer HL (1969) *Rhabdochona beatriceinsleyae* n. sp. (Nematoda: Spriruridea: Rhabdochonidae), from the Antarctic zoarcid, *Rhigophila dearborni*. Trans Am Microsc Soc 88:460–471

Holloway HL, Klewer HL, Husain A (1967) Notes on the genus *Ascarophis* Beneden, 1871, in Antarctic fishes. Proc Helminthol Soc Wash 34:222–227

Holloway HL, Spence JA (1980) Ecology of animal parasites in McMurdo Sound, Antarctica. Comp Physiol Ecol 5:262–284

Howie CA, Jones NV, Williams IC (1968) A report on the death of sheathbill, *Chionis alba* (Gmelin) at Signy Island, South Orkney Islands during the winter of 1965. Br Antarct Surv Bull 18:79–83

Ivanov VA (2006) *Guidus* n. gen. (Cestoda: Tetraphyllidea), with description of a new species and emendation of the generic diagnosis of *Marsupiobothrium*. J Parasitol 92:832–840

Ivanov VA, Campbell RA (2002) *Notomegarhynchus navonae* n. gen. and n. sp. (Eucestoda: Tetraphyllidea), from skates (Rajidae: Arhynchobatinae) in the southern hemisphere. J Parasitol 88:340–349

Jarecka L (1984) Development of *Hymenolepis arctowskii* Jarecka et Ostas, 1984 (Cestoda, Hymenolepididae) in the intermediate host *Brachinecta gaini* Daday (Branchiopoda) of the Antarctic. Acta Parasitol Pol 29:337–342

Jarecka L, Ostas J (1984) *Hymenolepis arctowskii* sp. n. (Cestoda, Hymenolepididae) from *Larus dominicanus* Licht. of the Antarctic. Acta Parasitol Pol 29:189–196

Jensen K, Bullard SA (2010) Characterization of a diversity of tetraphyllidean and rhinebothriidean cestode larval types, with comments on host associations and life- cycles. Int J Parasitol 40:889–910

Johnston TH (1937) Parasitic Nematoda. Australasian Antarctic Expedition 1911–14. Sci Rep Ser C 10:74

Johnston TH, Mawson PM (1945) Parasitic nematodes. Reports B.A.N.Z. Antarctic Research Expedition, Series B, 8:73–160

Jones NV, Williams IC (1967) The cestode parasites of the sheathbill, *Chionis alba* (Gmelin), from Signy Island, South Orkneys Islands. J Helminthol 41:151–160

Jones NV, Williams IC (1969) The nematode and acanthocephalan parasites of the sheathbill, *Chionis alba* (Gmelin) at Signy Island, South Orkney Islands and a summary of host parasite relationships in the sheathbill. J Helminthol 43:59–67

Kleinertz S, Christmann S, Silva LM, Hirzmann J, Hermosilla C, Taubert A (2014) Gastrointestinal parasite fauna of Emperor Penguins (*Aptenodytes forsteri*) at the Atka Bay, Antarctica. Parasitol Res 113:4133–4139

Kuchta R, Scholz T, Bray RA (2008) Revision of the order Bothriocephalidea Kuchta, Scholz, Brabec and Bray, 2008 (Eucestoda) with amended generic diagnoses and keys to families and genera. Syst Parasitol 71:81–136

Laskowski Z, Rocka A, Zdzitowiecki K, Ghigliotti L, Pisano E (2005) New data on the occurrence of internal parasitic worms in the *Gymnodraco acuticeps* and *Cygnodraco mawsoni* (Bathydraconidae) fish in the Ross Sea, Antarctica. Pol Polar Res 26:37–40

Laskowski Z, Rocka A, Zdzitowiecki K, Ozouf-Costaz C (2007) Occurrence of endoparasitic worms in dusky notothen, *Trematomus newnesi* (Actinopterygii Nototheniidae), at Adélie Land, Antarctica. Pol Polar Res 28:3742

Laskowski Z, Rocka A (2014) Molecular identification larvae of *Onchobothrium antarcticum* (Cestoda: Tetraphyllidea) from marbled rockcod, *Notothenia rossi*, in Admiralty Bay (King George Island, Antarctica). Acta Parasitol 59:767–772

Leiper RT, Atkinson EL (1914) Helminthes of the British Antarctic Expedition, 1910–1913. Proc R Soc Lond 1:222–226

Leiper RT, Atkinson EL (1915) Parasitic worms, with a note on a free-living nematode. Br Antarctic Terra Nova Exp Nat His Rep Zool 2:19–60

Lyadov VN (1985) Zoogeographical characteristics of the helminths of fishes from the Antarctic zone of the World Ocean. In: Hargis WJ Jr (ed) Parasitology and pathology of marine organisms of the World Ocean. NOAA Technical Reports NMFS 25:41–43

Mawson PM (1953) Parasitic Nematoda collected by the Australian National Antarctic Research Expedition: Heard Island and Macquarie Island, 1948–1951. Parasitology 43:291–297

Moravec F (1987) Revision of capillariid nematodes (subfamily Capillariinae) parasitic in fishes. Studie CSAV 3 Academia Praha:141 pp

Moser M, Cowen RK (1991) The effects of periodic eutrophication on parasitism and stock identification of *Trematomus bernacchii* (Pisces: Nototheniidae) in McMurdo Sound, Antarctica. J Parasitol 77:551–556

Mozgovoy AA (1953) Ascaridata of animals and man, and the diseases caused by them. Osnovy nematodologii. II. Izdatielstvo AN SSSR, Moskva:616 pp (in Russian)

Odening K (1982) Cestoden aus Flugvogeln der Sudshetlands (Antarktis) und der Falklandinseln (Malwinen). Angew Parasitol 23:202–223

Palm HW, Reimann N, Spindler M, Plötz J (1998) The role of the rock cod *Notothenia coriiceps* Richardson, 1844 in the life-cycle of Antarctic parasites. Polar Biol 19:399–406

Parukhin AM (1989) *Parasitic worms of bottom fishes of southern seas*. Naukova Dumka, Kiev:pp145 (in Russian)

Parukhin AM, Lyadov VN (1982) Helminth fauna of food Nototheniidae fishes from Kerguelen region. Ekologiya Morya 10:49–56 (in Russian)

Parukhin AM, Sysa VN (1975) The question of infection of fish of the suborder Notothenioidei of subantarctic waters. Trudy Nauchnoj Konferencii Parazitologov USSR 2, Naukova Dumka, Kiev:97–99 (in Russian)

Parukhin AM, Zaitsev AK (1984) The infection by helminths of different age *Notothenia squamifrons* in the subantarctic part of the Indian Ocean. Nauchnye Doki Vyss Shkoly Biol Nauki 10:34–37 (in Russian)

Petter AJ (1974) Essai de classification de la famille des Cucullanidae. Bull Mus Nat Hist Paris 255, Zoologie 177:1469–1491

Prudhoe S (1969) Cestodes from fish, birds and whales. Rep B.A.N.Z. Antarctic Res Exped 8B:171–193

Railliet A, Henry A (1912) Helminthes recueillis par l'Expédition antarctique française du Pourquoi-Pas. I. Cestodes d'Oiseaux. Bull Mus Nat Hist Paris 18:35–39

Reimer LV (1987) Helminthen von Fischen der Antarktis. Fischerei-Forschung, Rostock 25:36–40

Rennie J, Reid A (1912) The Cestoda of the Scottish Antarctic Expedition. Trans R Soc Edinb 48:441–453

Rocka A (1999a) The tetraphyllidean cercoids from teleosts occurring in the Weddell Sea (Antarctic). Acta Parasitol 44:115–118

Rocka A (1999b) Biometrical variability and occurrence of *Ascarophis nototheniae* Nematoda, Cystidicolidae), a parasitic nematode of Antarctic and subantarctic fishes. Acta Parasitol 44:188–192

Rocka A (2002) Nematodes of fishes in the Weddell Sea (Antarctic). Acta Parasitol 47:294–299

Rocka A (2003) Cestodes of the Antarctic fishes. Pol Polar Res 24:261–276

Rocka A (2004) Nematodes of the Antarctic fishes. Pol Polar Res 25:135–152

Rocka A (2006) Helminths of Antarctic fishes: life cycle biology, specificity and geographical distribution. Acta Parasitol 51:26–35

Rocka A, Zdzitowiecki K (1998) Cestodes in fishes of the Weddell Sea. Acta Parasitol 43:64–70

Rodjuk GN (1985) Parasitic fauna of the fishes of the Atlantic part of the Antarctic (South Georgia Island and South Shetland Isles). In: W.J. Hargis Jr (ed) Parasitology and pathology of marine organisms of the World Ocean. NOAA Technical Reports NMFS 25:31–32

Ropert-Coudert Y et al (2014) Chapter 8. Biogeographic patterns of birds and mammals. In: De Broyer C, Koubbi P, Griffiths HJ, Raymond B, Udekem d'Acoz C d' et al (eds) Biogeographic Atlas of the Southern Ocean. Scientific Committee on Antarctic Research, Cambridge, pp 364–387

Ruhnke TR (2011) Tapeworms of elasmobranchs (Part III). A monograph on the Phyllobothriidae (Platyhelminthes, Cestoda). Bull Univ Nebr State Mus 25:1–208

Ruhnke TR, Caira JN, Cox A (2015) The cestode order Rhinebothriidea no longer family- less: a molecular phylogenetic investigation with erection of two new families and description of eight new species of *Anthocephalum*. Zootaxa 3904:51–81

Schmidt H (1965) Tetrameres wetzeli n. sp., a new Tetrameres species from the Rock-hopper Penguin Eudyptes chrysocome Forst. Z Parasitenk:71–81

Southwell T (1925) A monograph on the Tetraphyllidea with notes on related cestodes. Mem Liverpool Sch Trop Med 2:368

Szidat L. Graefe G (1968) The parasites of *Parachaenichthys charcoti* an Antarctic fish, in relation to problems of zoogeography. In: Haeffer W (ed) Symposium on Antarctic Oceanography, Santiago –Chile, 13–16 September 1966. Scott Polar Research Institute and SCAR, Cambridge, pp 169–170

Temirova SI, Skrjabin AS (1978) Tetrabotriaty i mezocestoidaty. In: Ryzikov KM (ed) Osnovy cestodologii IX. Izdatielstvo Nauka, Moscow, pp 5–112

Vidal V, Ortiz J, Diaz JI et al (2012) Gastrointestinal parasites in chinstrap penguins from Deception Island, South Shetlands, Antarctica. Parasitol Res 111:723–727

Wojciechowska A (1990a) *Onchobothrium antarcticum* sp.n. (Tetraphyllidea) from *Bathyraja eatonii* (Günther, 1876) and a plerocercoid from Notothenioidea (South Shetlands, Antarctic). Acta Parasitol Pol 35:113–117

Wojciechowska A (1990b) *Pseudanthobothrium shetlandicum* sp.n. and *P. notogeorgianum* sp.n. (Tetraphyllidea) from rays in the regions of the South Shetlands and South Georgia (Antarctic). Acta Parasitol Pol 35:181–186

Wojciechowska A (1991a) New species of the genus *Phyllobothrium* (Cestoda, Tetraphyllidea) from Antarctic batoid fishes. Acta Parasitol Pol 36:63–68

Wojciechowska A (1991b) Some tetraphyllidean and diphyllidean cestodes from Antarctic batoid fishes. Acta Parasitol Pol 36:69–74

Wojciechowska A (1993a) The tetraphyllidean and tetrabothriid cercoids from Antarctic bony fishes. I. Morphology. Identification with adult forms. Acta Parasitol 38:15–22

Wojciechowska A (1993b) The tetraphyllidean and tetrabothriid cercoids from Antarctic bony fishes. II. Occurrence of cercoids in various fish species. Acta Parasitol 38:113–118

Wojciechowska A, Pisano E, Zdzitowiecki K (1995) Cestodes in fishes at the Heard Island (Subantarctic). Pol Polar Res 16:205–212

Wojciechowska A, Zdzitowiecki K, Pisano E, Vacchi M (1994) The tetraphyllidean cercoids from bony fishes occurring in the Ross Sea (Antarctic). Acta Parasitol 39:13–15

Zdzitowiecki K (1978) Wstępne badania nad pasożytami ryb okolic Południowych Szetlandów i Południowej Georgii. Kosmos 6:651–659

Zdzitowiecki K (2001a) New data on the occurrence of fish endoparasitic worms off Adelie Land, Antarctica. Pol Polar Res 22:159–165

Zdzitowiecki K (2001b) Occurrence of endoparasitic worms in a fish, *Parachaenichthys charcoti* (Bathydraconidae), off the South Shetland Islands (Antarctica). Acta Parasitol 46:18–23

Zdzitowiecki K, Cielecka D (1996) Morphology and occurrence of *Dichelyne* (*Cucullanellus*) *fraseri* (Baylis, 1929), a parasitic nematode of Antarctic and sub-Antarctic fishes. Acta Parasitol 41:30–37

Zdzitowiecki K, Dróżdż J (1980) Redescription of *Stegophorus macronectes* (Johnston et Mawson, 1942) and description of *Stegophorus arctowskii* sp. n. (Nematoda, Spirurida) from birds of South Shetlands (the Antarctic). Acta Parasitol Pol 27:205–212

Zdzitowiecki K, Rocka A, Pisano E, Ozouf-Costaz C (1998) A list of fish parasitic worms collected off Adelie Land (Antarctic). Acta Parasitol 43:71–74

Zdzitowiecki K, Szelenbaum-Cielecka D (1984) *Anomotaenia dominicana* (Railliet et Henry, 1912) (Cestoda, Dilepididae) from the Dominican gull *Larus dominicanus* Licht. of the Antarctic. Acta Parasitol Pol 29:49–58

Zdzitowiecki K, White M, Rocka A (1997) Digenean, monogenean and cestode infection of inshore fish at the South Orkney Islands. Acta Parasitol 42:18–22

Zdzitowiecki K, Zadróżny T (1999) Endoparasitic worms of *Harpagifer antarcticus* Nybelin, 1947 off the South Shetland Islands (Antarctic). Acta Parasitol 44:125–130

Chapter 7
Inventorying Biodiversity of Anisakid Nematodes from the Austral Region: A Hotspot of Genetic Diversity?

Simonetta Mattiucci, Michela Paoletti, Paolo Cipriani, Stephen C. Webb, Juan T. Timi, and Giuseppe Nascetti

7.1 Introduction

Inventorying of anisakid nematode biodiversity is the discovering, surveying, quantifying and mapping of species, populations and their genetic diversity and variability. Biodiversity cannot be investigated without first having a systematic foundation. However, any biodiversity assessment of anisakid nematodes inferred only from morphology is incomplete at best (Mattiucci and Nascetti 2008), despite the recent

S. Mattiucci (✉)
Department of Public Health and Infectious Diseases, Section of Parasitology, "Sapienza - University of Rome", P.le Aldo Moro, 5, 00185 Rome, Italy
e-mail: simonetta.mattiucci@uniroma1.it

M. Paoletti • G. Nascetti
Department of Ecological and Biological Sciences, Tuscia University, University Road s/n, 01100 Viterbo, Italy
e-mail: m.paoletti@yahoo.it; nascetti@unitus.it

P. Cipriani
Department of Ecological and Biological Sciences, Tuscia University, University Road s/n, 01100 Viterbo, Italy

Department of Public Health and Infectious Diseases, Section of Parasitology, "Sapienza - University of Rome", P.le Aldo Moro, 5, 00185 Rome, Italy
e-mail: paolo.cipr@gmail.com

S.C. Webb
Cawthron Institute, Private Bag 2, Nelson 7042, New Zealand

J.T. Timi
Laboratorio de Ictioparasitología, Instituto de Investigaciones Marinas y Costeras (IIMyC), Consejo Nacional de Investigaciones Científicas y Técnicas (CONICET)-Facultad de Ciencias Exactas y Naturales, Universidad Nacional de Mar del Plata, Funes 3350, 7600 Mar del Plata, Buenos Aires, Argentina

S. Klimpel et al. (eds.), *Biodiversity and Evolution of Parasitic Life in the Southern Ocean*, Parasitology Research Monographs, Vol. 9,
DOI 10.1007/978-3-319-46343-8_7

finding that morphological analysis and historical hypotheses often share with molecular-based taxonomic assessment a considerable congruence (Mattiucci et al. 2014). This highlights the importance of the detection and delimitation of cryptic species of anisakid nematodes inferred from molecular-based assessments, which allows more accurate assessment of biodiversity. This, in turn, permits elucidation of patterns and process in their evolution and ecology, including biogeography, host-parasite association and co-evolution. In addition, a true picture of anisakids and their genetic diversity facilitates understanding of their temporal and spatial distribution also related to their hosts demographic changes and marine ecosystem food webs.

This is of fundamental importance across the Boreal and Austral Regions, leading to greater understanding of the variation of biodiversity as a result of global change. On the other hand, it has been recently suggested that the parasitic abundance and genetic variability values of anisakid nematodes could be used for monitoring of the status of the marine trophic webs (Mattiucci and Nascetti 2008; Mattiucci et al. 2015a; Zarlenga et al. 2014). Indeed, anisakid nematodes of the genera *Anisakis* Dujardin, 1845, *Pseudoterranova* Krabbe, 1878, and *Contracaecum* Railliet et Henry, 1913, parasites of the alimentary tract of aquatic vertebrates, display indirect life-cycles in aquatic ecosystems and involve various hosts at different levels in the food webs. Marine mammals (cetaceans and pinnipeds) serve as definitive hosts; fish and squid serve as intermediate/paratenic hosts, and crustaceans serve as first intermediate hosts (Fig. 7.1). Thus, integrating molecular systematics of anisakid nematodes with ecological data will allow description of their global biodiversity and patterns of temporal and spatial partitions that influence their biodiversity. Findings suggest that anthropogenic change is one of those influences (Mattiucci and Nascetti 2008; Zarlenga et al. 2014).

The aim of this review is to provide an inventorying of the biodiversity, at species and gene level, of those anisakid species so far discovered belonging to the genera *Anisakis*, *Pseudoterranova* and *Contracaecum*, from the Austral Region, including: (1) taxa recognized as "biological species" based on the application of different molecular genetic markers; (2) current molecular/genetic approaches to identify them at any life-history stage; (3) ecological data relating to the geographical distribution, definitive host-association and host-preferences; (4) estimates of genetic variability values inferred from nuclear and mitochondrial genes as a possible indicator of the integrity of marine food webs; (5) collecting data so far reported concerning their possible zoonotic role to humans.

7.2 How Many Anisakid Species Are There?

"Cryptic" or "sibling" species (Nadler and Pérez-Ponce de León 2011) are ubiquitous among the anisakid nematodes (Mattiucci and Nascetti 2008). Their discovery has been bolstered in the last two decades by large-scale surveys from both Boreal and Austral Regions (Mattiucci and Nascetti 2008; Klimpel and Palm 2011; Shamsi

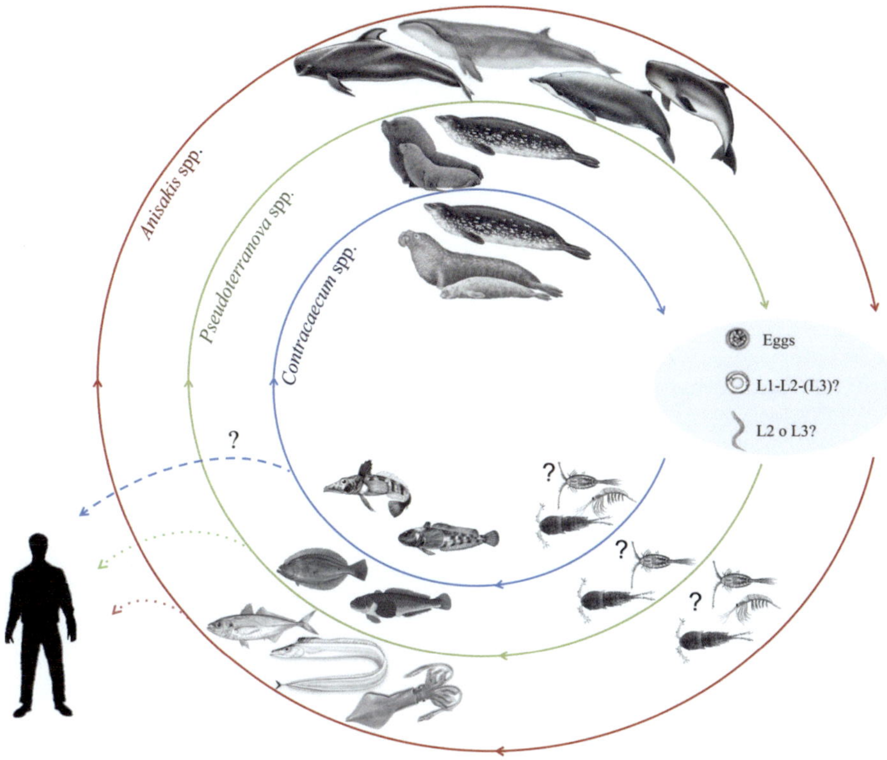

Fig. 7.1 Schematic representations of the hypothetic life-cycles of species of the genera *Anisakis*, *Pseudoterranova* and *Contracaecum* from the Austral Region, with their definitive and intermediate hosts, including real and hypothetical zoonotic role in human infections

2014) with resulting inventories based on molecular/genetic methodologies. Disclosure of cryptic biodiversity in anisakid nematodes starts with observations of: (1) considerable variability in morphological characters in a nominal anisakid species and (2) broad host range, often including species of different ecology and belonging to different families. Further steps include the demonstration that the single nominal species is indeed a complex of "biological species". This has been facilitated by the use of genetic/molecular markers, able to demonstrate the reproductive isolation between sympatric and allopatric populations of anisakids (Mattiucci and Nascetti 2008).

The present section summarizes the current taxonomy of anisakid species of the genera *Anisakis*, *Pseudoterranova* and *Contracaecum* (here considering only those species maturing in pinnipeds) which have been genetically characterized to date, infecting definitive and intermediate/paratenic hosts from the Austral Region (Tables 7.1, 7.2, 7.3 and 7.4 and Fig. 7.2).

A synopsis of each recognized anisakid species from the Austral Region, including data on both the definitive and intermediate hosts and the geographical range, is also provided in the present review.

Table 7.1 Definitive hosts of the *Anisakis* spp. from the Austral Region, identified by molecular/genetic markers

	A. pegreffii	*A. berlandi*	*A. ziphidarum*	*A. nascettii*	*A. brevispiculata*	*A. paggiae*
Cetaceans						
Delphinidae						
Cephalorhynchus hectori	NZ	–	–	–	–	–
Globicephala melas	NZ, CHI, SSI	NZ, SA, CHI	–	–	–	–
Tursiops truncatus	SA	–	–	–	–	–
Grampus griseus	NZ	NZ	–	–	–	–
Kogiidae						
Kogia breviceps	–	NZ	–	–	SA	SA
Kogia sima	–	AU	–	–	–	SA
Neobalaenidae						
Caperea marginata	SA	–	–	–	–	–
Ziphiidae						
Mesoplodon bowdoini	–	NZ	NZ	NZ	–	–
Mesoplodon densirostris	–	–	SA	–	–	–
Mesoplodon grayi	–	–	SA	NZ, SA	–	–
Mesoplodon layardii	–	–	SA	NZ	–	–
Mesoplodon mirus	–	–	SA	NZ, SA	–	–
Ziphius cavirostris	–	–	CHI, SA	–	–	–
Pinnipeds						
Phocidae						
Mirounga leonina	–	SSI	–	–	–	–

Data from: Mattiucci and Nascetti (2008), Mattiucci et al. (2009, 2014a, 2014b), Shamsi (2014), and Mattiucci, unpublished data

Sampling locality codes: CHI Chilean coast, NZ New Zealand, SA South Africa, SSI South Shetland Islands, AU Australian waters

Table 7.2 Intermediate/paratenic hosts for the *Anisakis* spp. sampled in fish and squids from the Southern Hemisphere, identified by molecular/genetic markers

	A. pegreffii	*A. berlandi*	*A. ziphidarum*	*A. nascettii*
Cephalopods				
Ommastrephidae				
Nototodarus sloanii	NZ	–	–	–
Ommastrephes angolensis	SA	SA	–	–
Todaropsis eblanae	SA	–	–	–
Onychoteuthidae				
Moroteuthis ingens	–	–	–	TA
Fishes				
Bramidae				
Brama brama	SA	–	–	–
Carangidae				
Trachurus trachurus	NZ	–	–	–
Trachurus capensis	SA	–	–	–
Dussumieridae				
Etrumeus whiteheadi	–	SA	–	–
Emmelichthyidae				
Emmelichthys nitidus nitidus	SA	–	–	–
Gempylidae				
Thyrsites atun	–	SA	–	–
Lophiidae				
Lophius vomerinus	SA	–	–	–
Merlucciidae				
Macruronus novazelandiae	–	NZ	–	–
Merluccius capensis	SA	–	–	–
Moridae				
Pseudophycis bachus	NZ	NZ	–	–
Myctophidae				
Electrona carlsbergi	–	SSI	–	–
Gymnoscopelus nicholsi	SSI	SSI	–	–
Ophidiidae				
Genypterus capensis	SA	–	–	–
Oreosomatidae				
Allocyttus niger	–	NZ	NZ	–
Pseudocyttus maculatus	–	NZ	–	–
Pinguipedidae				
Parapercis colias	NZ	NZ	–	–
Trachichthyidae				
Hoplostethus atlanticus	–	CHI, TA	–	–

(continued)

Table 7.2 (continued)

	A. pegreffii	*A. berlandi*	*A. ziphidarum*	*A. nascettii*
Trichiuridae				
Lepidopus caudatus	SA	–	–	–
Sebastidae				
Helicolenus dactylopterus	SA	–	–	–

Data from: Mattiucci and Nascetti (2008), Klimpel et al. (2010), Kuhn et al. (2011), Mattiucci et al. (2014a, b)
Sampling locality codes: CHI Chilean coast, NZ New Zealand, SA South Africa, SSI South Shetland Islands, TA Tasman Sea

Table 7.3 Definitive and intermediate/paratenic hosts of the *Pseudoterranova decipiens* species complex, so far reported from the Austral Region, identified by molecular/genetic markers

Host	*P. decipiens* sp. E	*P. cattani*
Pinnipeds		
Otaridae		
Otaria flavescens	–	CHI, ARG
Phocidae		
Leptonychotes weddellii	WS	–
Fishes		
Channichthydae		
Chaenocephalus aceratus	SSI	–
Merlucciidae		
Merluccius gayi	–	CHI
Notothenidae		
Notothenia coriiceps	SSI	–
Notothenia neglecta	SSI	–
Trematomus newnesi	SSI	–
Ophidiidae		
Genypterus maculatus	–	CHI
Paralichthydae		
Paralichthys patagonicus	–	ARG
Paralichthys microps	–	CHI
Paralichthys isosceles	–	ARG
Percophidae		
Percophis brasiliensis	–	ARG
Pinguipedidae		
Pseudopercis semifasciata	–	ARG
Serranidae		
Acanthistius patachonicus	–	ARG
Triglidae		
Prionotus nudigula	–	ARG

Data from Mattiucci and Nascetti (2008), Hernández-Orts et al. (2013), Timi et al. (2014)
Sampling locality codes: ARG Argentine waters, CHI Chilean coast, SSI South Shetland Islands, WS Weddell Sea

Table 7.4 Definitive and intermediate/paratenic hosts of the *Contracaecum* spp. from pinnipeds sampled from the Austral Region, identified by molecular/genetic markers

	C. osculatum sp. D	*C. osculatum* sp. E	*C. radiatum*	*C. miroungae*	*C. ogmorhini* (s. s.)
Pinnipeds					
Phocidae					
Mirounga leonina	–	–	–	SSI, ARG	ARG
Leptonychotes weddellii	RS, WS	RS, WS	RS, WS	–	–
Otariidae					
Arctocephalus australis	–	–	–	–	ARG
Arctocephalus pusillus	–	–	–	–	SA, AU
Fishes					
Bathydraconidae					
Gymnodraco acuticeps	RS	RS	–	–	–
Cygnodraco mawsoni	RS	RS	–	–	–
Channichthydae					
Cryodraco antarcticus	RS,WS	RS, WS	WS, RS	–	–
Chionodraco hamatus	RS,WS	RS, WS	WS, RS	–	–
Chionodraco myersi	–	–	WS, RS	–	–
Chaenodraco wilsoni	RS	RS	–	–	–
Pagetopsis macropterus	RS	RS	–	–	–
Notothenidae					
Notothenia neglecta	RS,WS	RS, WS	–	–	–
Trematomus centronotus	RS,WS	RS,WS			
Trematomus bernacchii	RS	RS	–	–	–
Trematomus hansoni	RS	RS			
Trematomus newnesi	RS	RS	–	–	–
Trematomus pennellii	RS	RS			

Data from: Mattiucci and Nascetti (2008), Mattiucci et al. (2015), and unpublished
Sampling locality codes: ARG Argentine waters, RS Ross Sea, SSI South Shetland Islands, WS Weddell Sea

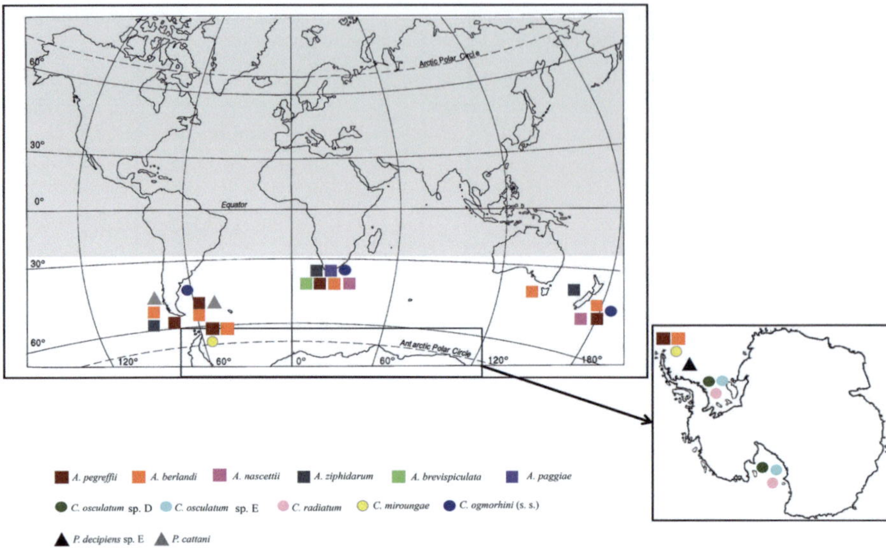

Fig. 7.2 World map highlighting the so far known distribution areas of the anisakid species of *Anisakis* (□), *Pseudoterranova* (Δ) and *Contracaecum* (○) in the Austral Region. The geographical areas indicated are those related to the sampling localities for their definitive and intermediate hosts (Data from Mattiucci and Nascetti (2008), Klimpel et al. (2010), Mattiucci et al. (2014, 2015a, 2015b)), Timi et al. (2014), (Shamsi 2014))

7.3 The Current Taxonomy of *Anisakis* spp. from the Austral Region

To date, nine species belonging to the genus *Anisakis* have been documented worldwide. Such species have been demonstrated to have distinct gene pools, to be characterized by distinct diagnostic genetic markers and to be reproductively isolated. The existence of those nine species as distinct phylogenetic units has been also demonstrated by various concatenated phylogenetic analyses, as inferred from nuclear and mitochondrial genes (Valentini et al. 2006; Mattiucci and Nascetti 2008; Cavallero et al. 2011; Mattiucci et al. 2014a). According to these phylogenetic analyses, four distinct clades within the genus *Anisakis* are clearly inferred, the existence of the three species *A. simplex* (s. s.), *A. pegreffii* and *A. berlandi* (= *A. simplex* sp. C) as distinct phylogenetic lineages (Mattiucci et al. 2014a). The topology of the Bayesian tree (Fig. 7.3) showed four main clades: first clade formed by (*A. berlandi* (*A. pegreffii* and *A. simplex* (s. s.)); a second clade formed by the two sibling species, *A. ziphidarum* and *A. nascettii*; a third clade formed by the species *A. physeteris*, *A. brevispiculata* and *A. paggiae*, with a support of 100 % posterior probability at the BI inference (Fig. 7.3). Finally, concatenated phylogenetic trees obtained from the combined nuclear and mitochondrial sequences depicted *A. typica* as a separate lineage; its position as the sister group to the other main clades received a posterior probability value of 100 % at the BI analysis (Mattiucci et al. 2014a).

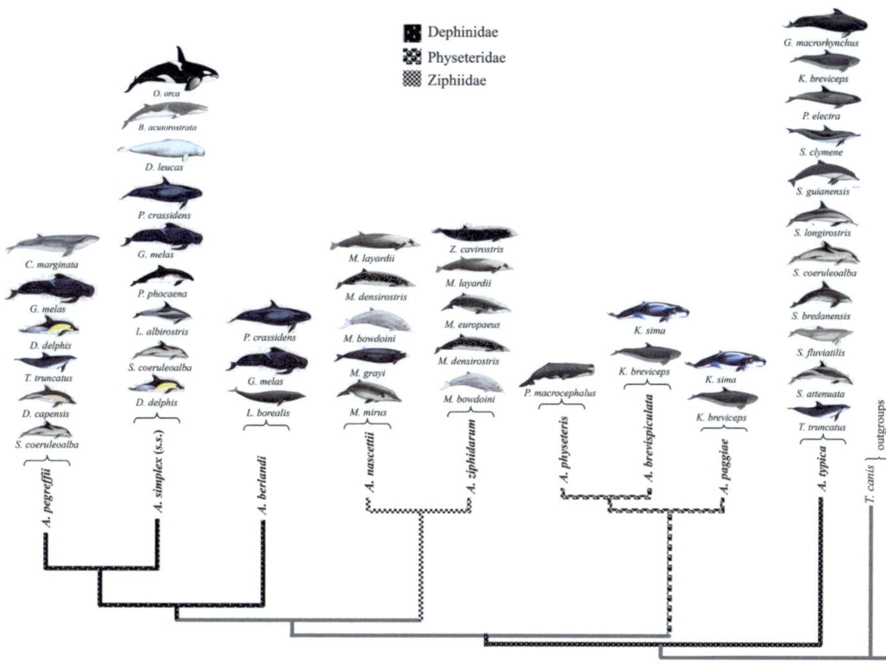

Fig. 7.3 Concatenated Bayesian inference (BI) tree obtained on the combined mtDNA *cox2*, *rrnS* and ITS region of rDNA sequences datasets of all the *Anisakis* so far genetically characterized (Data from Mattiucci et al. (2014a, b)), performed by MrBayes3.1 (Huelsenbeck and Ronquist 2005), using TrN+I+G model as selected by jModeltest 2.1 (Darriba et al. 2012) (with Akaike Information Criterion (AIC) calculation). The values of posterior probabilities, indicative of significant support, are given at the nodes. *T. canis* and *A. suum* were used as outgroups. The phylogenetic tree reported for all the *Anisakis* species, as representing distinct phylogenetic lineages, is mapped in association with families of cetaceans (Delphinidoidea, Ziphiidae, Physeteridae), as their main definitive hosts, so far reported in literature (Data from Mattiucci and Nascetti (2008), Mattiucci et al. (2009, 2014a), Shamsi (2014))

Among the nine species genetically characterized belonging to the genus *Anisakis*, those reported from the Austral Region are here reviewed with the sampling localities from where they have been identified, including the tables listing their definitive and intermediate/paratenic hosts (fish and squid) (Tables 7.1 and 7.2). They are the following:

Anisakis pegreffii Campana-Rouget and Biocca, 1955. Previously indicated as *A. simplex* A (see Nascetti et al. 1986), *A. pegreffii* occurs at both adult and larval stages in the Austral Region, between 30 and 60°S (Mattiucci et al. 2014). To date, it has been recorded at high abundance as an adult in dolphins, mainly belonging to the family Delphinoidea, and in a species of Neobalaenidae (*Caperea marginata*) from the south-east Atlantic coast (South African coast) (Table 7.1). It has been recorded, so far, in several species of fish and squids as a larva (Table 7.2). Among those hosts, three definitive and some intermediate/paratenic hosts are shared by *A. pegreffii* with *A. berlandi* in the Austral waters off New Zealand, the South African coast, Falkand Island and the southern Chilean coast (Table 7.2, Fig. 7.2).

Anisakis berlandi of Mattiucci et al. (2014a) [= *A. simplex* C of Mattiucci et al. (1997)] currently exhibits a discontinuous range (Mattiucci and Nascetti 2008). This includes, in the Austral Region: the Chilean Pacific, the South Shetland Islands, New Zealand waters and the South African Atlantic coast (Mattiucci and Nascetti 2008; Klimpel et al. 2010; Mattiucci et al. 2014 and unpublished). This species has been identified, at the adult stage, in sympatry and syntopy with *A. pegreffii* in *Globicephala melas* and *Grampus gryseus* from the New Zealand, and in *Globicephala melas* from south-west (South African coast) and south-east (Chilean coast) Pacific waters (Table 7.1). Very few specimens belonging to *A. berlandi* were identified in the pigmy sperm whale *Kogia sima* in south Pacific waters; in addition, it has been rarely identified also in *Mirounga leonina* from the sub-Antarctic area (South Shetland Islands) (Mattiucci and Nascetti 2008). Its Type I larvae were identified in several fish from Austral waters off New Zealand (Mattiucci and Nascetti 2008; Mattiucci et al. 2014), the South African coast (Mattiucci et al. 2014), Southern Shetland Islands (Klimpel et al. 2010) and the Southern Chilean coast (Table 7.2). Klimpel et al. (2010) stated that the occurrence of few larval specimens of *A. berlandi* (= *A. simplex* C) and *A. pegreffii* in myctophids from the southern waters of the Southern Ocean (i.e., South Shetland Islands) could be related to the introduction of those anisakid species from outside the Antarctica, through their migrating teleosts intermediate hosts. Indeed, also the very low infection found in *M. leonina* from South Shetland Islands (Mattiucci and Nascetti 2008) could be retained as an accidental infection.

Anisakis ziphidarum Paggi, Nascetti, Webb, Mattiucci, Cianchi and Bullini, 1998, was first described, both genetically and morphologically, as an adult in the beaked whales *Mesoplodon layardii* and *Ziphius cavirostris* from the South Atlantic Ocean (off the South African coast). Since its first morphological description and genetic characterization (Paggi et al. 1998, it has been recently identified genetically as an adult in other species of beaked whales, such as *M. mirus* and *M. grayi*, in South Atlantic waters and in *Mesoplodon* sp. and *Z. cavirostris* from Chilean waters (Mattiucci and Nascetti 2008). Thus, its geographical range appears to be wide (Fig. 7.2) and mainly related to that of its definitive hosts. Scanty data are available concerning its infection at larval stage in fish and/or squid from the Austral region, but it occurs at low prevalence of infection in some fish species, such as *Allocyttus niger* (Table 7.2). However, it seems that this species may involve other intermediate hosts in its life-cycle, such as squid (Table 7.2), rather than fish, as these represent the main food source of beaked whales.

Anisakis nascettii Mattiucci, Paoletti, Webb, 2009, has been detected in the beaked whales *Mesoplodon mirus* and *M. grayi* from South African and New Zealand waters (Fig. 7.2, Table 7.1). The gene pool was found to be reproductively isolated from the sympatric species *A. ziphidarum* occurring in the same hosts and geographical region. It is genetically very distinct from the other species of *Anisakis* but is most closely related to *A. ziphidarum*. The third-stage larva of *A. nascettii* is apparently of Type I. It has been genetically identified, at the larval stage, infecting heavily the squid *Moroteuthis ingens* in Tasman Sea waters (Mattiucci et al. 2009). This appears to support the hypothesis that this species involves squids rather than fish in its life-cycle.

Anisakis brevispiculata Dollfus, 1966, was initially characterized genetically using allozymes based on material from a pygmy sperm whale, *Kogia breviceps*, stranded on the South African coast (Mattiucci et al. 2001). Its reproductive isolation from the morphologically closely related *A. physeteris* was demonstrated, establishing the validity of *A. brevispiculata* (see Mattiucci et al. 2001), which had been synonymized with *A. physeteris* by Davey (1971). Later, its mitochondrial and nuclear sequences (Valentini et al. 2006; Mattiucci et al. 2014) established its genetic relationship with respect to the other *Anisakis* spp., confirming that *A. brevispiculata* clusters well with those *Anisakis* species forming the second clade.

Anisakis paggiae Mattiucci, Nascetti, Dailey, Webb, Barros, Cianchi and Bullini, 2005, clusters with *A. physeteris* and *A. brevispiculata*. This third species was first demonstrated by allozymes (Mattiucci et al. 2005) and mtDNA *cox2* sequence analysis (Valentini et al. 2006). It was also first described morphologically as an adult parasite of the pygmy sperm whale, *Kogia breviceps*, and the dwarf sperm whale, *K. sima*, from the South African Atlantic coast (Mattiucci et al. 2005). Scanty data are available regarding the identification of the intermediate hosts in the life-cycle of *A. paggiae* from Austral waters. Several larvae of Type II have been identified as belonging to this species in fish from Atlantic waters (in *Xiphias gladius*) (Mattiucci et al. 2015a), thus suggesting that other hosts, not yet detected, are involved in the life-cycle of this *Anisakis* species.

The high genetic heterogeneity of the *Anisakis* spp. is now also supported by some differential morphological features. The major clades can be delineated as follows: the clade including the species of the *A. simplex* complex (i.e., *A. simplex* (s. s.), *A pegreffii* and *A. berlandi*) has the following characteristics: (a) the ventriculus, in the adult stage, is longer than broad and often sigmoid in shape; (b) male spicules are long and often unequal (Mattiucci et al. 2014); (c) larval Type I morphology (*sensu* Berland 1961). Whereas, the clade including the species *A. ziphidarum* and *A. nascettii* shows (a) the ventriculus, in the adult stage, is longer than broad and often sigmoid in shape; (b) male spicules equal (see (Mattiucci et al. 2009); (c) larval Type I morphology (*sensu* Berland 1961). Finally, the clade encompassing the species *A. physeteris*, *A. brevispiculata* and *A. paggiae* shows (a) the ventriculus, in the adult stage, is short, never sigmoid and broader than long; (b) male spicules that are short, stout and of similar length (Mattiucci et al. 2005); (c) Type II larval morphology (*sensu* Berland 1961).

In addition, some morphological and morphometric characters are so far known which help in distinguishing the sibling species of the *A. simplex* complex (i.e., *A pegreffii*, *A. simplex* (s. s.) and *A. berlandi*) (Mattiucci et al. 2014). Furthermore, some morphological features, of diagnostic value, available in male and female adult specimens, were used to help in distinguishing *A. paggiae* from *A. physeteris* and *A. brevispiculata* (see Mattiucci et al. 2005), and *A. ziphidarum* from *A. nascettii* (see Mattiucci et al. 2009).

7.4 The Current Taxonomy of *Pseudoterranova* spp. from the Austral Region

To date, six biological species are recorded in the *Pseudoterranova decipiens* complex. Indeed, population genetic analysis, performed at first by allozyme markers on specimens of *P. decipiens* (*s. l.*) recovered from fish and seal species, collected at several locations in the North Atlantic Ocean, demonstrated the existence of a remarkable genetic heterogeneity with striking variation in allele frequencies among the samples (Paggi et al. 1991). Three distinct biological species occurred sympatrically in the samples of *P. decipiens* (*s. l.*) collected in seal hosts from those geographical areas, with no gene flow between them. The three taxa genetically recognized were thus provisionally designated as *P. decipiens* sp. A, *P. decipiens* sp. B and *P. decipiens* sp. C (Paggi et al. 1991). Morphological analysis carried out on male specimens identified by allozyme markers as *P. decipiens* A and B allowed the detection of significant differences in a number of characters between these two members; on the basis of such differences the nomenclature designation for *P. decipiens* sp. A and *P. decipiens* sp. B was proposed (see Paggi et al. 2000). Thus, the names *Pseudoterranova krabbei* Paggi, Mattiucci, Gibson, Berland, Nascetti, Cianchi and Bullini, 2000, and *P. decipiens* (*s. s.*) were proposed, respectively, for the species A and B, and a formal description of the two taxa was provided (see Paggi et al. 2000). Later on, the name of *P. bulbosa* (Cobb, 1888) was proposed for the taxon *P. decipiens* sp. C (see Mattiucci et al. 1998), as the latter taxon was demonstrated to correspond morphologically with *Ascaris bulbosa* described by Cobb (1888) from the bearded seal, *Erignathus barbatus*, at Spitzbergen (NE Atlantic Ocean). A further taxon, provisionally designated as *P. decipiens* sp. D (Mattiucci et al. 1998), was later included in the *P. decipiens* complex; this was detected by exhibiting several fixed differences at allozyme loci with respect to the other cryptic species. It was found to occur sympatrically with *P. bulbosa* in the same geographical areas (Japanese waters) and occasionally in the same definitive host, the bearded seal *Erignathus barbatus*, from which it was demonstrated to be reproductively isolated (Mattiucci et al. 1998). *Pseudoterranova decipiens* sp. D was found to correspond to the measurements and tail drawing of *Porrocaecum azarasi* Yamaguti and Arima, 1942, based on specimens recovered in the ribbon seal *Phoca (= Histriophoca) fasciata* on the islands of Sakhalin and Hokkaido. This taxon was synonymized by Margolis (1956) with "*Phocanema decipiens*". Therefore, Mattiucci et al. (1998) proposed the name *Pseudoterranova azarasi* (Yamaguti and Arima, 1942) n. comb. for the species *P. decipiens* sp. D.

In the Austral Hemisphere, using allozyme markers on larval and adult populations of *P. decipiens* (*s. l.*) collected from four fish species and the southern sea lion, *Otaria byronia* (= *Otaria flavescens*), in the SE Pacific Ocean, a further member of the *P. decipiens* complex has been shown to exist (George-Nascimento and Llanos 1995). In its formal description, this taxon was named *P. cattani* (George-Nascimento and Urrutia 2000). As stated above, this species was found as an adult in *O. byronia* on the Chilean coast. Using molecular markers in the internal transcribed spacers of ribosomal DNA (ITS rDNA), this species was previously shown to cluster with the *P. decipiens* complex (Zhu et al. 2002).

Finally, *P. decipiens* sp. E of Bullini, Arduino, Cianchi, Nascetti, D'Amelio, Mattiucci, Paggi Orecchia, Plötz, Smith and Brattey, 1997, was rarely genetically detected in the Antarctic Weddell seal, *Leptonychotes weddellii* (see Bullini et al. 1997).

A genetic identification and morphological characterization of larval *Pseudoterranova* spp. from three fish species sampled from Argentine waters (i.e.,

Acanthistius patachonicus and *Pseudopercis semifasciata*) and from *Notothenia coriiceps* from Antarctic waters was carried out by Timi et al. (2014). Larvae were sequenced for their genetic/molecular identification, including the mitochondrial cytochrome *c* oxidase subunit II (mtDNA *cox2*), the first (ITS-1) and the second (ITS-2) internal transcribed spacers of the nuclear ribosomal DNA, and compared with all species of the *P. decipiens (sensu lato)* species complex. Further, adults of *Pseudoterranova* spp. from the definitive host, the southern sea lion, *Otaria flavescens*, from Argentine and Chilean coasts were sequenced at the same genes. The sequences obtained at the ITS-1 and ITS-2 genes from all the larvae examined from fish of Argentine waters, as well as the adult worms, matched 100 % the sequences for the species *P. cattani*. While, the sequences obtained at mtDNA *cox2* gene for Antarctic larvae matched 99 % those available in GenBank for the sibling *P. decipiens* sp. E. In the same paper (Timi et al. 2014), phylogenetic analysis strongly supported *P. cattani* and *P. decipiens* sp. E as two distinct phylogenetic lineages and depicted the species *P. decipiens* sp. E as a sister taxon to the remaining taxa of the *P. decipiens* complex. In addition, larval morphometry was similar between specimens of *P. cattani* from Argentina, but significantly different from those of *P. decipiens* sp. E, indicating that larval forms can be distinguished based on their morphology (Timi et al. 2014).

Pseudoterranova cattani is common and abundant in a variety of fish species from Chile, whereas few host species harbour these larvae in Argentina where, on the contrary, they show low levels of infection. Finally, that study revealed that the life-cycle of *P. cattani* involves mainly demersal and benthic organisms, with a marked preference in large-sized benthophagous fish (Table 7.3). Those studies indicate that members of the genus *Pseudoterranova* in the areas of Pacific and Atlantic coasts of South America are representatives of a unique species, namely *P. cattani*, whose distribution mirrors that of its definitive host, *O. flavescens* (Timi et al. 2014). In addition, some of the records of *Pseudoterranova* spp. from that region could be erroneous or in need of validation by using molecular/genetic markers for their identification. The distribution of definitive hosts has been postulated as one of the most important biotic factors determining the distribution of *P. decipiens* (*s. l.*) (McClelland 2002). *Otaria flavescens* is distributed over a broad latitudinal range along the South American coastline, from Peru in the Pacific to Brazil in the Atlantic (Vaz-Ferreira 1982). This species is apparently the only suitable definitive host for *P. cattani* (George-Nascimento and Llanos 1995). Indeed, the South American fur seal *Arctocephalus australis*, sympatric with the sea lions in Uruguay and in some localities of the Argentine coasts, has been reported as harbouring only larval stages in Patagonia (Hernández-Orts et al. 2012). It is likely that the lower densities of sea lions in the northern Argentine coasts could be responsible also for a small population of *P. cattani* present in that geographic area. In contrast, higher levels of parasitism have been recorded in *O. flavescens* in Chile (prevalence = 100 %, mean abundance = 131.1 ± 125.5) (George-Nascimento 1991). Furthermore, environmental conditions can also affect the distribution and/or abundance of suitable previous invertebrate hosts, still unknown for *P. cattani*. A combination of factors seems to drive the population size of *P. cattani* in the northern boundary of distribution of *O. flavescens* in the Atlantic, including the environmental conditions (warm waters with low salinity), the density and dietary preferences (or prey availability) of definitive hosts and the life-cycle pathways of the parasite (Timi et al. 2014).

7.5 The Current Taxonomy of *Contracaecum* spp. from the Austral Region

7.5.1 The Contracaecum osculatum (s. l.) *Complex of Species*

First genetic studies based on allozyme markers on this species complex have demonstrated the reproductive isolation and the absence of gene flow among sympatric and allopatric populations of *C. osculatum* (*s. l.*) hosted by pinnipeds from Arctic and Antarctic regions (Mattiucci and Nascetti 2008). Those genetic markers have proved the existence of several biological species within *C. osculatum* (*s. l.*), considered previously as a cosmopolitan species and parasitic in various definitive seal hosts. These nematode species are often very similar morphologically but reproductively isolated (sibling species). Actually, they are the Arctic sibling species named as *C. osculatum* sp. A, *C. osculatum* sp. B and *C. osculatum* (*s. s.*) (see Nascetti et al. 1993; Mattiucci et al. 1998), and the two Antarctic members named *C. osculatum* sp. D and *C. osculatum* sp. E (see Orecchia et al. 1994). Later, those species of the *C. osculatum* (*s. l.*) complex have been genetically characterized on the basis of other genetic/molecular markers, such as the sequences analysis of the internal transcribed spacers of ribosomal DNA (ITS rDNA) (Nadler et al. 2005) and the mitochondrial *cox2* gene sequences analysis (Mattiucci et al. 2008). Further, the single strand conformation polymorphism (SSCP) analysis of the ITS rDNA was performed to screen for sequence variation within and among individuals of the *C. osculatum* (*s. l.*) species complex (Zhu et al. 2000; Hu et al. 2001). Inter-taxon differences in SSCP profiles were detected between those *Contracaecum* taxa. A reliable genetic differentiation of the sibling species from one another, revealed at the ITS rDNA sequences analysis, was recorded, except in the case of the two Antarctic members, i.e., *C. osculatum* sp. D and *C. osculatum* sp. E, which exhibited identical ITS rDNA sequences and SSCP profiles (Zhu et al. 2000). Similarly, SSCP-based analyses of three mitochondrial DNA (mtDNA) regions, namely cytochrome c oxidase subunit I (*cox1*) and the small and the large subunit of ribosomal RNA (*ssrRNA* and *lsrRNA*), respectively, in the Arctic and Antarctic members of *C. osculatum* (*s. l.*) (Hu et al. 2001), detected nucleotide differences considered diagnostic among all the sibling species of *C. osculatum* (*s. l.*) from the Arctic Boreal region, with the exception of the two Antarctic members, for which those markers failed to distinguish *C. osculatum* sp. D and sp. E (Hu et al. 2001).

In contrast, reproductive isolation and fixed alternative alleles were found at some diagnostic loci between the two sympatric sibling species from the Antarctic Ocean by the use of multilocus allozymes electrophoresis (MAE) (Orecchia et al. 1994). In more recent years, sequences analysis of the mitochondrial *cox2* gene of specimens belonging to *C. osculatum* sp. D and *C. osculatum* sp. E, previously identified by allozymes, was able to support the existence of the two Antarctic members of *C. osculatum* (*s. l.*) as two distinct phylogenetic units (Mattiucci et al. 2008, 2015).

Contracaecum osculatum sp. D and *C. osculatum* sp. E of Orecchia, Mattiucci, D'Amelio, Paggi, Plotz, Cianchi, Nascetti, Arduino and Bullini, 1994, occur sympatrically in the same definitive host, the Weddell seal, *Leptonychotes weddellii*, and have so far been reported from both the Weddell and the Ross Seas (Antarctica) (Orecchia et al. 1994). The larval stages of the two sibling species have been identified by diagnostic allozyme markers and sequences analysis of the mtDNA *cox2* gene, from several fish species belonging to the families Channicthydae, Bathydraconidae and Nototheniidae, in which a differential distribution of the two sibling species is reported (Mattiucci and Nascetti 2007; Mattiucci et al. 2015a) (Table 7.4, Fig. 7.1). The two species, *C. osculatum* sp. D and *C. osculatum* sp E, were found in the same individual fish hosts, showing a strict sympatry and syntopy. They showed also differences in the host infection site: the relative proportion of *C. osculatum* sp. D was significantly higher in the fish liver (Mattiucci et al. 2015a). Moreover, a significant statistical difference in the relative proportions by *C. osculatum* sp. D and *C. osculatum* sp. E in the fish species was observed (Mattiucci et al. 2015a). This finding could be related to the ecological and feeding habits of the fish host species. Thus, considering the relative frequencies observed of the two species of *Contracaecum* occurring in the different host species, each one characterized by its feeding ecology and diets, some conclusion has been drawn regarding the possible life-cycles of *C. osculatum* sp. D and *C. osculatum* sp. E in the Antarctic food web (Mattiucci et al. 2015a). *C. osculatum* sp. D seems to be mainly associated with fishes characterized by bentho-pelagic habits, and feeding above all other small fishes and Antarctic krill (presumably *Euphasia crystallorophias*, the euphasid present in the Ross sea), like *Chionodraco hamatus* and *Trematomus hansoni*. Thus, *C. osculatum* sp. D could include, in its biological cycle, a planktonic intermediate host, such as *E. crystallorophias* in the Ross Sea, and probably *E. superba* in other Antarctic areas. Instead, *C. osculatum* sp. E showed higher frequencies in the fish species, such as *T. bernacchii*, specialized in predation of strictly benthic organisms. This observation indicates that a possible first intermediate invertebrate host could be represented by an amphipod/polychaete/isopod, with a benthic life-cycle – habit (Mattiucci et al. 2015a).

7.5.2 *The* Contracaecum ogmorhini *Species Complex*

The pinniped parasite *Contracaecum ogmorhini* Johnston and Mawson, 1941, first described from the leopard seal, *Hydrurga leptonyx*, in South Australian waters, was later synonymized with *C. osculatum* (see Johnston and Mawson 1945). However, it was considered valid by Fagerholm and Gibson (1987). The species was found to be genetically heterogeneous using allozyme markers (18 enzyme loci), indicating the existence of two reproductively isolated taxa (sibling species) included within the morphospecies. A formal description of the two taxa was given by Mattiucci et al. (2003), and they were named *C. ogmorhini* Johnston and Mawson, 1941 (*sensu stricto*) and *C. margolisi* Mattiucci, Cianchi, Nascetti, Paggi, Sardella, Timi, Webb, Bastida, Rodriguez and Bullini, 2003. A morphological description of *C. ogmorhini* (s. s.) from *Arctocephalus australis* was given by Timi et al. (2003). *Contracaecum ogmorhini* (s. s.) has been detected as an adult in the otariid seals *Arctocephalus pusillus pusillus*, *A. pusillus doriferus* and *A. australis* in the Austral region (Mattiucci et al. 2003; Timi et al. 2003; Mattiucci and Nascetti 2008).

Contracaecum radiatum (v. Linstow, 1907) Baylis, 1920: the taxonomic status of this species was confirmed genetically by Arduino et al. (1995) on the basis of 24 enzyme loci. Several allozymes were found to be diagnostic between *C. radiatum* and the other taxa so far characterized as belonging to *Contracaecum* species from seals ((Arduino et al. 1995; Mattiucci et al. 2008), and unpublished data). Reproductive isolation from the two Antarctic members of the *C. osculatum* complex (i.e., *C. osculatum* sp. D and *C. osculatum* sp. E), occurring sympatrically in the same definitive hosts (the Weddell seal), was proved by the lack of F1 hybrids and recombinant or introgressed individuals between the Antarctic taxa in the sympatric areas of the Weddell and Ross Seas (Arduino et al. 1995). The genetic relationships between *C. radiatum* and other congeneric taxa were later inferred from LSU rDNA sequences (Nadler et al. 2000) and mtDNA *cox2* sequence analyses (Mattiucci et al. 2008). Morphological distinction between *C. radiatum* and *C. osculatum* (*s. l.*) was established by Klöser and Plötz (1992). *Contracaecum radiatum* has been genetically identified as an adult in *Leptonychotes weddellii* and as a larva in the pelagic channichthyd fishes *Chionodraco hamatus* and *Criodraco antarcticus* (see (Arduino et al. 1995)). This finding supports a previous report by Klöser et al. (1992), according to which *C. radiatum* has become adapted to a pelagic food web. Other definitive hosts recorded for this species in Antarctic waters are the leopard seal, *Hydrurga leptonyx*, and the Ross seal, *Ommatophoca rossi* (see Baylis 1937; Dailey 1975). Genetic investigations on this parasite of Antarctic seals are needed in order to determine any host preference of *C. radiatum* in the Antarctic waters.

*Contracaecum mirounga*e Nikolskii, 1974: the taxonomic status of the species was confirmed genetically by allozyme markers (20 enzyme loci) (Mattiucci et al. 2008). It was detected genetically as an adult in *Mirounga leonina* from the Antarctic and sub-Antarctic areas (Mattiucci et al. 2008) and also in the otariid *Arctocephalus australis* (Mattiucci and Nascetti 2008). There is reproductive isolation between the two Antarctic members of the *C. osculatum* complex (i.e., *C. osculatum* sp. D and *C. osculatum* sp. E) occurring sympatrically in the same definitive host (the Weddell seal). The genetic relationships between *C. miroungae* and other congeneric taxa were later inferred from LSU rDNA sequences (Nadler et al. 2000) and from the mtDNA *cox2* sequences analyses (Mattiucci et al. 2008). No data of genetically identified larvae of this species are available so far.

7.6 Current Methods for the Identification of Anisakid Nematodes from the Austral Region

In the last two decades, the reported diversity of anisakid species has increased due to the detection by genetic markers of several morphologically very similar sibling species, which thus showed reproductively isolated gene pools that certify them as "biological species". There are now morphospecies, or species complexes, based on previously recognized cosmopolitan species (*sensu lato*), that may actually comprise several recognized species. This genetic approach has solved one of the major problems in the systematics of anisakid nematodes: the occurrence of the parallelism and convergence of morphological features, which confound the systematic value of some morphological criteria and often accompany a high genetic and ecological divergence between the species. The lack of morphological differences in these parasites may be due to various factors, such as similar selection pressures causing the conservation of a common adaptive morphology. Consequently, some morphological characters have little or no taxonomic value because of the evolutionary coadaptation of these endoparasites to the stable habitat represented by their localization in definitive hosts. Indeed, morphospecies may appear to have multiple host species, i.e., parasite populations isolated in their hosts have diverged genetically but have conserved morphological features. Moreover, species identification based on morphological characters makes identification very difficult and speculative, especially for larval stages as they lack reliable diagnostic features at the species level.

The assessment of anisakid nematode biodiversity based on molecular genetic markers represents the preferable method, so far, for specific diagnosis. This prospect gains importance when the unambiguous identification of those anisakids with a zoonotic potential is an essential requirement for a proper epidemiological survey.

The species concept (BSC) (Mayr 1963) was well supported by the application of allozyme markers within certain anisakid morphospecies, such as those of *Anisakis*, *Pseudoterranova* and *Contracaecum* (Mattiucci and Nascetti 2008; Mattiucci et al. 2014). Reproductive isolation and absence of gene flow were demonstrated by these allozyme loci between sympatric and allopatric sibling species, establishing their specific status (Paggi et al. 1991; Nascetti et al. 1993; Mattiucci et al. 1997, 2001, 2003, 2005). Allozyme markers have allowed: (1) genetic characterization of different species of anisakid nematodes, (2) estimation of their genetic differentiation, (3) establishment of their genetic relationships, (4) identification of their larval stages which lack diagnostic morphological characters and (5) clarification of hybridization phenomena between very close sibling species (Mattiucci et al. 2016). Later on, the introduction of the polymerase chain reaction (PCR)-derived molecular methodologies confirmed taxonomic decisions involving anisakid species previously based on allozyme markers. Reference individuals, initially characterized by allozymes, have been used to develop DNA-based approaches for species identification, such as direct sequencing of nuclear and mitochondrial DNA genes (Mattiucci et al. 2014, 2016). Thus, phylogenetic analysis provided a new perspective for the delimitation of anisakid sibling species, including hierarchical relatedness and relative rates of evolution. An evolutionary perspective provides a conceptual approach to view species as independent evolutionary lineages and provides another approach for delimiting species (Nadler et al. 2005). Indeed, based on phylogenetic DNA analysis, sibling anisakid species have been confirmed by methods that can test the hypothesis of lineage independence by analysing many individual specimens and sometimes detecting new genotypes and species (Nadler et al. 2000, 2005; Valentini et al. 2006; Mattiucci and Nascetti 2008; Cavallero et al. 2011; Mattiucci et al. 2014).

Based on allozyme diagnostic loci for different anisakid taxa, easy and rapid identification of large numbers of individuals can be performed; this method is particularly valuable for identifying larval individuals collected from several intermediate/paratenic hosts, and often occurring in mixed infections. Accordingly, such species identifications have been demonstrated to be very informative tools for answering epidemiological questions involving geographical range, host preference and life-cycles of these parasites. Moreover, because numerous allozymes analyses have been applied to thousands of individuals, they have contributed greatly to our knowledge of the genetic diversity of anisakid populations collected from various ecosystems in the Boreal and Austral Regions. However, allozymes tool is limited to frozen-preserved or fresh individuals. This preservation constraint has now been resolved by DNA-based diagnostic techniques, which have the advantage of also being able to use alcohol-preserved specimens. In contrast with allozymes, the DNA-based techniques have increased our ability to study phylogenetic relationships between related anisakids based on the evolutionary lineage concept. However, the PCR-DNA molecular derived methodologies, so far applied to the systematics of anisakid nematodes, include nowadays, the application of a multigene approach in order to have a robust identification of the considered taxa. They are the sequences analysis of mitochondrial genes, such as the cytochrome oxidase 2 (mtDNA *cox2*) (Valentini et al. 2006; Mattiucci and Nascetti 2006, 2008; Mattiucci et al. 2014), and the small subunit of rRNA (*rrnS*) (Mattiucci et al. 2014) and of nuclear genes, such as the ITS region of rDNA (Nadler et al. 2005); the elongation factor–1 alpha 1 nDNA (EF1 $\alpha-1$ nDNA region) (Mattiucci et al. 2016); PCR-restriction fragment length polymorphism (PCR-RFLPs) of ITS region of rDNA (D'Amelio et al. 2000; Pontes et al. 2005).

7.7 Host Preference in Anisakid Nematodes and Host-Parasite Co-phylogenetic Pathways

As described above, the phylogenetic relationships between *Anisakis* spp. shows presence of distinct main clades, as inferred from the multigene sequences analysis (Mattiucci et al. 2014a). Phylogenetic relationships between *Anisakis* spp. is supported also by ecological data and specific host-parasite relationships (Fig. 7.3).

Those have been suggested to “mirror”, in several host-parasite associations (Mattiucci and Nascetti 2008), the phylogenetic relationships so far proposed for their definitive hosts (Milinkovitch 1995; Cassens et al. 2000; Nikaido et al. 2001; Arnason et al. 2004). Indeed, Mattiucci et al. (2014a) provided support for the existence of host specificity among *A. simplex* (s. s.), *A. pegreffii* and *A. berlandi* for “oceanic dolphins” and whales, as suggested by Mattiucci and Nascetti (2008) and others (e.g., Klimpel et al. 2008; Cavallero et al. 2011). The three species have been identified as the only species of *Anisakis* parasitizing striped dolphins, pilot whales and minke whales. Interestingly, *G. melas* was found to host all three species, depending on its locality, in relation to the geographical ranges reported for *A. pegreffii*, *A. simplex* (s. s.) and *A. berlandi*. Notably, for pilot whales in South Pacific waters, *A. pegreffii* and *A. berlandi* were detected in sympatry and in syntopy in the same individual host. In contrast, *A. pegreffii* has been identified, based on molecular markers (sequences data of the ITS-1 and ITS-2 rDNA) and described morphologically, from the short-beaked common dolphin, *Delphinus delphis*, and the common bottlenose dolphin, *Tursiops truncatus*, from south eastern Australian waters; similarly, larval stages belonging to the species *A. berlandi* were found in a dwarf sperm whale from the same geographical area (Shamsi 2014). All these findings appear to confirm that the three species share, in different geographical areas, the same definitive hosts, and they involve in their life-cycles different pelagic and demersal fish hosts in their respective ranges (Mattiucci and Nascetti 2008).On the other hand, the same definitive host, the pilot whale, has previously been found in Spanish Atlantic waters to be parasitized by adults of *A. pegreffii* and *A. simplex* (s. s.) in sympatry (Mattiucci et al. 2014a). Interestingly, it has been suggested that two subspecies of pilot whales exist (Rice 1998), with subspecies *G. melas melas* in the Boreal region and subspecies *G. melas edwardii* (Smith) in the Austral region. The occurrence of *A. simplex* (s. s.) in Boreal individuals of pilot whales (see (Mattiucci and Nascetti 2008)) and the detection of *A. berlandi* in Austral specimens of pilot whales (Mattiucci et al. 2014a) seem to support this hypothesis, and the possible use of *Anisakis* spp. for gathering information also on the migration routes and population structure of their definitive hosts.

In addition, the odontocetes *Physeter catodon*, *Kogia breviceps* and *K. sima* are the main definitive hosts for *A. physeteris*, *A. brevispiculata* and *A. paggiae*, respectively (Mattiucci et al. 2001, 2005; Mattiucci and Nascetti 2006) (Fig. 7.3, Table 7.1). The beaked whales *Ziphius cavirostris* and several species belonging to the genus *Mesoplodon* are hosts of *A. ziphidarum* and *A. nascettii*, which are partitioned in the distinct clade I in the *Anisakis*-parasite phylogenetic tree (Fig. 7.3). According to the phylogenetic hypothesis proposed by Arnason et al. (2004), the Cetacea group splits into monophyletic Mysticeti (baleen whales) and monophyletic Odontoceti (toothed whales). The Odontoceti diverged into the four extant lineages, Physeteridae (sperm whales: represented by the sperm whales), Ziphiidae (beaked whales), Platanistidae (Indian river dolphins) and Delphinoidea (encompassing the families Iniidae, Monodontidae, Phocoenidae and Delphinidae). Phylogenetic trees provided by Nikaido et al. (2001) and Arnason et al. (2004) were congruent in depicting the branching order of the extant cetacean lineages, where the families Physeteridae

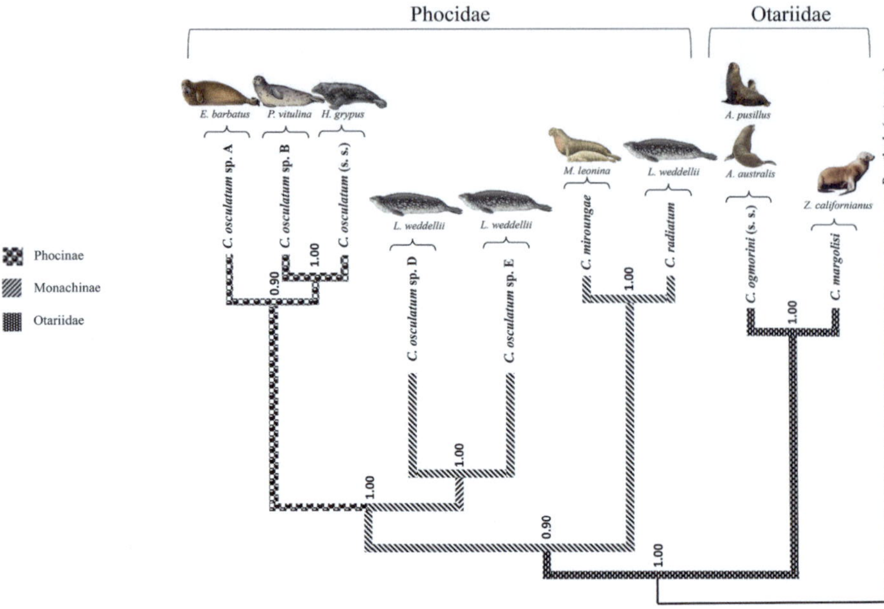

Fig. 7.4 Bayesian inference (BI) tree obtained from mtDNA *cox2* sequences analysis of *Contracaecum* spp., performed by MrBayes3.1 (Huelsenbeck and Ronquist 2005), using TrN+G model as selected by jModeltest2.1 (Darriba et al. 2012) (with Akaike Information Criterion (AIC) calculation). The values of posterior probabilities, indicative of significant support, are given at the nodes. *Pseudoterranova ceticola* was used as outgroup. The phylogenetic tree reported for *Contracaecum* species, as representing distinct phylogenetic lineages, is mapped in association with families of pinnipeds (Phocidae, Otariidae), as their main definitive hosts

and Kogiidae represent basal taxa, followed by the Ziphidae and the freshwater and marine dolphins as the most derived. In accordance with those analyses, the branching order so far proposed for the *Anisakis* taxa showed that nematodes from the sperm whale and pygmy sperm whales (i.e., *A. physeteris*, *A. brevispiculata* and *A. paggiae*) always occupy a basal and well-supported lineage, followed by those parasitizing the beaked whales (*A. ziphidarum* and *A. nascettii*). The species of the *A. simplex* complex and *A. typica*, parasites of delphinoids, are the most derived (Mattiucci and Nascetti 2008) (Fig. 7.3).

Similar investigations have been carried out into host-parasite associations between *Contracaecum* taxa and their definitive hosts, the pinnipeds of the Families Phocidae and Otariidae. The presence of the two main clades, as presented in the phylogenetic relationships among *Contracaecum* spp. (Fig. 7.4), is supported also by the ecological data concerning host preference (Nascetti et al. 1993) and specific host-parasite relationships (Mattiucci and Nascetti 2008). Phylogenetic relationships so far proposed, and here reviewed for species of *Contracaecum*, parallel that reported for their definitive hosts based on molecular data (Arnason et al. 1995; Deméré et al. 2003). Several phocid seals (true seals) in the Phocinae are hosts of the species of the *C. osculatum* complex (i.e., *C. osculatum* sp. A, *C. osculatum* sp. B, *C. osculatum* (s. s.), *C. osculatum* sp. D, *C. osculatum* sp. E). Moreover, seals in the Monachinae are the main hosts for *C. miroungae* and *C. radiatum* in the sub-Antarctic and Antarctic region (Table 7.4, Fig. 7.4). These anisakids are included in clade I in the parasite phylogenetic tree. Whereas, the otariid species in the Otariinae (sea lions), *Zalophus californianus*, and in the Arctocephalinae (fur seals) *Arctocephalus* spp. are hosts of the *C. ogmorhini* species complex (*C. margolisi* and *C. ogmorhini* (s. s.)) (Fig. 7.4). These anisakids are included in a distinct clade in the *Contracaecum*-parasite phylogenetic tree. Although a complete species-level phylogeny for pinnipeds, including fossil and extant taxa, is yet unavailable, a molecular assessment of pinniped relationships was performed by Arnason et al. (1995) using the complete sequences of the mitochondrial cytochrome b gene (mtDNA *cytb*) of the Phocidae, Odobenidae and Otariidae. Later, Deméré et al. (2003) used a composite tree inferred from the basic topology of generic level, morphological and molecular data, fossil taxa and consensus phylogeny of the phocid subfamilies to propose an integrated hypothesis for pinniped evolutionary biogeography. According to that data elaboration, the Pinnipedia includes three major monophyletic clades: (1) the Otariidae (fur seals and sea lions), (2) the Odobenidae and (3) the Phocidae (true seals), *plus* the extinct desmatophocids. In this combined tree, the fur seals and sea lions comprising the Otariinae (*Zalophus californianus*) and the Arctocephalinae (*Arctocephalus* spp. from the Southern Hemisphere) are represented as well-supported basal groups (Deméré et al. 2003; Arnason et al. 1995). In accordance with that analysis, the branching order so far proposed for the *Contracaecum* taxa showed that nematodes from the Otariidae (i.e., *C. ogmorhini* (s. s.) from *Arctocephalus* spp. and *C. margolisi* from *Zalophus californianus*) always occupy a basal lineage of the parasite phylogenetic tree, with the species of the *C. osculatum* complex from the Phocinae (true seals) as the most derived (Fig. 7.4).

According to Mattiucci and Nascetti (2008) speciation of the members of *C. osculatum* complex is apparently related to their geographical isolation, through that of their hosts, as well as to a rapid host-parasite adaptation and co-evolution. Such processes apparently occurred in different times during the Plio-Pleistocene, when extreme climatic variation took place. The genetic relationships found between the members of the *C. osculatum* complex suggest that the evolutionary divergence of the most differentiated species [*C. osculatum* (s. s.)] started more than three million years ago, in a Pleistocene refuge (the Baltic Sea). As to the other *C. osculatum* species, their evolutionary divergence probably took place during the Pleistocene, when the complex achieved a distribution over both polar regions. This process involved two distinct colonizations of the marine Antarctic region by ancestors of the Northern Hemisphere, giving rise to *C. osculatum* sp. D and *C. osculatum* sp. E, both parasites of *Leptonychotes weddellii*. This hypothesis seems to fit with the evolutionary biogeography of a pinnipedimorph hypothesis based on both dispersal and vicariant events in the context of a species-level phylogenetic framework proposed by Deméré et al. (2003). This hypothesis supports an eastern North Pacific origin during the late Oligocene coincident with start of glaciation in Antarctica. During the late Miocene, pinnipedimorphs remained restricted to the eastern North Pacific, where they began to diversify. Fur seals remained restricted to the North Pacific until the late Pliocene, with a dispersal and rapid speciation in the Southern Ocean during the Pleistocene. The phocine seal diversification took place in the Arctic and North Atlantic during the late Miocene with a subsequent dispersal into the Paratethys and Pacific during the Pleistocene. Finally, the monachine seals, including *Mirounga leonina* and *Leptonychotes weddellii*, seem to have the Southern Hemisphere as the centre of diversification (Deméré et al. 2003).

The mode of speciation that apparently fits well with the anisakid nematodes is the *peripatric model* proposed by Mayr (1963, 1976). This involves the geographical isolation of small populations, whose genetic structure begins to differ from the parental one by different genetic mechanisms. In the case of the *Anisakis* spp., the *C. osculatum* and *P. decipiens* species complexes, molecular genetic data strongly suggest that adaptation to different hosts and speciation is related to the geographical isolation of the hosts. Such processes apparently occurred in different times from the lower Miocene to Pliocene, and Pleistocene, when extreme climatic variation took place. During glacial maxima (a period also of lowest sea level), smaller populations of hosts and their endoparasites could have remained isolated in marine refuges, promoting genetic divergence and coadaptation. Then, during interglacial periods, geographical ranges might have expanded, favouring host range expansion (Bullini et al. 1997; Mattiucci and Nascetti 2008).

7.8 Anisakids as Indicators of Trophic Web Stability and Habitat Disturbance of Marine Ecosystems from the Austral Region

Food webs are networks of trophic relationships, which map the location of energy flow in a community. The transmission pathways of parasites with indirect life-cycles are fully included in food webs of aquatic ecosystems. In other words, just as food webs have exerted strong selective pressure on the evolution of parasite transmission strategies, parasites are now shaping some of the ecological properties of existing food webs. The transmission routes of anisakid nematodes follow closely the trophic relationships among their successive hosts, and, thus, they are parasites embedded in food webs. As a consequence, the completion of such life-cycles, as complicated as those of anisakid nematodes, requires stable trophic webs. As a result, the life-cycle of anisakid nematodes in marine ecosystems characterized by various degrees of habitat disturbance could be affected by changes in host population size. Indeed, when the population size of the hosts participating in the life-cycle of these parasites is reduced, due to different causes (pollution, by-catch of marine mammals, viral diseases of marine mammals, overfishing, etc.), the population size of their anisakid endoparasites could also be reduced. This would result in a higher probability of genetic drift in the parasite gene pools and, consequently, a decrease in their genetic variability values (Mattiucci and Nascetti 2008). In this context, quantifying population density and estimating the genetic diversity of those parasites, whose life-cycle is embedded in a marine ecosystem food webs, could be an indirect analysis of the demographic reductions and population bottlenecks (due to anthropogenic causes such as habitat fragmentation, and over-exploitation) of those definitive and intermediate/paratenic hosts which are involved in their life-cycle. Indeed, there is a general understanding that the quantity and quality of genetic diversity of natural populations may influence their viability (Frankham 2010). It is sometimes predicted that reductions in natural population sizes, among the other effects, could negatively impact their genetic diversity. This loss of genetic diversity is a result of increased genetic drift in small populations. Because genetic drift acts more rapidly in small populations, overall genetic diversity is expected to be roughly proportional to the size of a population.

It has been shown (Mattiucci and Nascetti 2008) that the distribution of the genetic variability of anisakid nematode populations in geographical areas with different levels of environmental stress is likely to reflect the influence of a range of factors that could promote their genetic diversity. These include a large effective parasite population size, the wide range, availability, and population size of their hosts, and the stability of marine trophic webs. The values of the genetic variability [estimated at the parameters of: percentage of polymorphic loci (*P*); mean number of alleles per locus (*A*); and expected heterozygosity per locus (*He*)], obtained at 19 allozyme loci, were compared among 53 populations of nematodes belonging to 20 species of *Anisakis*, *Pseudoterranova* and *Contracaecum* from several hosts in the Boreal and Austral Regions (Mattiucci and Nascetti 2007, 2008). Austral populations of species belonging to these three genera exhibited significantly higher genetic variability values than those from the Boreal regions [expected mean of heterozygosity per locus, $He=0.19$ (in Austral populations) and $He=0.09$ (in Boreal populations) ($P<0.01$)] (Mattiucci and Nascetti 2007, 2008). A more remarkable difference in their genetic variability values was observed when only Antarctic and sub-Antarctic populations were compared directly with Arctic and sub-Arctic populations [$He=0.23$ and $He=0.07$ ($P<0.001$), respectively] (Mattiucci and Nascetti 2008). One conclusion is that the observed values of genetic variability could be related to extreme latitudes, a parameter often considered as relevant (Nevo et al. 1984). However, the data suggested that a significantly higher level of genetic diversity found in the Antarctic members considered (i.e., *C. osculatum* sp. D, *C.osculatum* sp. E, *C. radiatum*, *P. decipiens* sp. E, *A. berlandi* and *A. pegreffii* populations from sub-Antarctic regions) coincide with a lower degree of habitat disturbance (e.g., overfishing, by-catch of cetaceans, hunting and diseases mortality of seals, sea water pollution and acidification). This would allow host species to reach higher population sizes, resulting in higher anisakid population sizes, with a reduced probability of genetic drift phenomena in the parasite gene pools. Consequently, a higher level of genetic diversity in the Antarctic populations of these nematodes was observed (Mattiucci et al. 2015). Likewise, a much higher abundance and intensity of infection was observed in the Antarctic populations and species of anisakid nematodes (Mattiucci and Nascetti 2008; Mattiucci et al. 2015a). The data

Table 7.5 Genetic diversity values so far observed in populations and species of the genera *Anisakis*, Pseudoterranova and Contracaecum from Austral hemisphere, as estimated from the mtDNA *cox2* sequences analysis. *N* number of sequences, *Nh* ± *sd* number of haplotypes and standard deviation, *hd* haplotype diversity, π ± *sd* nucleotide diversity and standard deviation, *S* number of polymorphic sites, *K* average number of differences

Species	Nh	π±s.d.	*h*±s.d.	*K*	*S*
Anisakis berlandi	96	0053±0.006	0.998±0.002	33.58	433
A. pegreffii	90	0.020±0.004	0.996±0.002	12.58	146
A. nascettii	12	0.009±0.002	0.987±0.035	5.79	31
A. brevispiculata	4	0.020±0.004	0.900±0.161	12.60	29
A. paggiae	6	0.041±0.005	0.00±0.096	25.73	59
Pseudoterranova cattani	17	0.017±0.003	0.993±0.021	8.72	32
P. decipiens sp. E	30	0.007±0.003	0.953±0.022	3.61	36
Contracaecum osculatum sp. D	288	0.026±0.004	0.991±0.001	13.65	181
C. osculatum sp. E	138	0.020±0.002	0.992±0.004	10.75	141
C. miroungae	16	0.023±0.004	0.987±0.023	11.96	53
C. radiatum	4	0.025±0.005	1.000±0.177	12.83	24

Data from Mattiucci et al. (2014, 2015a), and unpublished

are consistent with biotic factors, such as the host density of those suitable definitive and intermediate hosts for the anisakid nematodes in the Antarctic waters, that contribute in maintaining the high genetic diversity in the anisakid gene pools. Large populations of anisakid nematodes, such as those from the Antarctic, show higher levels of genetic diversity. It is likely that elevated host density in the Antarctic and sub-Antarctic areas will lead to an increase in anisakid parasite prevalence and abundance, in both suitable definitive and intermediate hosts.

The high levels of parasitic infection reported in the two cryptic species *C. osculatum* sp. D and *C. osculat*um sp. E in Antarctic fish species, which are prey for the Weddell seal, are consistent with the high integrity and stability of the food webs in this pristine marine ecosystem. This, in turn, facilitates the completion of the life-cycles of Antarctic and sub-Antarctic anisakid nematodes (Mattiucci and Nascetti 2007). The parasitic infection levels by *C. osculatum* sp. D and sp. E and their estimates of genetic variability showed no statistically significant variation, over a temporal scale, 1994 *vs* 2012), thus suggesting that the low habitat disturbance of the Antarctic region permits, despite the "extreme" ecological conditions of marine ecosystems, the maintenance of stable trophic webs (Mattiucci et al. 2015).

So far, existing results on the genetic variability of anisakid nematodes, at both nuclear (Mattiucci and Nascetti 2008) and mitochondrial level (Table 7.5, Fig. 7.5), have shown that the genetic diversity (variability) estimates of the host-parasite systems formed by anisakid populations of the genera *Anisakis*, *Pseudoterranova* and *Contracaecum*, from fish and marine mammals, are higher in the Austral Region (i.e., Antarctic and sub-Antarctic regions) than in other geographical areas of the Boreal Region, from where populations of members belonging to those genera, are completing their life-cycles. Particularly, in the populations from the Austral

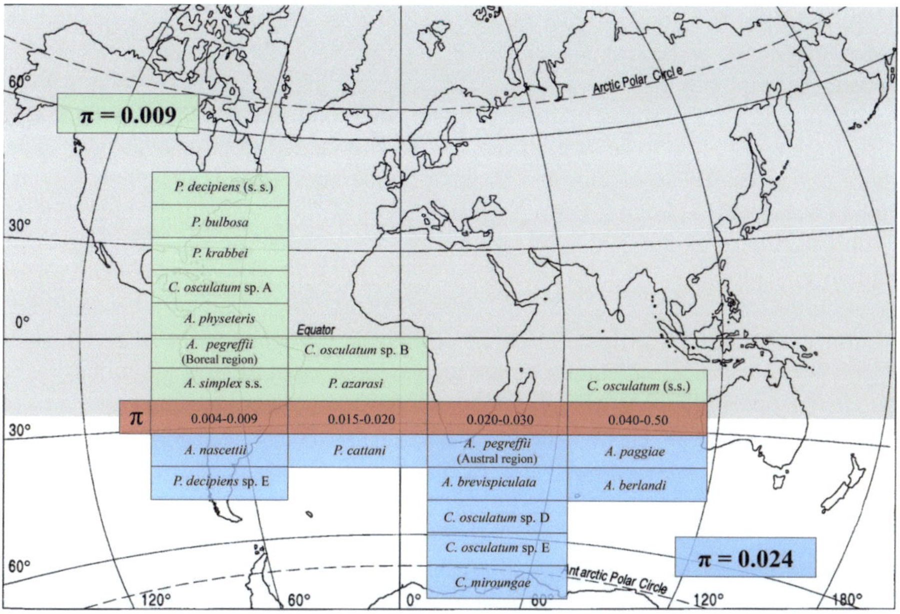

Fig. 7.5 Distribution of the nucleotide diversity values (on average, π) (*reddish coloured cells*), as estimated from the mtDNA *cox2* sequences analysis (see also Table 7.5), among anisakid populations so far considered from Austral Regions of the Southern Hemisphere (*blue coloured cells*), in comparison with those so far calculated in populations/species of anisakids from Boreal Regions of the Northern Hemisphere (*green coloured cells*) (Data from Mattiucci et al. (2014, 2015a), Timi et al. (2014), and unpublished)

Region, a high level of nucleotide diversity was observed at mitochondrial level (on average, π=0.024) ((Mattiucci et al. 2015a) and unpublished data); whereas, a very low level of genetic variability in the mtDNA *cox2* gene was found, so far, in populations from the Boreal region (on average, π=0.009) (Mattiucci et al. unpublished data) (Fig. 7.5). Such differences can be explained by the lower habitat disturbance of the Austral Region, which permits the maintenance of more stable trophic webs in these ecosystems (Mattiucci and Nascetti 2008).

7.9 Are Those Anisakids Species from the Austral Region of Zoonotic Importance to Humans?

Human infection occurs when raw, undercooked or marinated fish, containing living larvae, is eaten. The L3 stage of anisakids infecting the flesh of sea fish or squid can be ingested alive by humans, causing the zoonotic disease known as anisakidosis. Therefore, the study of anisakids is relevant for medicine, veterinary, food inspection, hygiene, legislation and fishery industry.

Anisakiasis, the zoonotic disease due to *Anisakis simplex* (*s. l.*) has gained an increasing health and economic relevance, in particular in those countries where the consumption of raw fish and squid is frequent and human cases are increasingly reported in many European countries (Spain, Italy and France). A number of fish dishes are considered to be of high risk for the contraction of human anisakiasis in those countries. These include the Spanish boquerones and marinated anchovies, Italian marinated anchovies, Latin American "ceviche", etc. Among the nine species of *Anisakis* described above, only two are reported, so far, as causative agents of human anisakiasis: *A. simplex* (s. s.) and *A. pegreffii* (Stallone et al. 1996; D'Amelio et al. 1999; Moschella et al. 2004; Umehara et al. 2007; Fumarola et al. 2009; Mattiucci et al. 2011, 2013). However, despite *A. pegreffii* occurs in several fish species of economic/commercial importance in the Austral Region, no human cases due to that species have been so far documented from this geographic area. In contrast, several cases of gastric, intestinal and gastro-allergic anisakiasis have been molecularly identified as attributable to *A. pegreffii* in patients after eating raw or poorly cooked fish originating from sea waters of the Boreal Region (Mattiucci et al. 2011, 2013). Similarly, no data concerning the possible infectiveness to humans of the species *A. berlandi*, often co-infecting with *A. pegreffii* the same fish species from the Austral Region, are so far available.

Larvae of the *Pseudoterranova decipiens* species complex, known as "sealworms" or "codworms", are the second most common pathogen among anisakids reported from humans, after the species of the *Anisakis simplex* complex. Nematodes of the genus *Pseudoterranova* have proven to be not only a costly problem for seafood processors, but a risk for human health, due to the severe pathology they can cause when consumed with raw or undercooked fish (McClelland 2002; Zhu et al. 2002; Mattiucci et al. 2013), provoking in humans the fish-borne zoonotic disease named "pseudoterranoviasis" or "pseudoterranovosis". The first reports of human infections by *P. decipiens* (*s. l.*) were from North America (Margolis 1977; Lee et al. 1985), followed by cases described as transient infections in California. In Korea, human pseudoterranovosis was first described by Lee et al. (1985), and more recently a human case due to the species *P. azarasi* was reported in Japan (Arizono et al. 2011) and in Italy; the last likely due to imported fish (Cavallero et al. 2016). No data are so far available on documented cases of pseudoterranovosis from the Austral Region that includes the molecular identification of the zoonotic species. On the other hand, for instance, Timi et al. (2014) documented that the musculature of most of the studied fish species obtained from Argentine waters was free of this parasite, with the exception of *A. brasilianus* and *P. patagonicus*, both showing high parasite burdens. Consequently, these fish species constitute the most potentially hazardous threat for human health, if consumed raw or undercooked. Because of the scattered distributions of *Pseudoterranova* spp. and their geographic patterns of pathogenicity, it has been suggested that their pathological effects in humans could differ among anisakid species (Arizono et al. 2011; Mattiucci et al. 2013). In addition some regional effect has been found in the degree of pathology caused by larval *Pseudoterranova*. In Japan, most patients have severe pathology caused by penetration of the alimentary tract, whereas most cases diagnosed in Europe and Chile have been classified as "transient luminal" and asymptomatic, with worms being expelled by coughing, vomiting or defecation (Smith 1999; McClelland 2002). Consequently, although no human cases have been reported in Argentina, the risk for human health should be expected to be similar to that recorded in Chile. On the other hand, the absence of human cases in Argentina is likely due to the fact that the culinary tradition involves mainly well-cooked fish (Timi et al. 2014).

As concerning the zoonotic role of those *Contracecum* spp. occurring in fish species from the Austral Region, a human case, due to Contracaecum, was reported by Shamsi and Butcher (2011). However, it is also true that some fish species, such as those from Antarctic area, are not of commercial value, nor they have found to be infected in the fish musculature (Mattiucci et al. 2015a). In other cases, as stated above, very scanty data have so far been collected on intermediate hosts involved in the life-cycles of other *Contracaecum* spp., such as *C. miroungae* and *C. ogmorhini.*

It is clear that studies on the zoonotic potential of these nematodes should be also extended to the geographical areas of the Austral Region. This may contribute to more correct diagnosis of anisakidosis, which may presently be overlooked in these less anisakid-aware regions.

7.10 Conclusions and Future Prospects

Detecting and delimiting cryptic parasites species is vital to our understanding of their responses to perturbation and variation in physiological tolerances that may determine their geographic distributions, potential host associations and patterns of disease. Molecular/genetic analysis of anisakid nematodes has provided essential tools for their basic species recognition and their ecology. However, despite the extensive literature on the occurrence and description of anisakid nematodes from the Austral Region, we need more information about the full extent of their geographical distribution, life-cycles, host range and epidemiology of the species of *Anisakis*, *Pseudoterranova* and *Contracaecum*. This will also allow to clarify the possible transmission of the disease (anisakidosis) to humans.

In addition, molecular characterization of biodiversity can be useful to address scale phenomena that are critical to understanding temporal and spatial distributions in some geographic regions where cryptic biodiversity in term of species is now being revealed (Hoberg et al. 2015). In this matter, the discovery of cryptic anisakid species in the Austral Region has allowed an assessment of local biodiversity at both species and gene level. Given the growing evidence that biodiversity could be increasingly affected by human influence, measuring and monitoring the global biodiversity of those parasites would be of great importance. Two complementary strategies have been suggested to examine the effects of habitat disturbance on the genetic variability of parasite populations: (*1*) comparison of different datasets of populations inhabiting disrupted ecosystems (spatial scale) and/or (*2*) comparison of particular datasets of populations through time, from the same geographical area (temporal scale) (Mattiucci and Nascetti 2008). In this scenario, at the spatial scale, genetic diversity and parasite density (abundance) of anisakid populations of the genera *Anisakis*, *Contracaecum* and *Pseudoterranova* from the Southern Ocean have been found to reach high values (Mattiucci and Nascetti 2008). While at the temporal scale, the genetic variability estimates of the two cryptic species, i.e., *C. osculatum* sp. D and *C. osculatum* sp. E, over a period of almost 20 years in the Antarctic ecosystem, showed no statistically significant differences in their parasite mean intensity values and their genetic variability estimates (at both mitochondrial and nuclear level) from Antarctic fish species (Mattiucci et al. 2015a). Furthermore, no statistically significant difference was observed in the relative frequencies of the two species, *C. osculatum* sp. D and *C. osculatum* sp. E in the fish species here considered, over a time. On the other hand, in the Antarctic, the two species, *C. osculatum* sp. D and *C. osculatum* sp. E, share the same definitive host, the Weddell seal *Leptonychotes weddellii*, in which they occur – even syntopically – at very high parasitic burden, with several thousands of specimens collected from a single seal host (Mattiucci and Nascetti 2008). These findings seem to support the hypothesis that the low level of habitat disturbance (pollution, overfishing, mortality by disease and hunting of seals) of the Antarctic region permits the maintenance of more stable trophic webs in this ecosystem. This seems to support the evidence that Antarctic ecosystem is still a "pristine" ecosystem. This same level of ecosystem stability would allow definitive and intermediate/paratenic host species involved in the life-cycles of the two Antarctic species of parasites to reach higher population sizes. This will result, as a consequence, in the observation of high and stable density of parasite populations, with high and stable genetic variability values, over a temporal scale. In other words, monitoring the demography of anisakid parasites and their genetic diversity (variability) values, also by the use of suitable molecular/genetic data generated from DNA microsatellites, SNPs, and next generation sequencing, will be future tools for monitoring the cryptic biodiversity, at both species and gene level, of anisakids from the Southern Ocean.

Acknowledgements Part of the research reported in this review on anisakid nematodes from Austral Region was funded by MIUR-PNRA (Piano Nazionale Ricerche in Antartide) 2013 AZ01/09, and by a Research Grant from "Sapienza- University of Rome" (year 2015).

References

Arduino P, Nascetti G, Cianchi R, Plötz J, Mattiucci S et al (1995) Isozyme variation and taxonomic rank of *Contracaecum radiatum* (v. Linstow, 1907) from the Antarctic Ocean (Nematoda, Ascaridoidea). Syst Parasitol 30:1–9

Arizono N, Miura T, Yamada M, Tegoshi T, Onishi K (2011) Human infection with *Pseudoterranova azarasi* roundworm. Emerg Infect Dis 17:555–556

Arnason U, Bodin K, Gullberg A, Ledje C, Mouchaty S (1995) A molecular view of pinniped relationships with particular emphasis on the true seals. J Molecular Evol 40:78–85

Arnason U, Cullberg A, Janke A (2004) Mitogenomic analysis provide new insights into cetacean origin and evolution. Gene 26:27–34

Baylis HA (1937) On the ascarids parasitic in seals, with special reference to the genus *Contracaecum*. Parasitology 29:121–130

Berland B (1961) Nematodes from some Norwegian marine fishcs. Sarsia 2:1–50

Bullini L, Arduino P, Cianchi R, Nascetti G, D'Amelio S et al (1997) Genetic and ecological research on Anisakid endoparasites of fish and marine mammals in the Antarctic and Arctic-Boreal Regions. In: Battaglia B, Valencia J, Walton DW (eds) Antarctic communities: species, structure and survival. Cambridge University Press, Cambridge/New York, p 464

Cassens I, Vicario S, Waddell VG, Balchowsky H, Van Belle D et al (2000) Independent adaptation to riverine habitats allowed survival of ancient cetacean lineages. Proc Natl Acad Sci U S A 97:11343–11347

Cavallero S, Nadler SA, Paggi L, Barros NB, D'Amelio S (2011) Molecular characterization and phylogeny of anisakid nematodes from cetaceans from Southeastern Atlantic coasts of USA, Gulf of Mexico, and Caribbean Sea. Parasitol Res 108:781–792

Cavellero S, Scrivano D, D'Amelio S (2016) First case report of invasive pseudoterranoviasis in Italy. Parasitol Int 65:488–490

D'Amelio S, Mathiopoulos KD, Brandonisio O, Lucarelli G, Doronzo F, Paggi L (1999) Diagnosis of a case of gastric anisakidosis by PCR-based restriction fragment length polymorphism analysis. Parassitologia 41:591–593

D'Amelio S, Mathiopoulos KD, Santos CP, Pugachev ON, Webb SC, Picanco M, Paggi L (2000) Genetic markers in ribosomal DNA for the identification of members of the genus *Anisakis* (Nematoda: Ascaridoidea) defined by polymerase chain reaction-based restriction fragment length polymorphism. Int J Parasitol 30:223–226

Dailey MD (1975) The distribution and intraspecific variation of helminth parasites in pinnipeds. Rapp P-V Reun Cons Int Explor Mer 169:338–352

Darriba D, Taboada GL, Doallo R, Posada D (2012) jModelTest 2: more models, new heuristics and parallel computing. Nat Methods 9:772

Davey JT (1971) A revision of the genus *Anisakis* Dujardin, 1845 (Nematoda: Ascaridata). J Helminthol 45:51–72

Deméré TA, Berta A, Adam PJ (2003) Pinnipedimorph evolutionary biogeography. In: Flynn LJ (ed). Vertebrate fossils and their context: contributions in Honor of Richard H. Tedford. Bull Am Mus Nat Hist 279:32–76

Fagerholm HP, Gibson DI (1987) A redescription of the pinniped parasite *Contracaecum ogmorhini* (Nematoda, Ascaroidea), with an assessment of its antiboreal circumpolar distribution. Zool Scr 16:19–24

Frankham R (2010) Challenges and opportunities of genetic approaches to biological conservation. Biol Conserv 143:1919–1927

Fumarola L, Monno R, Ierardi E, Rizzo G, Giannelli G, Lalle M, Pozio E (2009) *Anisakis pegreffii* etiological agent of gastric infections in two Italian women. Foodborne Pathog Dis 6:1157–1159

George-Nascimento M (1991). La estructura de los ensambles comunitarios de parásitos metazoos de vertebrados marinos: un acercamiento a distintos niveles jerárquicos. Doctoral Thesis. Pont Univ Cat Chile. pp 332

George-Nascimento M, Llanos A (1995) Micro-evolutionary implications of allozymic and morphometric variations in sealworms *Pseudoterranova* sp. (Ascaridoidea: Anisakidae) among sympatric hosts from the Southeastern Pacific Ocean. Int J Parasitol 25:1163–1171

George-Nascimento M, Urrutia X (2000) *Pseudoterranova cattani* sp. nov. (Ascaridoidea: Anisakidae), a parasite of the South American sea lion *Otaria byronia* De Blainville from Chile. Rev Chil Hist Nat 73:93–98

Hernández-Orts JS, Montero FE, Juan-García A, García NA, Crespo EA et al (2012) Intestinal helminth fauna of the South American sea lion *Otaria flavescens* and fur seal *Arctocephalus australis* from northern Patagonia, Argentina. J Helminthol doi:10.1017/S0022149X12000454

Hernández-Orts JS, Aznar FJ, Blasco-Costa I, García NA, Víllora-Montero M et al (2013) Description, microhabitat selection and infection patterns of sealworm larvae (*Pseudoterranova decipiens* species complex, Nematoda: Ascaridoidea) in fishes from Patagonia, Argentina. Parasit Vectors 6:252

Hoberg EP, Agosta SJ, Boeger WA, Brooks DR (2015) An integrated parasitology: revealing the elephant through tradition and invention. Trends Parasitol 31:128–133

Hu M, D'Amelio S, Zhu XQ, Paggi L, Gasser R (2001) Mutation scanning for sequence variation in three mitochondrial DNA regions for members of the *Contracaecum osculatum* (Nematoda: Ascaridoidea) complex. Electrophoresis 22:1069–1075

Huelsenbeck JP, Ronquist R (2005) Bayesian analysis of molecular evolution using Mr Bayes. In: Stat Methods Mol Evol, Nielsen R. (ed). Springer, New York. pp 183–232

Johnston TH, Mawson PM (1945) Parasitic Nematodes. Reports of the British, New Zealand and Australian Antartic. Research Expedition 5:74–159

Klimpel S, Palm HW (2011) Anisakid nematode (Ascaridoidea) life cycles and distribution: increasing zoonotic potential in the time of climate change? Progress in parasitology. Springer, Berlin/Heidelberg, pp 201–222

Klimpel S, Busch M, Kuhn T, Rohde A, Palm HW (2010) The *Anisakis simplex* complex off the South Shetland Islands (Antarctica): endemic populations versus introduction through migratory hosts. Mar Ecol Prog Ser 403:1–11

Klöser H, Plötz J (1992) Morphological distinction between adult *Contracaecum radiatum* and *Contracaecum osculatum* (Nematoda Anisakidae) form the Weddel seal (*Leptonychotes weddelli*). Zool Scr 21:129–132

Klöser H, Plötz J, Palm H, Bartsch A, Hubold G (1992) Adjustment of anisakid nematode life cycles to the high Antarctic food web as shown by *Contracaecum radiatum* and *C. osculatum* in the Weddell sea. Antarct Sci 4:171–178

Kuhn T, García-Márquez J, Klimpel S (2011) Adaptive radiation within Marine Anisakid Nematodes: a zoogeographical modeling of cosmopolitan, zoonotic parasites. PLoS ONE 1–6, e28642. doi:10.1371/journal.pone.0028642

Lee AH, Kim SM, Choi KY (1985) A case of human infection with the larva of *Terranova* type A. Korean J Pathol 19:463–467

Margolis L (1956) Parasitic helminths and arthropods from pinnipeds of the Canadian Pacific Coast. J Fish Res Board Can 13:489–505

Margolis L (1977) Public health aspects of "codworm" infection: a review. J Fish Res Board Can 34:887–898

Mattiucci S, Nascetti G (2006) Molecular systematics, phylogeny and ecology of anisakid nematodes of the genus *Anisakis* Dujardin, 1845: an update. Parasite 13:99–113

Mattiucci S, Nascetti G (2007) Genetic diversity and infection levels of anisakid nematodes parasitic in fish and marine mammals from Boreal and Austral hemispheres. Vet Parasitol 148:43–57

Mattiucci S, Nascetti G (2008) Advances and trends in the molecular systematics of anisakid nematodes, with implications for their evolutionary ecology and host-parasite co-evolutionary processes. Adv Parasitol 66:47–148

Mattiucci S, Nascetti G, Cianchi R, Paggi L, Arduino P et al (1997) Genetic and ecological data on the *Anisakis simplex* complex with evidence for a new species (Nematoda, Ascaridoidea, Anisakidae). J Parasitol 83:401–416

Mattiucci S, Paggi L, Nascetti G, Ishikura H, Kikuchi K et al (1998) Allozyme and morphological identification of *Anisakis, Contracaecum* and *Pseudoterranova* from Japanese waters (Nematoda, Ascaridoidea). Syst Parasitol 40:81–92

Mattiucci S, Paggi L, Nascetti G, Abollo E, Webb SC et al (2001) Genetic divergence and reproductive isolation between *Anisakis brevispiculata* and *Anisakis physeteris* (Nematoda: Anisakidae). Int J Parasitol 31:9–14

Mattiucci S, Cianchi R, Nascetti G, Paggi L, Sardella N et al (2003) Genetic evidence for two sibling species within *Contracaecum ogmorhini* Johnston & Mawson (1941) (Nematoda: Anisakidae) from otariid seals of boreal and austral regions. Syst Parasitol 54:13–23

Mattiucci S, Nascetti G, Dailey M, Webb SC, Barros N, Cianchi R, Bullini L (2005) Evidence for a new species of *Anisakis* Dujardin, 1845: morphological description and genetic relationships between congeners (Nematoda: Anisakidae). Syst Parasitol 61:157–171

Mattiucci S, Paoletti M, Webb SC, Sardella N, Timi JT et al (2008) Genetic relationships among species of *Contracaecum* Railliet & Henry, 1912 and *Phocascaris* Host, 1932 (Nematoda: Anisakidae) from pinnipeds inferred from mitochondrial *cox2* sequences, and congruence with allozyme data. Parasite 15:408–419

Mattiucci S, Paoletti M, Webb SC (2009) *Anisakis nascettii* n. sp. (Nematoda: Anisakidae) from beaked whales of the southern hemisphere: morphological description, genetic relationships between congeners and ecological data. Syst Parasitol 74:199–217

Mattiucci S, Paoletti M, Borrini F, Palumbo M, Palmieri RM et al (2011) First molecular identification of the zoonotic parasite *Anisakis pegreffii* (Nematoda: Anisakidae) in a paraffin-embedded granuloma taken from a case of human intestinal anisakiasis in Italy. BMC Infect Dis 11:82

Mattiucci S, Fazii P, De Rosa A, Paoletti M, Salomone Megna A et al (2013) Anisakiasis and gastroallergic reactions sssociated with *Anisakis pegreffii* infection, Italy. Em Infect Dis 19:496–499

Mattiucci S, Cipriani P, Webb SC, Paoletti M, Marcer F et al (2014) Genetic and morphological approaches distinguish the three sibling species of the *Anisakis simplex* species complex, with a species designation as *Anisakis berlandi* n. sp. for *A. simplex* sp. C (Nematoda: Anisakidae). J Parasitol 100:199–214

Mattiucci S, Cipriani P, Paoletti M, Nardi V, Santoro M et al (2015a) Temporal stability of parasite distribution and genetic variability values of *Contracaecum osculatum* sp. D and *C. osculatum* sp. E (Nematoda: Anisakidae) from fish of the Ross Sea (Antarctica). Int J Parasitol Parasites Wildl 4:356–367

Mattiucci S, Cimmaruta R, Cipriani P, Abaunza P, Bellisario B, Nascetti G (2015b) Integrating parasite data and host genetic structure in the frame of an holistic approach for stock identification in Mediterranean Sea fish species. Parasitology 142:90–108

Mattiucci S, Acerra V, Paoletti M, Cipriani P, Levsen A et al (2016) No more time to stay 'single' in the detection of *Anisakis pegreffii*, *A. simplex* (*s. s.*) and hybridization events between them: a multi-marker nuclear genotyping approach. Parasitology 143:998–1011

Mayr E (1963) Animal species and evolution. Belknap Press/Harvard University Press, Cambridge, Massachusetts

Mayr E (1976) Evolution and the diversity of life. Belknap Press/Harvard University Press, Cambridge, Massachusetts

McClelland G (2002) The trouble with sealworms (*Pseudoterranova decipiens* species complex, Nematoda): a review. Parasitology 124:183–203

Milinkovitch MC (1995) Molecular phylogeny of cetaceans prompts revision of morphological transformations. Trends Ecol Evol 10:328–334

Moschella CM, Mattiucci S, Mingazzini P, De Angelis G, Assenza M et al (2004) Intestinal anisakiasis in Italy: case report. J Helminthol 78:271–273

Nadler SA, Pérez-Ponce de León G (2011) Integrating molecular and morphological approaches for characterizing parasite cryptic species: implications for parasitology. Parasitology 138:1688–1709

Nadler SA, D'Amelio S, Fagerholm HP, Berland B, Paggi L (2000) Phylogenetic relationships among species of *Contracaecum* Railliet & Henry, 1912 and *Phocascaris* Host, 1932 (Nematoda: Ascaridoidea) based on nuclear rDNA sequence data. Parasitology 121:455–463

Nadler SA, D'Amelio S, Dailey MD, Paggi L, Siu S et al (2005) Molecular phylogenetics and diagnosis of *Anisakis, Pseudoterranova*, and *Contracaecum* from northern pacific marine mammals. J Parasitol 91:1413–1429

Nascetti G, Paggi L, Orecchia P, Smith JW, Mattiucci S, Bullini L (1986) Electrophoretic studies on *Anisakis simplex* complex (Ascaridida: Anisakidae) from the Mediterranean and North East Atlantic. Int J Parasitol 16:633–640

Nascetti G, Cianchi R, Mattiucci S, D'Amelio S, Orecchia P et al (1993) Three sibling species within *Contracaecum osculatum* (Nematoda, Ascaridida, Ascaridoidea) from the Atlantic Arctic-boreal region: reproductive isolation and host preferences. Int J Parasitol 23:105–120

Nevo E, Beiles A, Ben-Shlomo R (1984) The evolutionary significance of genetic diversity: ecological, demographic and life history correlates. Evol Dyn Genet Divers 53:14–213

Nikaido M, Matsuno F, Hamilton H, Brownell RL Jr, Cao Y et al (2001) Retroposon analysis of major cetacean lineages: the monophyly of toothed whales and the paraphyly of river dolphins. Proc Natl Acad Sci U S A 98:7384–7389

Orecchia P, Mattiucci S, D'Amelio S, Paggi L, Plotz J, Cianchi R, Nascetti G et al (1994) Two new members in the *Contracaecum osculatum* complex (Nematoda, Ascaridoidea) from the Antarctic. Int J Parasitol 24:367–377

Paggi L, Nascetti G, Cianchi R, Orecchia P, Mattiucci S et al (1991) Genetic evidence for three species within *Pseudoterranova decipiens* (Nematoda, Ascaridida, Ascaridoidea) in the North Atlantic and Norwegian and Barents Seas. Int J Parasitol 21:195–212

Paggi L, Nascetti G, Webb SC, Mattiucci S, Cianchi R, Bullini L (1998) A new species of *Anisakis* Dujardin, 1845 (Nematoda: Anisakidae) from beaked whale (Ziphiidae): allozyme and morphological evidence. Syst Parasitol 40:161–174

Paggi L, Mattiucci S, Gibson DI, Berland B, Nascetti G et al (2000) *Pseudoterranova decipiens* species A and B (Nematoda: Ascaridoidea): nomenclatural designation, morphological diagnostic characters and genetic markers. Syst Parasitol 45:185–197

Pontes T, D'Amelio S, Costa G, Paggi L (2005) Molecular characterization of larval anisakid nematodes from marine fishes of Madeira by a PCR-based approach, with evidence for a new species. J Parasitol 91:1430–1434

Rice DW (1998) Marine mammals of the world: systematics and distribution, 4th edn. Soc Mar. Mamm Spec Publ, Lawrence

Shamsi S (2014) Recent advances in our knowledge of Australian anisakid nematodes. Int J Parasitol Parasites Wildl 3:178–187

Shamsi S, Butcher AR (2011) First report of human anisakidosis in Australia. Med J Aust 194:199–200

Smith JW (1999) Ascaridoid nematodes and pathology of the alimentary tract and its associated organs in vertebrates, including man: a literature review. Helminthol Abstr 68:49–96

Stallone O, Paggi L, Balestrazzi A, Mattiucci S, Montinari M (1996) Gastric anisakiasis in Italy: case report. Med J Sur Med 4:13–16

Timi JT, Sardella NH, Mattiucci S (2003) *Contracaecum ogmorhini* Johnston et Mawson, 1941 (Nematoda: Anisakidae), parasite of *Arctocephalus australis* (Zimmermann, 1783) off the Argentinean coasts. Helminthologia 40:27–31

Timi JT, Paoletti M, Cimmaruta R, Lanfranchi AL, Alarcos AJ, Garbin L et al (2014) Molecular identification, morphological characterization and new insights into the ecology of larval *Pseudoterranova cattani* in fishes from the Argentine coast with its differentiation from the Antarctic species, *P. decipiens* sp. E (Nematoda: Anisakidae). Vet Parasitol 199:59–72

Umehara A, Kawakami Y, Araki J, Uchida A (2007) Molecular identification of the etiological agent of the human anisakiasis in Japan. Parasitol Int 56:211–215

Valentini A, Mattiucci S, Bondanelli P, Webb SC, Mignucci-Giannone A, Colom-Llavina MM, Nascetti G (2006) Genetic relationships among *Anisakis* species (Nematoda: Anisakidae) inferred from mitochondrial *cox-2* sequences, and comparison with allozyme data. J Parasitol 92:156–166

Vaz-Ferreira R (1982) Otaria flavescens (Shaw), South American sea lion. Mammals in the Seas. F. A. O., Fisheries Series 6:447–495

Zarlenga DS, Hoberg E, Rosenthal B, Mattiucci S, Nascetti G (2014) Anthropogenics: Human influence on global and genetic homogenization of parasite populations. J Parasitol 100:756–772

Zhu XQ, D'Amelio S, Paggi L, Gasser RB (2000) Assessing sequence variation in the internal transcribed spacers of ribosomal DNA within and among members of the *Contracaecum osculatum* complex (Nematoda: Ascaridoidea: Anisakidae). Parasitol Res 86:677–683

Zhu XQ, D'Amelio S, Palm HW, Paggi L, George–Nascimento M, Gasser RB (2002) SSCP–based identification of members within the Pseudoterranova decipiens complex (Nematoda:Ascaridoidea: Anisakidae) using genetic markers in the internal transcribed spacers of ribosomal DNA. Parasitology 124:615–623

Chapter 8
Acanthocephalans in Sub-Antarctic and Antarctic

Zdzisław Laskowski and Krzysztof Zdzitowiecki

8.1 Introduction

Acanthocephalans (spiny head worms) are a medium-sized phylum (about 1000 species have been described) of usually small (few mm to over 1m) vertebrate intestinal parasites. They are pseudocoelomates with bilateral symmetry and usually cylindrical bodies. The sexes are separate, with females usually larger than males. The body consists of a proboscis, neck, and trunk. The proboscis, neck, and internal organs connected with them (proboscis receptacle and lemnisci) form the forebody. In some cases, the trunk may be divided into two parts of different shape: fore-trunk and hind-trunk. The proboscis is armed with recurved hooks. The hooks consist of two parts: blade (thorn) and root, both usually directed posteriorly. Hooks situated at the base of the proboscis (basal hooks) are usually rootless. The proboscis (usually retractable) may be invaginated into the proboscis receptacle. The latter contains a cerebral ganglion. Two lemnisci lie parallel to the proboscis receptacle. The trunk may be unarmed, or armed with spines. This armament is usually restricted to the anterior part of the trunk, but sometimes reaches the posterior end of the body. The genital pore may be subterminal or terminal. Spines surrounding the genital pore are often separated from the other ones by a bare zone. In such cases, the armament of the trunk is divided into somatic and genital spines. Ligaments (one or two) run along the trunk, and sexual organs are attached to them. The male reproductive system consists of 2 testes, cement glands (4–8 in number in Antarctic species), seminal ducts, cement ducts and reservoirs, Säfftigen's pouch, penis and the copulatory bursa (retracted or everted). The female reproductive system consists of ovarian balls, a uterine bell (an organ for selection of immature and

Z. Laskowski (✉) • K. Zdzitowiecki
Witold Stefański Institute of Parasitology, Polish Academy of Sciences,
Twarda 51/55, 00-818 Warsaw, Poland
e-mail: laskowz@twarda.pan.pl

S. Klimpel et al. (eds.), *Biodiversity and Evolution of Parasitic Life in the Southern Ocean*, Parasitology Research Monographs, Vol. 9,
DOI 10.1007/978-3-319-46343-8_8

mature eggs), a uterus, and a vagina, with a single or a double sphincter. Ovarian balls are enclosed in ligament sacs in juvenile females and are liberated during maturation. Eggs mature in the pseudocoelom of a female. In fact, in mature females these are not eggs, but the first larval stage (acanthors) enclosed in 3–4 envelopes. More correct terms are "shelled acanthors" and "embryophores", but these are rarely used. Acanthocephalans have reduced the muscular, nervous, circulatory, and excretory systems and complete loss of the digestive system. Absorption and excretion take place through the tegument. The latter contains a system of canals known as the lacular system. The number and arrangement of main lacular canals are of fundamental value in the classification of higher taxa (classes). Excretion is by diffusion except in Oligacanthorhynchidae (with two protonephridial organs). The life cycles involve an arthropod (intermediate host) and a vertebrate definitive or paratenic host. Eggs are shed with the host's faeces, when the definitive host, the appropriate intermediate host ingests them, and the acanthor is liberated and pierces the gut wall. In the arthropod body cavity, the acanthor develops into an acanthella and then into an infective cystacanth, which matures to adulthood in the gut of the definitive host, following ingestion of the infected arthropod (Amin 1987; Zdzitowiecki 1991).

One of the present authors (Zdzitowiecki) published in 1991 the monograph of Antarctic Acanthocephala, this chapter contains new data of this parasites.

The phylum includes four classes: Archiacanthocephala, Eoacanthocephala, Palaeacanthocephala, and Polyacanthocephala

Representatives of two orders of Palaeacanthocephala (Echinorhynchida and Polymorphida) occur in notothenioid fishes (Zdzitowiecki 1991). Echinorhynchida use fishes as definitive hosts and occur in the lumen of the alimentary tract. Fishes become infected by feeding on crustaceans (intermediate hosts), or in cases of Polymorphida also small infected fishes which play a role as paratenic hosts of Polymorphida localized in cysts in the body cavity. Crustaceans of the order Amphipoda were recorded as intermediate hosts of two echinorhynchid species, *Aspersentis megarhynchus* (Linstow 1892) and *Metacanthocephalus johnstoni* Zdzitowiecki 1983, and three polymorphids, *Corynosoma bullosum* (Linstow 1892), *C. hamanni* (Linstow 1892), and *C. pseudohamanni* Zdzitowiecki, 1984, in Antarctica (Hoberg 1986; Zdzitowiecki 2001; Zdzitowiecki and Presler 2001; Laskowski et al. 2008). Definitive hosts of Antarctic polymorphids are marine mammals and birds. The infective stage, the cystacanth, is similar to the mature worm, but differs from the latter in the size of the trunk and degree of development of the sexual organs (Zdzitowiecki 1991). In cystacanths of the Polymorphidae Meyer, 1931 (with exceptions of Filicollis Lühe, 1911 and Profilicollis Meyer, 1931), the dimensions of the proboscis and the development and size of both the proboscis hooks and trunk spines are usually identical with those of adults. Cystacanths occur in intermediate and paratenic hosts in cysts and are contracted; this is especially so in that they have an introverted proboscis. Cystacanths should be collected alive, liberated from their cysts, and relaxed. Such material can be determined on the basis of most of the diagnostic morphological features useful for adults.

8.2 Checklist of the Antarctic and Sub-Antarctic Acanthocephala

Class **Palaeacanthocephala**

Order **Echinorhynchida**

Family **Heteracanthocephalidae;** Subfamily **Aspersentinae**

Genus ***Aspersentis***

Species:
Aspersentis megarhynchus (von Linstow 1892) (Fig. 8.1)
Aspersentis johni (Baylis 1929) (Fig. 8.2)
Aspersentis zanclorhynchi (Johnston and Best 1937) Smales 1996

Family **Arhythmacanthidae**

Genus ***Heterosentis***

Species:
Heterosentis heteracanthus Linstow 1896 (Fig. 8.3)
Heterosentis hirsutus Pichelin and Cribb 1999
Heterosentis zdzitowieckii (Kumar 1992)

Genus ***Hypoechinorhynchus***

Species:
Hypoechinorhynchus magellanicus Szidat 1950 (Fig. 8.4)

Family **Echinorhynchidae** Subfamily **Echinorhynchinae**

Genus ***Echinorhynchus***

Species:
Echinorhynchus petrotschenkoi Rodjuk 1984 (Fig. 8.5)
Echinorhynchus muraenolepisi Rodjuk 1984

Family **Rhadinorhynchidae** Subfamily **Gorgorhynchinae**

Genus ***Metacanthocephalus***

Species:
Metacanthocephalus campbelli (Leiper and Atkinson 1914)
Metacanthocephalus dalmori Zdzitowiecki, 1983
Metacanthocephalus johnstoni Zdzitowiecki, 1983 (Fig. 8.6)
Metacanthocephalus rennicki (Leiper and Atkinson 1914)

Order **Polymorphida**

Family **Polymorphidae**

Genus ***Profilicollis***

Species:
Profilicollis antarcticus Zdzitowiecki 1985 (Fig. 8.7)
Profilicollis novaezelandensis Brockerhoff and Smales, 2002

Genus ***Corynosoma***

Species:
Corynosoma arctocephali Zdzitowiecki, 1984 (Fig. 8.8)
Corynosoma australe Johnston, 1937
Corynosoma beaglense Laskowski, Jeżewski, Zdzitowiecki, 2008 (Fig. 8.9)
Corynosoma bullosum (Linstow 1892) (Fig. 8.10)
Corynosoma evae Zdzitowiecki, 1984 (Fig. 8.11)
Corynosoma gibsoni Zdzitowiecki, 1986 (Fig. 8.12)
Corynosoma hamanni Linstow 1892 (Fig. 8.13)
Corynosoma hannae Zdzitowiecki, 1984
Corynosoma pseudohamanni Zdzitowiecki, 1984 (Fig. 8.14)
Corynosoma shackletoni Zdzitowiecki, 1978

Genus ***Andracantha***

Species:
Andracantha baylisi (Zdzitowiecki 1986a, b, c, d, e, f, g) Zdzitowiecki, 1989 (Fig. 8.15)
Andracantha clavata (Goss 1940)

Genus: ***Bolbosoma***

Species:
Bolbosoma balaenae (Gmelin 1790)
Bolbosoma brevicolle (Malm 1867) (Fig. 8.16)
Bolbosoma hamiltoni Baylis 1929
Bolbosoma tuberculata Skryabin 1970
Bolbosoma turbinella australis Skryabin 1972

8.3 Representatives of Acanthocephalans Genera Occurring in Antarctica and Sub-Antarctica

(Zdzitowiecki 1991; Laskowski and Zdzitowiecki 2004, 2008; Laskowski et al. 2008, 2010)

Family **Heteracanthocephalidae** Petrotschenko 1956
Genus **Aspersentis** Van Cleave 1929

Diagnosis: Trunk spined. Proboscis cylindrical, relatively short. Ventral proboscis hooks larger than dorsal. Proboscic receptacle double-walled, ganglion in its posterior half. Neck short. Cement glands in males pyriform, six in number, forming compact group. Vaginal sphincter in females double. Eggs with polar prolongations of middle envelope. Parasites of fishes.

Aspersentis megarhynchus (Linstow 1892) (Fig. 8.1)

Synonyms: *A. austrinus* Van Cleave 1929, *Rhadinorhynchus wheeleri* Baylis 1929, *Heteracanthocephalus hureaui* Dollfus 1965.

Diagnosis (after Zdzitowiecki 1981): Proboscis hooks in 13–16 rows of 8–11. The largest hook is the third one counting from tip. Trunk spines conspicuous anteriorly (maximum length 35 μm), very small, and hardly visible at posterior trunk end. Lemnisci slightly longer than proboscis receptacle.

Male. Total dimensions 3.6–5.5 × 0.73–1.39 mm. Proboscis 0.47–0.63 × 0.20–0.31 mm. Maximum hook length 106–135 μm. Testes arranged in tandem to diagonally.

Female. Total dimensions 5.6–9.6 × 1.16–2.09 mm. Proboscis 0.51–0.73 × 0.29–0.35 mm. Maximum hook length 119–149 μm. Eggs 60–88 × 19–25 μm.

Suitable definitive hosts: fishes.

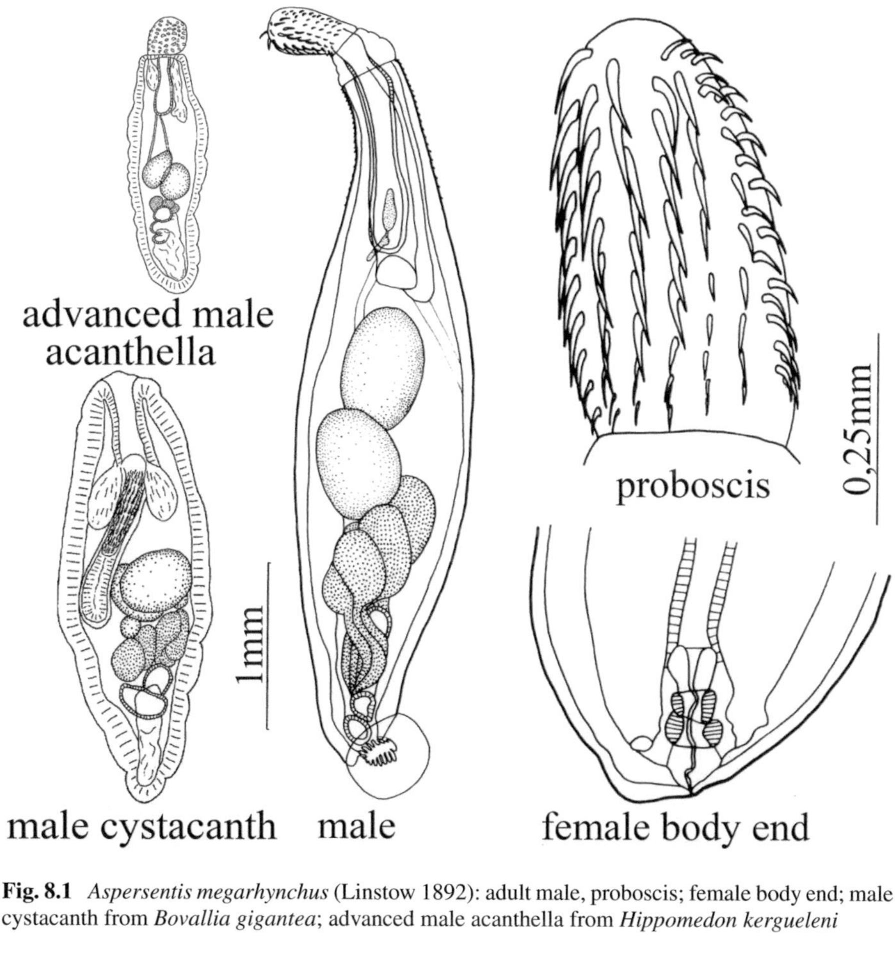

Fig. 8.1 *Aspersentis megarhynchus* (Linstow 1892): adult male, proboscis; female body end; male cystacanth from *Bovallia gigantea*; advanced male acanthella from *Hippomedon kergueleni*

Nototheniidae: *Notothenia acuta*, *N. coriiceps*, *N. cyanobrancha*, *N. rossii*, *Nototheniops mizops*, *Lindbergichthys nudifrons, Gobionotothen gibberifrons*, *Pagothenia bernacchii*, *P. hansoni*, *Trematomus newnesi*; Bathydraconidae: *Parachaenichthys charcoti, P. georgianus*; Channichthyidae: *Channichthys rhinoceratus*, *Chaenocephalus aceratus*; Harpagiferidae: *Harpagifer antarcticus.*

Intermediate hosts: amphipods. Eusiridae: *Bovallia gigantea*; Gammarellidae: *Gondogeneia antarctica*; Ischyroceridae: *Jassa ingens*; Lysianassoidea: *Hippomedon kergueleni* and *Orchomenella rotundifrons.*

Habitat: Males mainly in posterior half of small intestine, females mainly in large intestine. Few specimens in other parts of intestine.

Biology and ecology: According to Zdzitowiecki and Rokosz (1986), Zdzitowiecki (1990b), Zdzitowiecki and White (1996), Zdzitowiecki and Presler (2001), Zdzitowiecki and Laskowski (2004), Laskowski and Zdzitowiecki (2010), Laskowski et al. (2012), the species is associated with the inshore (fiord) environment, where infections of fishes take place. *N. coriiceps* and juvenile specimens of *N. rossii* living in Admiralty Bay (the South Shetland Islands) and *N. coriiceps* caught in the coastal zone at Signy Island (South Orkney Islands) are massively infected (prevalence 100%, maximum intensity of infection 180, 91, and 81, respectively). Other fishes are much less infected. *A. megarhynchus*, the dominant echinorhynchid species in the Admiralty Bay and South Orkney Islands, was extremely rare at the Vernadsky Station (Argentine Islands). Only two *N. coriiceps* specimens of 93 examined were infected by one and 14 parasites (prevalence 2 %). Adult specimens of *N. rossii* living in the open sea are also less infected, while other fishes living in the open sea at South Shetland Islands and at South Georgia are uninfected. The parasite occurs in fishes during the whole year, but infections of *N. coriiceps* and *N. rossii* in Admiralty Bay are more numerous in winter than in summer (incomplete seasonality). Cystacanths of *A. megarhynchus* were found in four sub-coastal host species belonging to four families of Amphipoda in the Admiralty Bay.

Distribution: Circumpolar species not far from the Antarctic convergence in the Sub-Antarctic (Kerguelen subregion); South Shetland Islands, South Orkney Islands, South Georgia, Heard, Kerguelen, Crozet, Macquarie, Ob Bank, and Argentine Islands (Linstow 1892; Van Cleave 1929; Baylis 1929; Joyeux and Baer 1954; Edmonds 1955, 1957; Dollfus 1965; Szidat and Graefe 1967; Golvan 1969; Parukhin and Sysa 1975; Parukhin and Lyadov 1982; Hoogesteger and White 1981; Zdzitowiecki 1981, 1987, 1990a, b; Zdzitowiecki and Rokosz 1986; Zdzitowiecki and Laskowski 2004; Rodjuk 1985; Reimer 1987).

Aspersentis johni (Baylis 1929) (Fig. 8.2)
Synonyms: *Rhadinorhynchus johni.*

Diagnosis (after Laskowski and Zdzitowiecki 2004): Proboscis almost cylindrical, relatively narrow (length/width ratio 2.16–3.22:1, mean 2.78:1), widest anteriorly, curved towards ventral side. Hooks normally arranged in 14 rows, rarely in 13 or 15 rows, of 10–13 hooks, either in same number in all rows around proboscis or with difference of one hook in neighbouring rows. Ventral hooks (with exceptions of 2–4 posterior-most) much larger than dorsal hooks. Number of large hooks gradually decreases in lateral rows; not less than 3 dorsal rows exclusively contain small hooks. Dimensions of large hooks decrease posteriorly; distal or sub-distal hooks are longest.

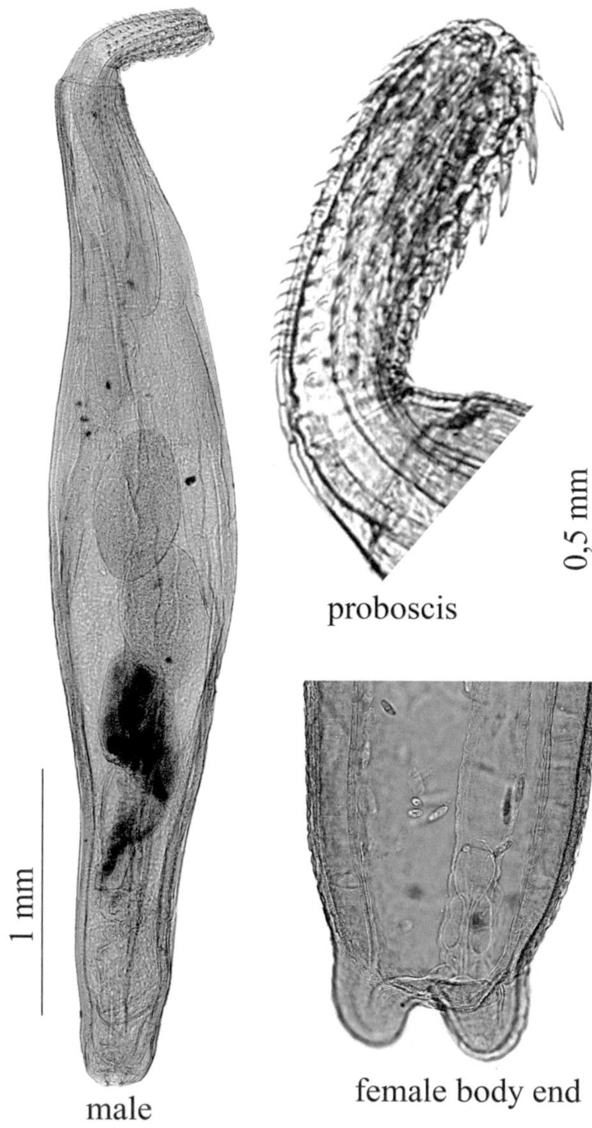

Fig. 8.2 *Aspersentis johni* (Baylis 1929): adult male, proboscis; female body end

All small hooks of similar length and with roots (roots of small hooks are hardly visible and often unmeasurable); blades and roots directed posteriorly. Roots of small hooks have process directed anteriorly. Blades are longer than roots, larger in females than males. Neck unarmed curved towards ventral side. Anterior trunk armed with spines, maximum length of spines on ventral side c.30 μm, smaller on dorsal side. Conspicuous ventral spines extend posteriorly over 9.7–13.9 (11.9)% of trunk length in males, 7.6–13.3 (10.7)% in females. Smaller spines of various dimensions are visible more posteriorly, especially near posterior end of trunk and on ventral side just beyond large spines. Proboscis receptacle extends posteriorly beyond range of large spines. Lemnisci longer and narrower than proboscis receptacle.

Male. Total dimensions 4.03–6.21×0.518–0.912 mm. Proboscis 0.419–0.580×0.156–0.226 mm. Maximum length of ventral hook 77–101 μm. Trunk spindle-shaped. Testes oval, tandem to oblique in mid-length of trunk. Cement glands pear-shaped, 6 in number, forming compact group. Posterior end of trunk oval; genital aperture shifted slightly to dorsal side. When everted genital bursa is bell-like.

Female. Total dimensions 6.12–8.54×0.71–1.16 mm. Proboscis 0.49–0.66×0.18–0.28 mm. Maximum length of ventral hook 84–108 μm. Eggs 87–102×20–26 μm. Trunk spindle-shaped, more elongate than in males, with 2 lateral lobes invariably present at posterior end. Uterine bell obscured by eggs. Vaginal sphincter double. Genital aperture in concavity between lateral lobes. Total length of female genital system (uterine bell, uterus, and vagina) was measurable approximately in 2 cases and reached 1.5 mm in immature specimen and 2.1 mm in mature specimen. Mature eggs elongate, with polar prolongations of middle envelope.

Definitive host: fishes.
Nototheniidae: *Patagonotothen longipes*; Merlucciidae: *Merluccius* sp.; Channichthyidae: *Champsocephalus esox*.

Habitat: intestine, large intestine (rectum).

Biology and ecology: (after Laskowski and Zdzitowiecki 2004, 2009): The infection of the *Patagonotothen longipes* and *Champsocephalus esox* at Beagle Channel (eastern mouth of the Beagle Channel): prevalence 85 and 25%, maximum intensity 18 and 4 parasites in one fish, respectively.

Distribution: Beagle Channel, Magellanic subregion of the Sub-Antarctic waters off the Falkland Islands (Baylis 1929; Laskowski and Zdzitowiecki 2004).

The only other representative of *Aspersentis* occurring in notothenioids is *A. megarhynchus* (Linstow 1892). Features useful to distinguish *A. johni* from *A. megarhynchus* are: 10–13 vs. 7–11 proboscis hooks in each row, the maximum length of the ventral hooks 77–108 vs. 106–149 μm, a narrower proboscis with a length/width ratio of 2.16–3.22:1 (mean 2.78:1) vs. 1.66–2.27:1 (mean 2.015:1), an egg length of 87–102 vs. 60–88 μm, and an unusual form of the posterior extremity of females (the presence of a terminal concavity between two lateral lobes). Another representative is *A. zanclorhynchi* (Johnston and Best 1937) Smales 1996, synonym Echinorhynchus sensu lato from *Zanclorhynchus spinifer* (Zdzitowiecki 1986a).

Family **Arhythmacanthidae** Yamaguti, 1935
Genus **Heterosentis** Van Cleave 1931

Diagnosis: Trunk spined anteriorly. Proboscis relatively short, cylindrical to globular. Two to three types of hooks along proboscis. Proboscis receptacle double-walled. Ganglion at base of proboscis receptacle. Neck short. Cement glands in males pyriform, six in number, forming compact group. Vaginal sphincter in females single. Eggs with polar prolongations of middle envelope. Parasites of fishes.

Heterosentis heteracanthus (Linstow 1896) (Fig. 8.3)

Diagnosis (after Zdzitowiecki 1984a): Proboscis short, narrowed at base. Hooks in 10 rows of 3–5. One large distal hook and 2–4 rootless basal hooks in every row.

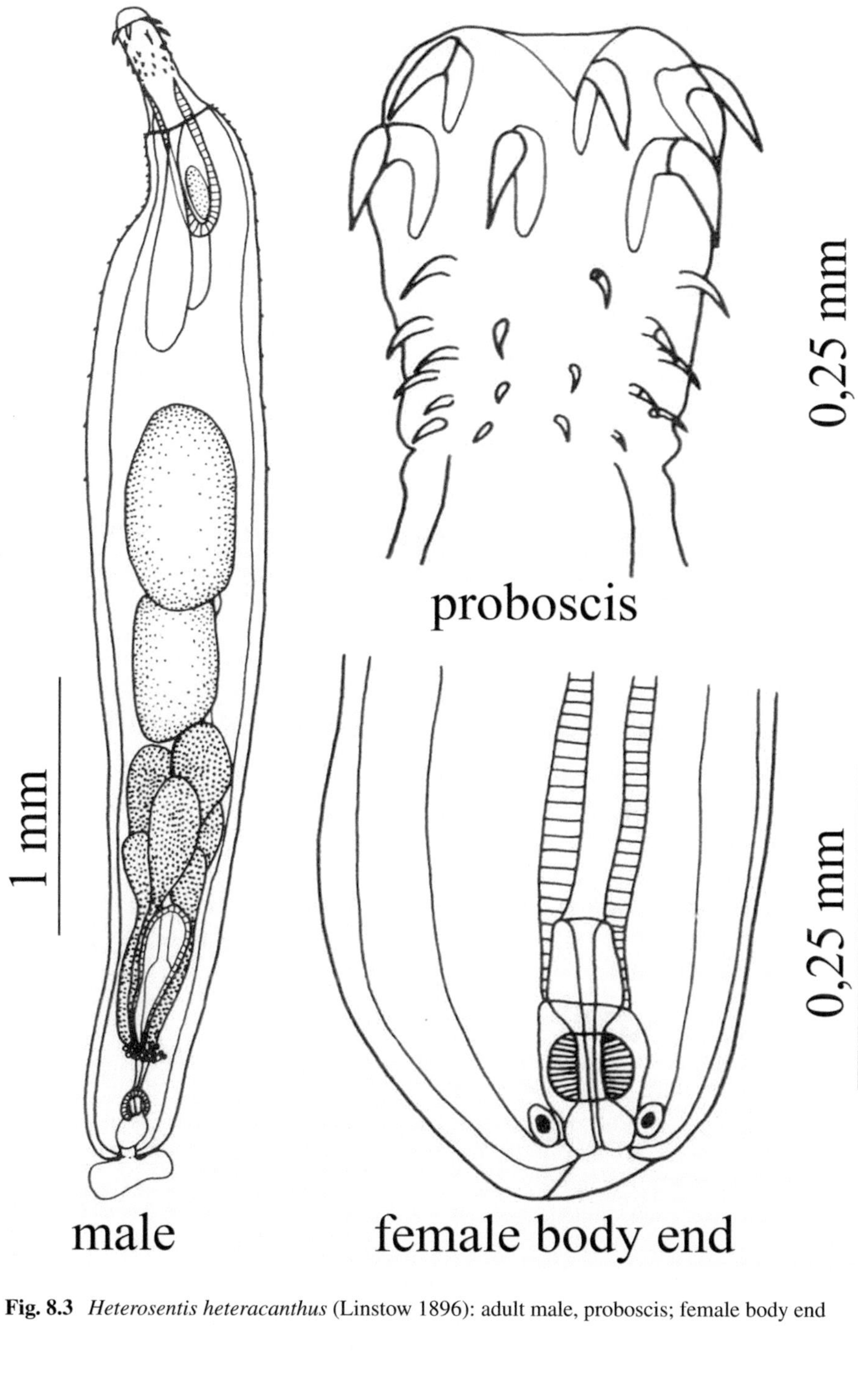

Fig. 8.3 *Heterosentis heteracanthus* (Linstow 1896): adult male, proboscis; female body end

Blade and root of distal hook similar in length. Lemnisci longer than proboscis receptacle.

Male. Total dimensions 3.6–4.5 × 0.43–0.58 mm. Proboscis 0.224–0.252 × 0.154–0.161 mm. Maximum length of distal hook 58–60 μm. Testes in tandem.

Female. Total dimensions 6.6 × 0.75 mm. Proboscis 0.264 × 0.195 mm. Maximum length of distal hook 76 μm.

Definitive hosts: fishes.

Atherinidae: *Chirostoma microlepidotus*; Nototheniidae: *N. coriiceps*, *N. rossii*, *N. squamifrons*, *Gobionotothen gibberifrons*, *N. nybelini*, *Lindbergichthys nudifrons*, *Patagonotothen longipes*, *P. tessellata*; Artedidraconidae: *Artedidraco mirus*; Bathydraconidae: *Parachaenichthys georgianus*: Channichthyidae: *Champsocephalus esox*.

Habitat: Mainly large intestine. Few specimens in posterior half of small intestine.

Biology and ecology: The species is rare in the Antarctic and it seems to be more frequent in fiords than in the open sea. Of the fish examined in the eastern mouth of the Beagle Channel, *Patagonotothen longipes* was the most infected (prevalence 50%, maximum intensity 25), *P. tessellata* and *Champsocephalus esox* were less infected (prevalence 15% and 10%, maximum intensity 17 and 1, respectively) (Laskowski and Zdzitowiecki 2009).

Distribution: Strait of Magellan (South America), Beagle Channel, South Shetland Islands, and South Georgia (Linstow 1896; Van Cleave 1931; Meyer 1931; Zdzitowiecki 1984a, 1986g, 1987, 1990b; Laskowski and Zdzitowiecki 2009).

Genus ***Hypoechinorhynchus*** Yamaguti, 1939

Diagnosis (after Pichelin and Cribb 1999) Hypoechinorhynchus have the characteristic abrupt transition from basal spines to apical hooks; they also possess longitudinal rows, which alternate in their possession of a middle spine. The middle and posterior spines are small, thin and without roots (or very reduced roots); the middle spine may be longer than the posterior spine. Each longitudinal row has at least one large hook with a root.

Hypoechinorhynchus magellanicus (Szidat 1950) (Fig. 8.4)

Diagnosis (after Laskowski and Zdzitowiecki 2008): Trunk with antero-dorsal curvature. Proboscis spherical (length/width ratio 0.89–1.22:1) slightly curved towards ventral side. Proboscis armature: 40 hooks, including 15 large hooks with root > c .50% length of blade and 25 rootless basal spines. Large hooks arranged in 10 alternating rows of 1 and 2 hooks; anterior hook of each pair slightly smaller than other hooks; each single large hook is followed in same row by 2 spines; pairs of large hooks are followed by single spines. Ten single spines are present at base of proboscis between rows. This arrangement of proboscis armature could be also interpreted as 3 transverse rows of 5 large hooks in each and 3 transverse rows of basal spines containing 5, 10, and 10 spines, respectively. Unarmed neck slightly curved towards ventral side. Trunk cylindrical, armed anteriorly with loosely

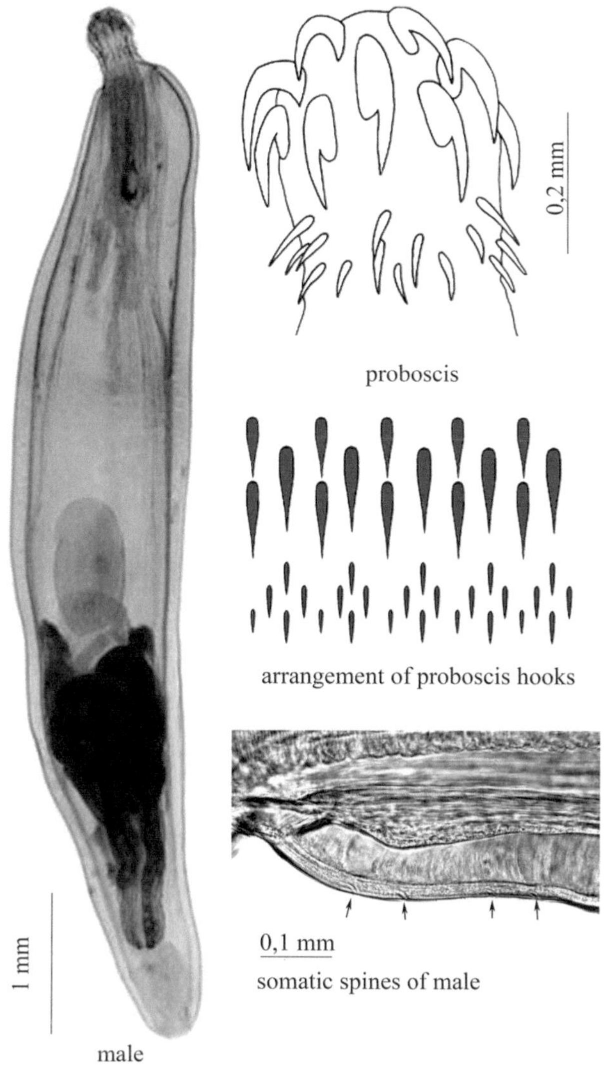

Fig. 8.4 *Hypoechinorhynchus magellanicus* (Szidat 1950): adult male, proboscis; arrangement of proboscis hooks; somatic spines of male

arranged small spines of 15–26×3–13 in size. Region of spination extends back 5.3–16.0% of trunk length. Proboscis receptacle double-walled, with ganglion at base. Lemnisci long, narrow, considerably longer than proboscis receptacle.

Male. Total length 5.90–7.81 mm. Proboscis 0.336–0.396×0.289–0.330 mm. Five large sub-apical hooks: blade length 109–138 µm, basal width 21–32 µm, root length c.60 µm in; posterior hook of pairs: blade length 132–170 µm, basal width 35–36 µm, root length 69–91 µm; single large hooks: blade length 145–184 µm, basal width 36–45 µm, root length 71–85 µm. Basal proboscis spines: blade length 37–94 µm, basal width 8–11 μm. Neck conical, 0.113–0.181 mm in length. Proboscis receptacle 0.875–1.072 × 0.237–0.318 mm. Length of lemnisci 1.800–2.137 mm. Trunk spindle-shaped, 5.435–7.230 × 1.187–1.387 mm; length/width ratio 4.58–5.21:1. Testes and cement glands form compact group far beyond proboscis receptacle. Testes more oblique than tandem; anterior testis 0.823–1.030 × 0.528–0.637 mm; posterior testis partly parallel with cement glands, 0.764–1.050 × 0.550–0.621 mm. Cement glands pear-shaped, 6 in number, form compact group. Säfftigen's pouch 0.990–1.157 × 0.221–0.238 mm. Genital pore terminal.

Female. Total length 9.54 mm and 7.78 mm. Proboscis 0.364 × 0.409 mm and 0.399–0.375 mm. Five sub-apical hooks: blade length 122–124 μm, basal width 32 μm, 61 μm root length; posterior hooks of pairs: blade length 155–162 μm, 36 μm basal width, root length 81 μm; single large hooks: blade length 173–175 μm, basal width 45 μm, root length 85 μm. Basal proboscis spines: blade length 50–94 μm, basal width 10–21 μm. Neck conical, 211–213 μm in length. Proboscis receptacle 0.953 × 0.294 and 1.053 × 0.346. Lemnisci, uterine bell, and uterus obscured by eggs. Trunk spindle-shaped, 8.975–1.820 mm and 7.167–1.658 mm. Trunk length/width ratio 4.93:1 and 4.32:1. Genital pore terminal. Vaginal sphincter single, c. 160 × 150 μm. Eggs with polar prolongations of middle envelope, 71–86 × 16–22 μm, mean 76 × 19 μm.

Suitable definitive hosts: fishes. Nototheniidae: *Eleginops maclovinus*; Channichthyidae: *Champsocephalus esox.*

Habitat: Large intestine.

Biology and ecology: Not known.

Distribution: Beagle Channel, Ushuaia (Tierra del Fuego, South America) (Szidat 1950; 1965 Laskowski and Zdzitowiecki 2008; Laskowski and Zdzitowiecki 2009).

Family **Echinorhynchidae** Cobbold, 1876
Genus **Echinorhynchus** Zoega in Müller, 1776

Diagnosis: Trunk cylindrical, not spined. Proboscis cylindrical. Neck short. Proboscis receptacle double-walled. Ganglion at half of length of proboscis receptacle. Lemnisci claviform. Cement glands in males spherical or oval, six in number, arranged either in a compact group or in line along the trunk. Testes in tandem. Vaginal sphincter in females single. Eggs elongated, with long polar prolongations of middle envelope. Parasites of fishes.

Echinorhynchus petrotschenkoi (Rodjuk 1984) (Fig. 8.5)

Synonyms: *Echinorhynchus* sp. Kagei et Watanuki, 1975, *E. nototheniae* Zdzitowiecki, 1986, *E. georgianus* Rodjuk 1986.

Diagnosis (after Zdzitowiecki 1989b): Proboscis hooks arranged in 14–20 rows of 9/10–14/15, including 1–2 basal ones. Blades of hooks longer than roots. Proboscis receptacle a little longer than lemnisci.

Male. Total dimensions 6.3–13.6 × 0.51–1.34 mm. Proboscis 0.766–1.015 × 0.218–0.303 mm. Maximum hook length 63–85 μm. Testes oval. Cement

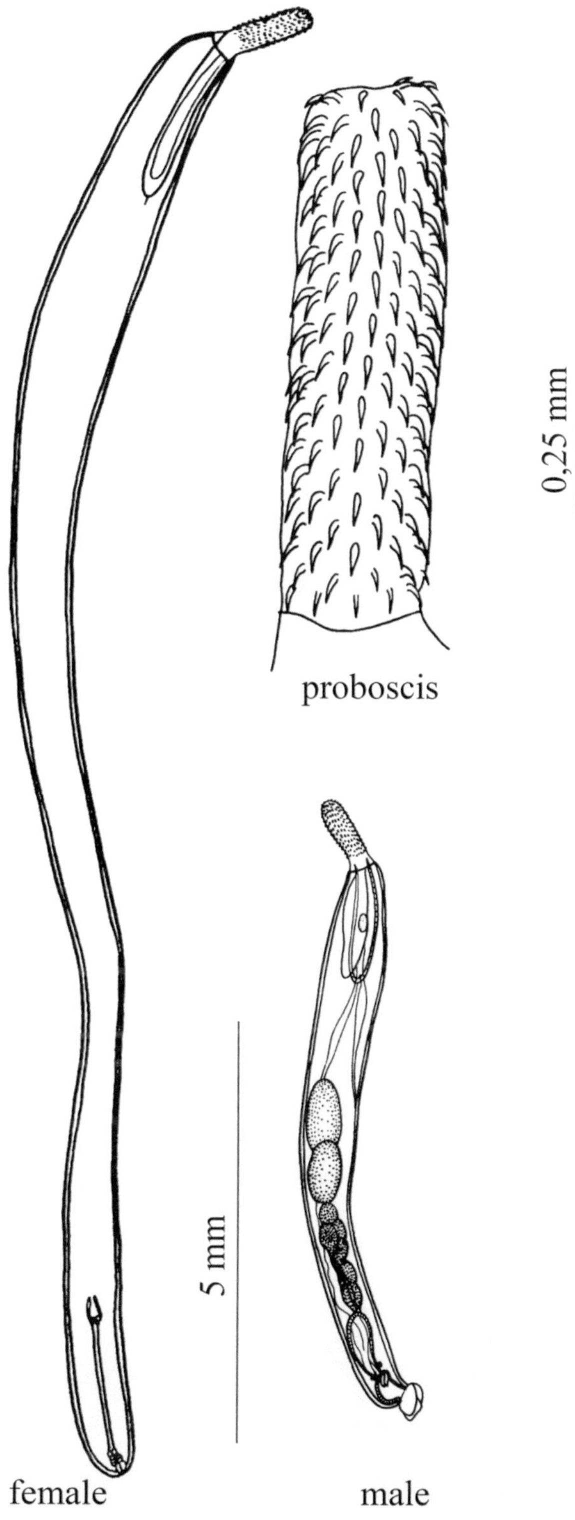

Fig. 8.5 *Echinorhynchus petrotschenkoi* (Rodjuk 1984): adult female and male, proboscis

glands arranged in principle along trunk, closely to each other. However, some cement glands can lie parallel forming one or two pairs.

Female. Total dimensions 12.2–30.0×0.75–1.48 mm. Proboscis 0.764–1.176×0.233–0.340 mm. Maximum hook length 64–87 µm. Eggs 89–121×19–25 µm.

Suitable definitive hosts: fishes. Muraenolepidae: *Muraenolepis microps*; Nototheniidae: *Dissostichus eleginoides*, *Pagothenia bernacchii*. Other (? unsuitable) definitive hosts: fishes. Nototheniidae: *Dissostichus mawsoni*, *Notothenia coriiceps*, *Nototheniops nybelini*, *Pagothenia hansoni*; Channichthyidae: *Chaenocephalus aceratus*, *Cryodraco antarcticus*.

Habitat: Small intestine.

Biology and ecology: The species is associated mainly with the open sea shelf environment (Zdzitowiecki 1990b). Prevalence of infection of the main definitive host, *M. microps*, at South Georgia 40%, maximum intensity 11.

Distribution: Probably circumpolar. Till now found at South Shetland Islands, South Georgia, and Syowa Station (Enderby Land) (Kagei and Watanuki 1975; Rodjuk 1984, 1986; Zdzitowiecki 1986d, g, 1989b, 1990b).

Family **Rhadinorhynchidae** Subfamily **Gorgorhynchinae**

Genus ***Metacanthocephalus*** Yamaguti, 1959

Diagnosis: Trunk not spined. Neck short. Proboscis cylindrical to ovoid. Ganglion in anterior half of proboscis receptacle. Proboscis receptacle double-walled. Lemnisci (in Antarctic species) longer than proboscis receptacle. Testes in tandem. Cement glands pyriform, eight in number, arranged in a compact group. Vaginal sphincter in females double. Eggs with polar prolongations of middle envelope. Parasites of fishes.

Metacanthocephalus johnstoni Zdzitowiecki, 1983 (Fig. 8.6)
Synonyms: *Leptorhynchoides campbelli* (1914) in Johnston and Best (1937) pro parte.

Diagnosis (after Zdzitowiecki 1983): Trunk oval or egg-shaped. Maximum width at half of its length. Proboscis cylindrical. Hooks arranged in 12–17 rows of 5–7/8 (usually 14–16×6–7). Blade of hook longer than root. Longest hook is the second or third one counting from base of proboscis.

Male. Total dimensions 3.60–7.37×0.60–2.10 mm. Proboscis 0.426–0.554×0.182–0.280 mm. Length: width ratio of proboscis 1.74–2.67: 1. Maximum hook length 71–86 µm.

Female. Total dimensions 6.06–8.66×1.94–3.22 mm. Proboscis 0.486–0.599×0.229–0.323 mm. Length: width ratio of proboscis 1.69–2.51: 1. Maximum hook length 79–96 µm. Length of female genital system 1.0–1.7 mm. Eggs 88–108×20–25 µm (mean 97×22 µm).

Suitable definitive hosts: fishes. Nototheniidae: *Notothenia coriiceps*, *N. rossii*, *Gobionotothen gibberifrons*, *Lindbergichthys nudifrons*, *Pagothenia bernacchii*, *P. hansoni*, *Trematomus eulepidotus*. *T. newnesi*; Bathydraconidae: *Parachaenichthys georgianus*.

Other definitive (? unsuitable) hosts: fishes. Channichthyidae: *Champsocephalus gunnari*.

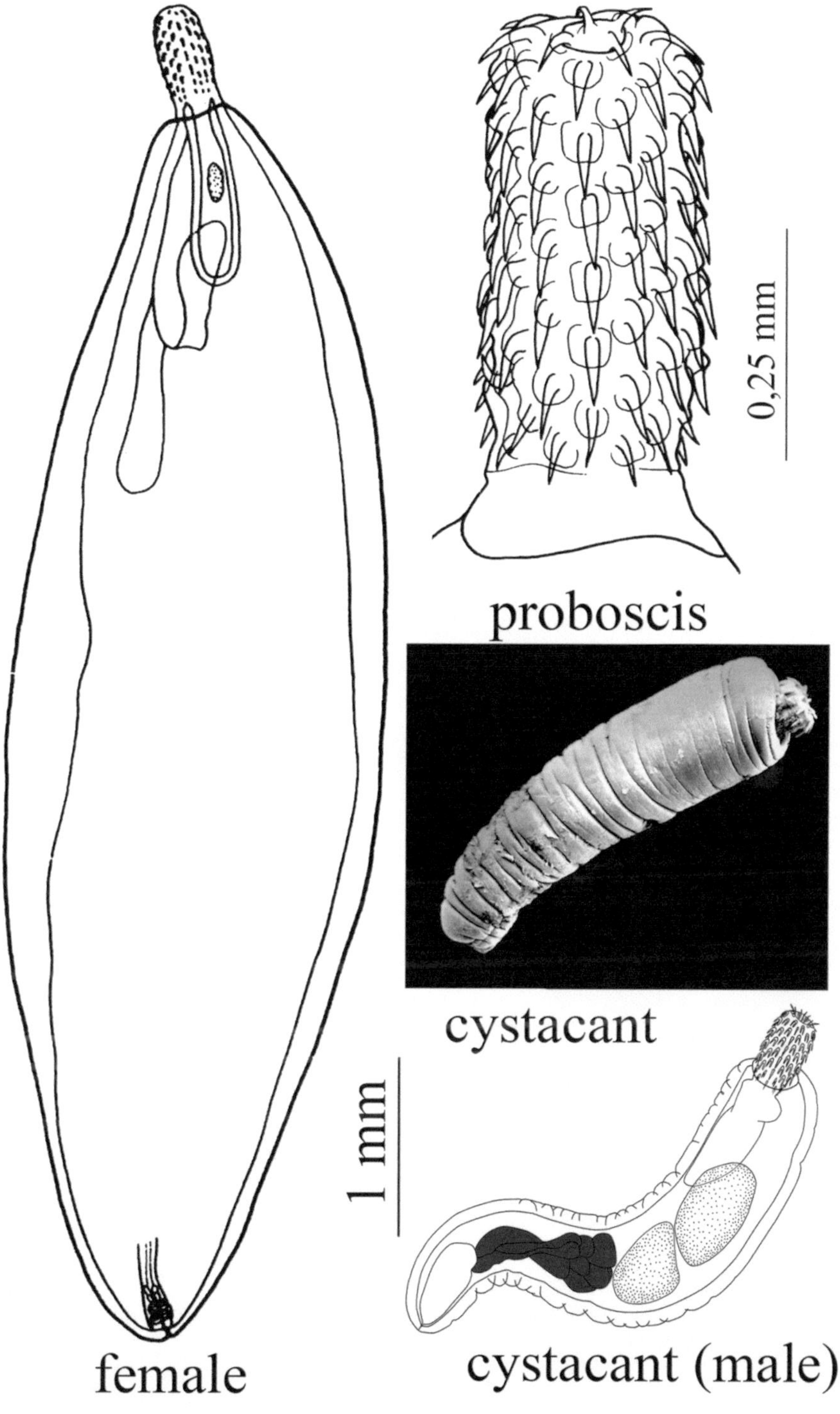

Fig. 8.6 *Metacanthocephalus johnstoni* Zdzitowiecki, 1983: adult female; proboscis and male cystacanth from *Cheirimedon femoratus*

Intermediate host: amphipods. Lysianassoidea: *Cheirimedon femoratus*.

Habitat: Mainly pyloric caeca and anterior half of small intestine. Few specimens occur more posteriorly in small and large intestine.

Biology and ecology: According to Zdzitowiecki 1986g, 1990b; Zdzitowiecki and Laskowski 2004; Laskowski et al. 2010, 2012, the species is associated with the inshore fiord environment and infections take place at a depth smaller than 50 m. *N. coriiceps* and juvenile *N. rossii* living in the Admiralty Bay (the South Shetland Islands) were heavily infected (prevalence 85 and 100%, maximum intensity 85 and 130 parasites in one fish). *M. johnstoni* infection of *N. coriiceps* were less abundant (prevalence was 74 % and maximum intensity 25). The species is rare in fishes living in the open sea at the South Shetland Islands and at South Georgia. It was found there almost exclusively in adults of *N. rossii.* The parasite occurs in fishes the whole year (lack of seasonality). Cystacanths of *Metacanthocephalus johnstoni* were found in the haemocoeloma of *C. femoratus* (5707 examined specimens) caught at the Galindez Island (Argentine Islands, Western Antarctica) with prevalence 0.51%. A total of 1416 specimens of *Cheirimedon femoratus* caught in the Admiralty Bay (South Shetland Islands) were found to be free of *M. johnstoni*

Distribution: The South Shetland Islands, South Georgia, Adelie Land, Argentine Islands (Johnston and Best 1937; Zdzitowiecki 1983; 1986g, 1987, 1990b; Zdzitowiecki and Laskowski 2004; Laskowski et al. 2007, 2010).

Order **Polymorphida**

Family **Polymorphidae**
Genus ***Profilicollis*** Meyer 1931

Diagnosis: Trunk cylindrical, spined in anterior half. Proboscis of both sexes spherical. Neck long. Proboscis receptacle long, double-walled. Lemnisci claviform. Testes in tandem. Cement glands tubular. Vaginal sphincter in females double. Eggs without polar prolongations of middle envelope. Parasites of birds.

Species:
Profilicollis antarcticus Zdzitowiecki 1985 (Fig. 8.7)

Diagnosis (after Zdzitowiecki 1985): Proboscis spherical, wider than long. Hooks relatively small, arranged in 18–22 rows of 7–8/9. Anterior 2–3 hooks solid, with short blades and long roots directed posteriorly. Posterior 4–5 hooks with long blades directed posteriorly and short roots directed anteriorly. Neck very long, constituting 15–22% of total body length. Anterior 16–24% of trunk covered with small spines. Lemnisci reaching more posteriorly than proboscis receptacle.

Male. Total dimensions 14.1–21.1 × 2.0–3.0 mm. Proboscis 0.86–1.56 × 1.06–1.98 mm. Maximum hook length 71–74 µm. Neck length 2.1–4.0 mm. Four cement glands.

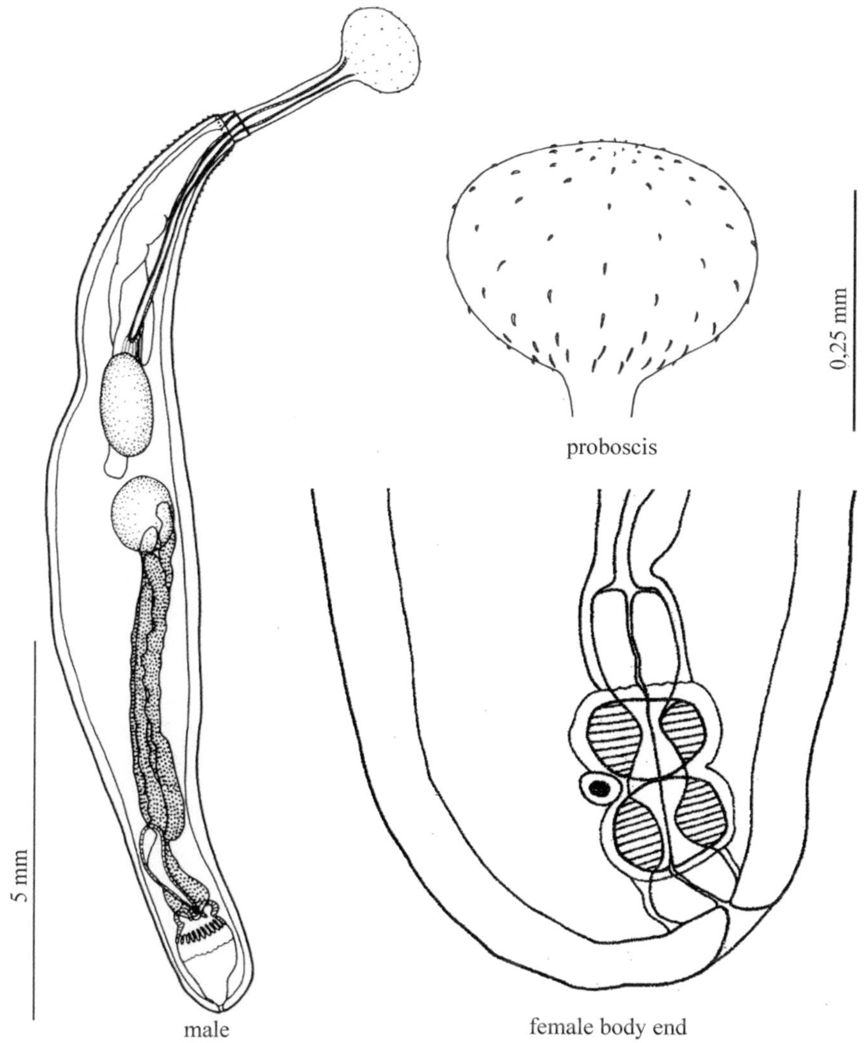

Fig. 8.7 *Profilicollis antarcticus* Zdzitowiecki 1985: adult male and proboscis

Female. One immature specimen was available. Total dimensions 11.1 × 1.2 mm. Proboscis 1.01 × 1.25 mm. Maximum hook length 80 μm. Neck length 2.46 mm. Genital pore subterminal. Eggs unknown.

Definitive hosts (probably unsuitable): birds (*Chionis alba*).

Habitat: Ileum and caecum.

Biology and ecology: Not known.

Genus ***Corynosoma***
Corynosoma bullosum (Linstow 1892) (Fig. 8.8)

Synonyms: *C. mirabilis* Skryabin 1966, *C. singularis* Skryabin et Nikolsky, 1971 pro parte.

Diagnosis (after Zdzitowiecki 1986c): Proboscis hooks in 16 (rarely 17 or 18) rows of 10/11–14/15, including 2–3/4 rootless basal ones. Distal hook the longest one. Hind-trunk cylindrical, considerably longer than fore-trunk. Genital armature separated from somatic one. Lemnisci flat, shorter than proboscis receptacle.

Male. Total dimensions 9.6–13.4 × 1.4–2.0 mm. Proboscis 0.91–1.35 × 0.31–0.37 mm. Maximum hook length 89–117 μm. Fore-trunk constitutes 30–40% of trunk length. Somatic armature covers 33–54% of trunk length on ventral side. Number of genital spines circa 80–250, usually 100–200. Cement glands tubular.

Female. Total dimensions 13.6–19.7 × 1.8–2.8 mm. Proboscis 1.11–1.33 × 0.34–0.40 mm. Maximum hook length 0.099–1.120 mm. Fore-trunk constitutes 20–32% of trunk length. Somatic armature covers 28–38% of trunk length on ventral side. Number of genital spines 3–120, usually 20–50. Genital pore terminal. Eggs 107–125 × 35–39 μm.

Suitable definitive hosts: elephant seals (*Mirounga leonina*, *M. angustirostris* (?)). Unsuitable definitive hosts: seals (*Hydrurga leptonyx*, *Leptonychotes weddelli*, *Lobodon arcinophagus*); whales (*Physeter catodon*). Juvenile specimens also in intestine of birds (*Phalacrocorax atriceps*, *Pygoscelis papua*).

Paratenic hosts: fishes. Nototheniidae: *Notothenia macrophthalma*, *N. coriiceps*, *N. rossii*, *N. squamifrons*, *N. nybelini*, *Nototheniops larseni*, *Gobionotothen gibberifrons*, *Lindbergichthys nudifrons*, *Dissostichus eleginoides*, *D. mawsoni*, *Pagothenia bernacchii*, *P. hansoni*, *Patagonotothen brevicauda guntheri*; Artedidraconidae: *Artedidraco mirus*, *Artedicraco* sp.; Bathydraconidae: *Parachaenichthys charcoti*, *P. georgianus*; Channichthyidae: *Chaenocephalus aceratus*, *Chionodraco rastrospinosus*, *Cryodraco antarcticus*, *Pseudochaenichthys georgianus*; Macrouridae: *Macrourus holotrachys*; Muraenolepidae: *Muraenolepis microps*; Liparidae: *Paraliparis* sp.

Intermediate hosts: amphipods. Lysianassoidea: *Waldeckia obesa*; Eusiridae: *Bovallia gigantea*.

Habitat: Small and large intestine.

Biology and ecology: According to Zdzitowiecki (1986b, g, 1990b), infections of paratenic hosts take place mainly in the open sea shelf environment, deeper than 100 m. Predatory fishes living at the South Shetland Islands and at South Georgia are massively infected, up to one thousand cystacanths in one host specimen (*D. eleginoides*).

Cystacanths of *C. bullosum* were found in amphipods (intermediate hosts) in Admiralty Bay (Zdzitowiecki 2001b; Zdzitowiecki and Presler 2001).

Fig. 8.8 *Corynosoma bullosum* (Linstow 1892): adult male and female; cystacanth from *Waldeckia obesa*

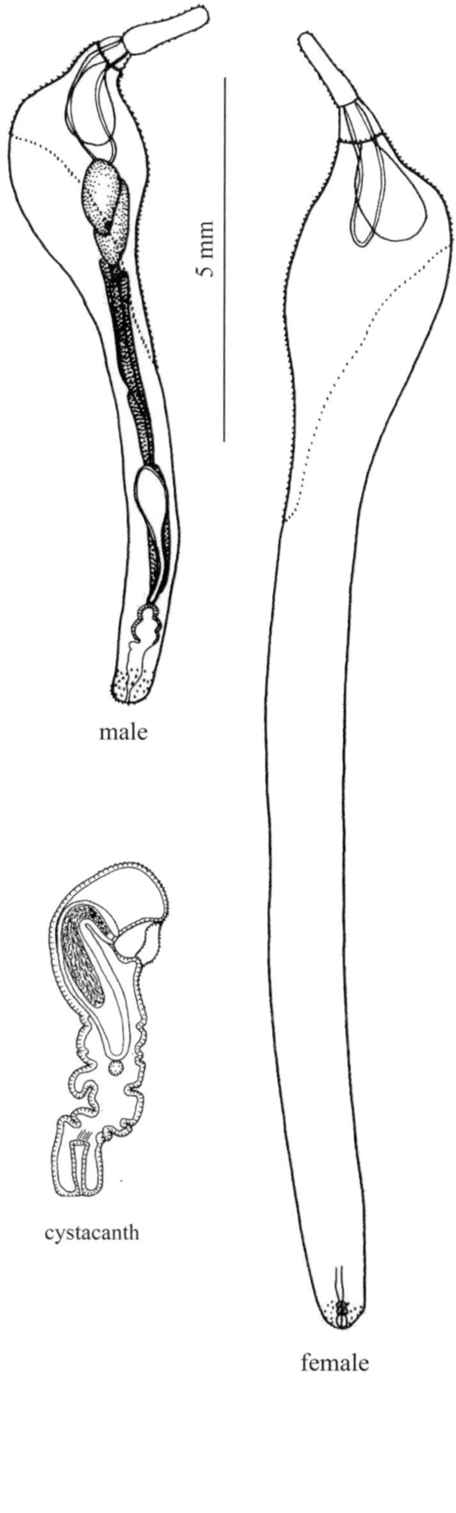

Three elephant seals examined on King George Island (the South Shetland Islands) harboured 2520–3753 parasites per host. Five elephant seals examined in the maritime Antarctic were less infected (Nikolsky 1974).

Corynosoma arctocephali Zdzitowiecki 1984b (Fig. 8.9)
Synonyms: *C. singularis* Skryabin et Nikolsky, 1971 pro parte.

Diagnosis (after Zdzitowiecki 1991): Proboscis hooks arranged in 19–22 rows of 10/11–13/14, including 3/4–4/5 rootless basal ones. Subdistal and prebasal hooks

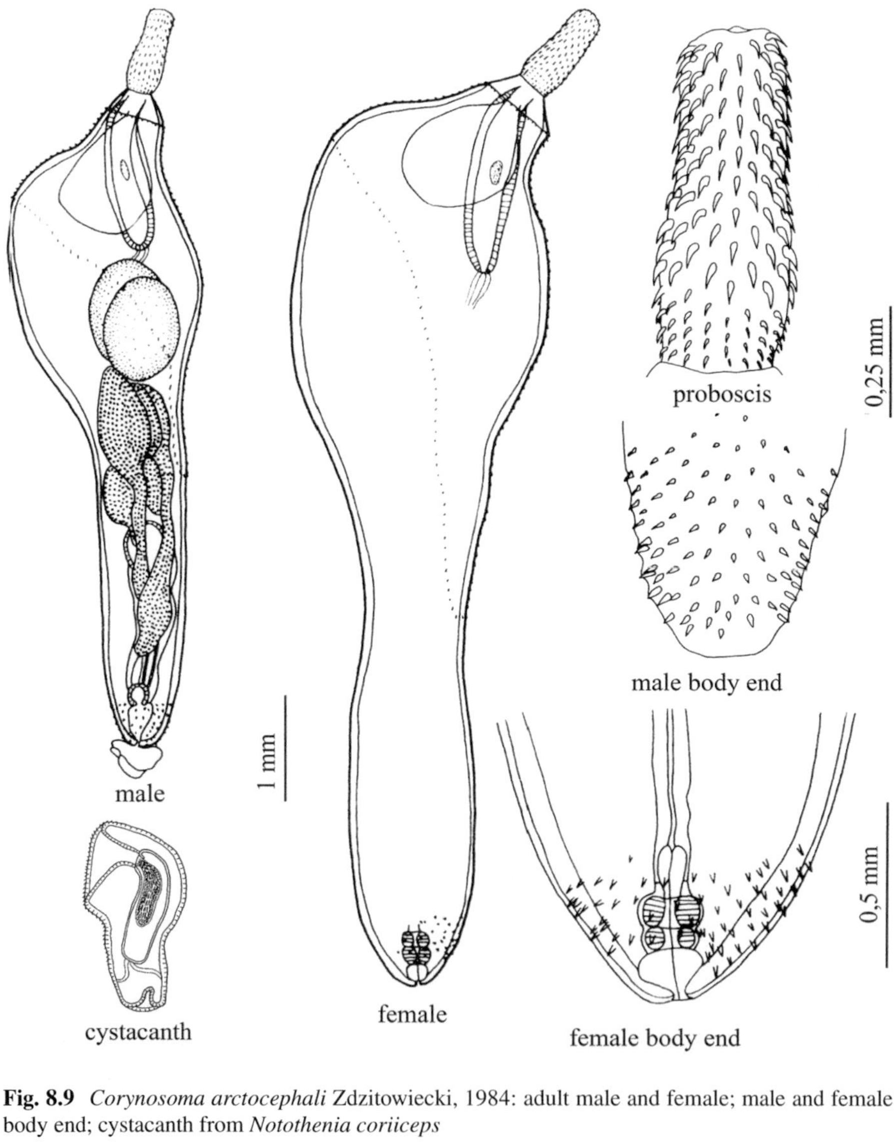

Fig. 8.9 *Corynosoma arctocephali* Zdzitowiecki, 1984: adult male and female; male and female body end; cystacanth from *Notothenia coriiceps*

the longest ones. Fore-trunk and hind-trunk of similar length. Hind-trunk cylindrical. Somatic spines cover about 60% of trunk length on ventral side. Genital spines (if present) separated from somatic ones. Lemnisci flat, shorter than proboscis receptacle.

Male. Total dimensions 6.9–7.7 × 1.4–2.0 mm. Proboscis 0.728–0.878 × 0.284–0.343 mm. Maximum hook length 66–76 μm. Genital spines, circa 150 in number, arranged in 8–9 irregular circles. Cement glands pyriform.

Female. Total dimensions 7.7–9.6 × 1.8–2.7 mm. Proboscis 0.821–1.001 × 0.313–0.343 mm. Maximum hook length 71–86 μm. Genital spines present (1–100) or absent. Genital pore terminal. Eggs 126–159 × 38–47 μm.

Suitable definitive hosts: seals (*Arctocephalus gazella*, *Hydrurga leptonyx*). Unsuitable definitive hosts: seals (*Lobodon carcinophagus*). Juvenile specimens also in intestine of birds (*Phalacrocorax atriceps*). Paratenic hosts: fishes. Nototheniidae: *Notothenia coriiceps*, *N. rossii*, *N. squamifrons*, *Lindbergichthys nudifrons*, *Dissostichus eleginoides*, *Patagonotothen brevicauda guntheri*; Bathydraconidae: *Parachaenichthys charcoti*, *P. georgianus*; Channichthyidae: *Chaenocephalus aceratus*, *Cryodraco antarcticus*; Muraenolepidae: *Muraenolepis microps*.

Habitat: Mainly posterior half of ileum. Few specimens in jejunum and large intestine.

Biology and ecology: According to Zdzitowiecki (1986b, g, 1990b), infections of paratenic hosts take place mainly in the fiord environment. The species was probably very rare at the beginning of the twentieth century, because its main definitive hosts, *A. gazella* (Antarctic fur seal), was almost completely exterminated. Thus *C. arctocephali* was absent in samples of cystacanths from fishes caught at South Georgia in 1925–1928 (Baylis 1929; Zdzitowiecki 1987). The population of fur seal increased under protection and so did the parasite population. Now, *C. arctocephali* is abundant in fishes of the fiord environment in the same area (Zdzitowiecki 1987, 1990b), in Admiralty Bay (Laskowski et al. 2012), at the South Orkney Islands (Zdzitowiecki and White 1996), and at the Argentine Islands (Zdzitowiecki and Laskowski 2004).

Numerical data concerning the occurrence of *C. arctocephali* in definitive hosts are limited. Maximum intensity found till now in fur seal was 65 acanthocephalans. The most heavily infected paratenic hosts: *N. rossii* at South Georgia (prevalence 91 %, maximum intensity 84 cystacanths) and *Notothenia coriiceps* at the South Orkney Islands (prevalence 100 %, maximum intensity 36 cystacanths).

Distribution: The South Shetland Islands, South Georgia, Antarctic Peninsula, Argentine Islands, Ross Sea (probably its northern part) (Skryabin and Nikolsky 1971; Nikolsky 1974; Zdzitowiecki 1978, 1984b, 1986b, c, 1987, 1990b; Rodjuk 1985; Hoberg 1986; Zdzitowiecki and Laskowski 2004).

Corynosoma hamanni (Linstow 1892) (Fig. 8.10)

Synonyms: *C. antarcticum* (Rennie 1906), *C. sipho* Railliet et Henry, 1907, *C. pacifica* Nikolsky 1974.

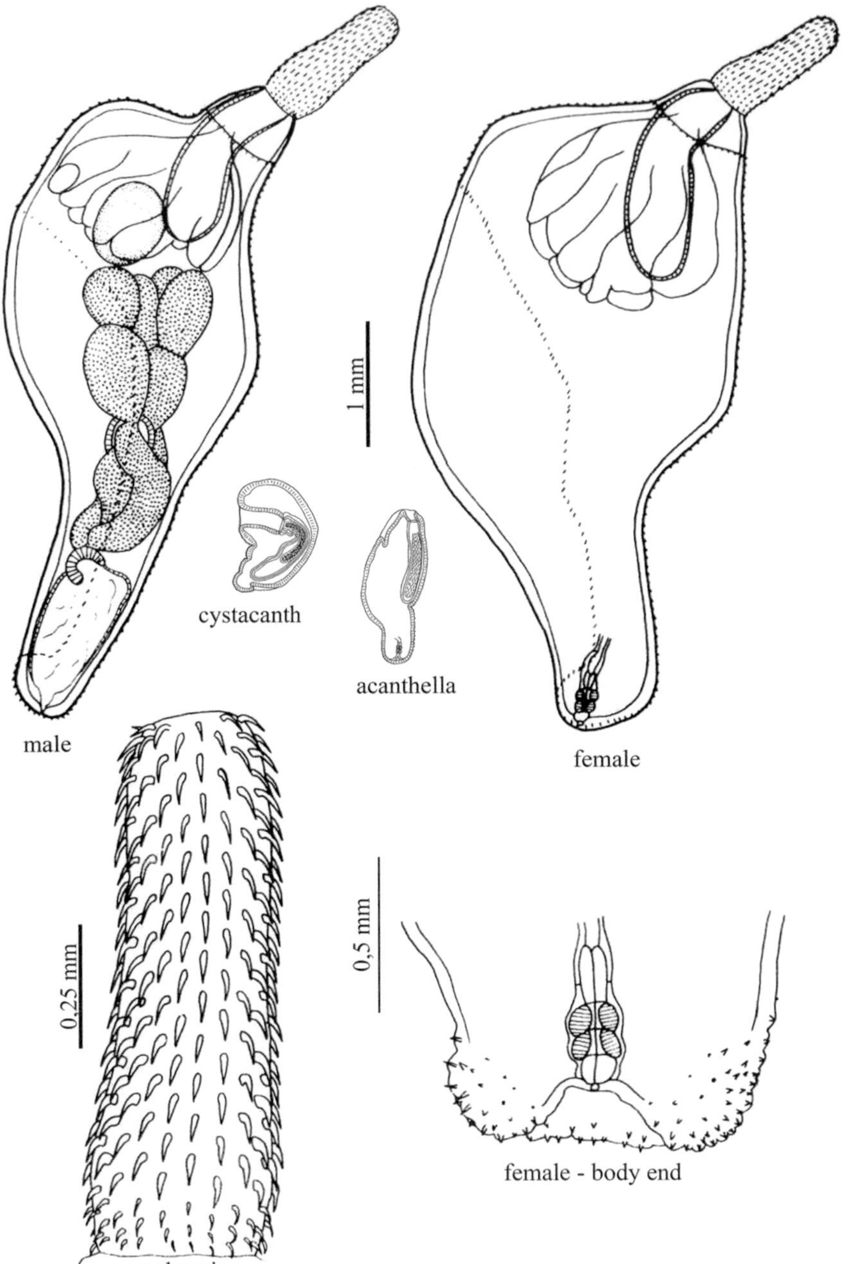

Fig. 8.10 *Corynosoma hamanni* (Linstow 1892): adult male and female; female body end; cystacanth from *Prostebbingia brevicornis*

Diagnosis (after Zdzitowiecki 1984c): Proboscis hooks arranged in 19–22 (usually 20) rows of 12/14–16, including 2/3–3/4 rootless basal ones. Subdistal (third to fifth) hooks the longest ones. Body shape depends from sex. Somatic and genital armature not separated. Lemnisci strongly folded, similar in length as proboscis receptacle.

Male. Total dimensions 5.2–7.1 × 1.7–2.5 mm. Proboscis 1.004–1.161 × 0.333–0.412 mm. Maximum hook length 77–98 μm. Fore-trunk constitutes 54–71 % of trunk length. Hind-trunk tapering posteriorly. Cement glands pyriform.

Female. Total dimensions 5.2–6.4 × 1.9–2.7 mm. Proboscis 1.072–1.278 × 0.339–0.410 mm. Maximum hook length 81–99 μm. Fore-trunk constitutes 59–80% of trunk length. Hind-trunk terminates with two lateral lobes. Slightly subterminal genital pore lies in concavity between lobes. Genital armature covers both lobes and ventral body side. Only narrow unarmed zone remains on mid-dorsal side at trunk end. Eggs 155–202 × 46–58 μm.

Suitable definitive hosts: seals (*Hydrurga leptonyx*, *Leptonychotes weddelli*). Unsuitable definitive hosts: seals (*Lobodon carcinophagus*). Juvenile specimens also in intestine of birds (*Chionis alba*, *Phalacrocorax atriceps*)

Paratenic hosts: fishes. Nototheniidae: *Notothenia coriiceps*, *N. rossii*, *Dissostichus mawsoni*, *Pagothenia bernacchii*, *P. hansoni*, *Trematomus newnesi*, *T. bernacchii*, *Lindbergichthys nudifrons*, *Gobionotothen gibberifrons*; Bathydraconidae: *Parachaenichthys charcoti*, *P. georgianus*; Channichthyidae: *Chaenocephalus aceratus*, *Chionodraco rastrospinosus*, *Cryodraco antarcticus*, *Pseudochaenichthys georgianus*; Harpagiferidae: *Harpagifer antarcticus*.

Intermediate host: amphipods. Eusiridae: *Prostebbingia brevicornis*.

Habitat: Pyloric part of stomach, duodenum, and anterior part of jejunum. Few specimens more posteriorly, in small and large intestine.

Biology and ecology: According to Zdzitowiecki (1986b, g, 1990b, Zdzitowiecki and White 1996, Zdzitowiecki and Presler 2001, Zdzitowiecki and Laskowski 2013, Laskowski and Zdzitowiecki 2010, Laskowski et al. 2012), infections of paratenic hosts take place in the fiord environment in the shallow water up to a depth of circa 50 m. Leopard seals, Weddell seals, and some paratenic hosts are massively infected, with up to several thousand parasites in one seal and over one hundred cystacanths in one fish. Probably all seals of both species mentioned above living in Admiralty Bay (the South Shetland Islands) are infected. *N. coriiceps*, *N. rossii*, and *Ch. aceratus* are the main paratenic hosts in this area (prevalence 96%, 100%, and 81%, maximum intensity 149, 166, and 123, respectively). *C. hamanni* found appears to be specific parasites of *Prostebbingia brevicornis*. Intermediate hosts occur mainly in sub-coastal waters (specimens examined were caught at the depth 5–15 m).

Distribution: Previous literature data concerning distribution and lists of hosts are partially doubtful and should be referred fully or partially to *Corynosoma pseudohamanni*. However, there are no doubts that the species occurs circumpolar: South Georgia, South Orkney Islands, South Shetland Islands, Antarctic Peninsula, Adelie Land, King George V Land, Argentine Islands, and maritime Antarctic (Linstow 1892; Rennie 1906; Railliet and Henry 1907; Baylis 1929; Johnston and

Best 1937; Markowski 1971; Nikolsky 1974; Zdzitowiecki 1978, 1984c, 1986a, b, 1987, 1990b; Rodjuk 1985; Hoberg 1986; Zdzitowiecki and White 1996; Zdzitowiecki and Laskowski 2004). Doubtful data: Enderby Land, Ongul Island, McMurdo Sound, Kerguelen, Crozet and Heard islands, Lena, Skiff, and Ob banks (Leiper and Atkinson 1914, 1915; Edmonds 1957; Golvan 1959; Nickol and Holloway 1968; Holloway and Nickol 1970; Kamegai and Ichihara 1973; Holloway and Spence 1980; Parukhin and Lyadov 1982).

Corynosoma pseudohamanni Zdzitowiecki 1984c (Fig. 8.11)

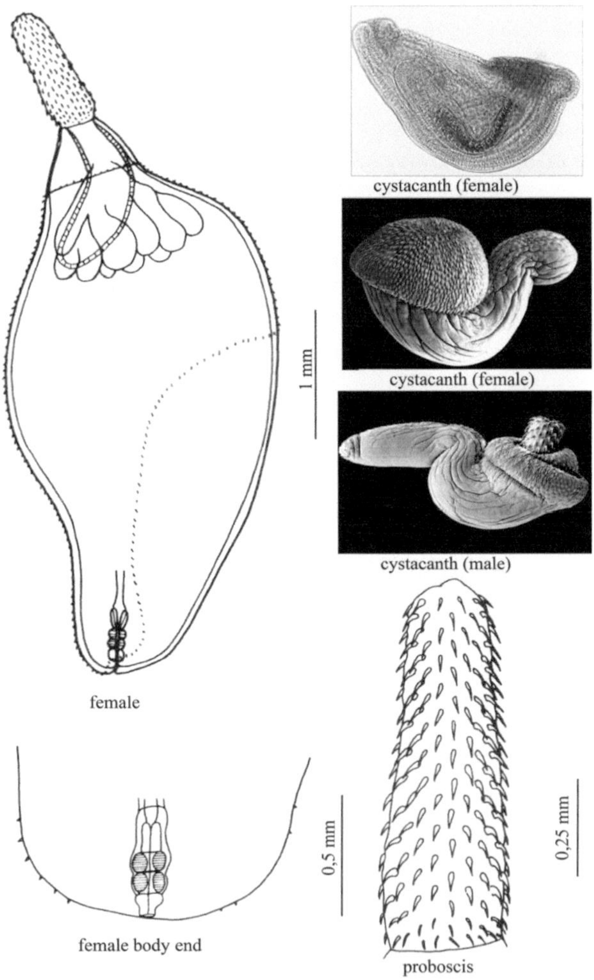

Fig. 8.11 *Corynosoma pseudohamanni* Zdzitowiecki, 1984: adult female; female body end; proboscis; cystacanth from *Cheirimedon femoratus*

Synonyms: *C. hamanni* of various authors nec Linstow (1892) pro parte, *C. antarcticum* of Johnston and Best (1937) nec Rennie (1906) pro parte.

Diagnosis (after Zdzitowiecki 1984c): Proboscis hooks in 18–22 rows of 10/11–14, including 1–2/3 rootless basal ones. Subdistal (second to fourth) hooks the longest ones. Body shape depends from sex. Somatic and genital armature not separated. Lemnisci strongly folded, similar in length as proboscis receptacle.

Male. Body shape similar to that of *C. hamanni*. Total dimensions 4.8–6.2 × 1.4–1.8 mm. Proboscis 0.799–0.929 × 0.258–0.325 mm. Maximum hook length 67–79 µm. Fore-trunk constitutes 56–69% of trunk length. Hind-trunk slightly tapering posteriorly. Cement glands pyriform.

Female. Total dimensions 3.9–5.3 × 1.3–2.1 mm. Proboscis 0.804–1.001 × 0.300–0.325 mm. Maximum hook length 64–81 µm. Fore-trunk constitutes 67–85% of trunk length. Hind-trunk slightly tapering posteriorly, with rounded end. Genital pore terminal. Genital armature ends just before genital pore on ventral side. Spines spread at sides before genital pore, but they never occur on dorsal side of hind-trunk. Eggs 92–120 × 29–40 µm.

Suitable definitive hosts: seals (*Leptonychotes weddelli*, *Hydrurga leptonyx*, *Lobodon carcinophagus*). Unsuitable definitive hosts: seals (*Arctocephalus gazella*, *Mirounga leonina*). Juvenile specimens also in intestine of birds (*Catharacta lonnbergi*, *Chionis alba*, *Larus dominicanus*, *Phalacrocorax atriceps*).

Paratenic hosts: fishes. Nototheniidae: *Notothenia coriiceps*, *N. rossii*, *N. nybelini*, *Lindbergichthys nudifrons*, *Gobionotothen gibberifrons*, *Dissostichus eleginoides*, *D. mawsoni*, *Pagothenia bernacchii*, *P. hansoni*, *Trematomus newnesi*; Bathydraconidae: *Parachaenichthys charcoti*; Channichthyidae: *Chaenocephalus aceratus*, *Champsocephalus gunnari*, *Chionodraco rastrospinosus*, *Cryodraco antarcticus*, *Gymnodraco acuticeps*; Harpagiferidae: *Harpagifer antarcticus*. Probably also further species of fishes listed by Holloway and Spence (1980) as paratenic hosts of *Corynosoma hamanni* in McMurdo Sound: Nototheniidae: *Pagothenia borchgrevinki*, *Trematomus centronotus*; Zoarcidae: *Lycodichthys dearborni*.

Intermediate hosts: amphipods. Eusiridae: *Pontogeneiella* sp.; Lysianassoidea: *Cheirimedon femoratus*.

Biology and ecology: According to Zdzitowiecki (1986b, g, 1990b; Zdzitowiecki and White 1996; Zdzitowiecki and Presler 2001; Zdzitowiecki and Laskowski 2004; Laskowski et al. 2007), infections of paratenic hosts take place in the fiord environment, but a little deeper than in the case of *Corynosoma hamanni*, at a depth of up to 100 m. Probably all Weddell seals living in the Admiralty Bay (South Shetland Islands) are infected; intensities of the infection sometimes exceed one thousand parasites per seal. *N. coriiceps*, *N. rossii*, *P. charcoti*, and *Ch. aceratus* are the main paratenic hosts in the same area (prevalence 99.6–100%, maximum intensity 856, 106, 219, and 263, respectively). At the Vernadsky Station (Argentine Islands) and at the South Orkney Islands, *N. coriiceps* was also heavily infected (prevalence 99% and 100%, maximum intensity 421 and 23, respectively). Cystacanths in intermediate hosts (*C. femoratus*) were found in Admiralty Bay and at Vernadsky Station.

Distribution: Circum-Antarctic: Antarctic Peninsula, Argentine Islands, South Shetland Islands, southern coasts of Weddell Sea, McMurdo Sound, Adelie Land, King George V Land, Enderby Land, South Orkney Islands, Ross Sea. Part of the material was originally referred to *C. hamanni*. It is here referred to *C. pseudohamanni* based on morphological data contained in papers of various authors (Leiper and Atkinson 1915; Johnston and Best 1937; Edmonds 1957; Golvan 1959; Nickol and Holloway 1968; Holloway and Nickol 1970; Holloway and Spence 1980; Zdzitowiecki 1978, 1984c, 1986a, b, 1990b; Hoberg 1986; Zdzitowiecki and White 1996; Zdzitowiecki and Laskowski 2004; Laskowski and Zdzitowiecki 2005, 2010). *C. pseudohamanni* is the only representative of the genus *Corynosoma* occurring without any doubts within the Antarctic Circle. The species was absent in the large sample of fishes examined at South Georgia (Zdzitowiecki 1990b).

Corynosoma beaglense Laskowski, Jeżewski, Zdzitowiecki, 2008 (Fig. 8.12)

Diagnosis (after Laskowski et al. 2008): Only juvenile specimens (cystacanths) of *Corynosoma beaglense* were found in *Champsocephalus esox* in Beagle Channel. It has an almost cylindrical proboscis (length 0.52–0.56 mm); a proboscis hook formula

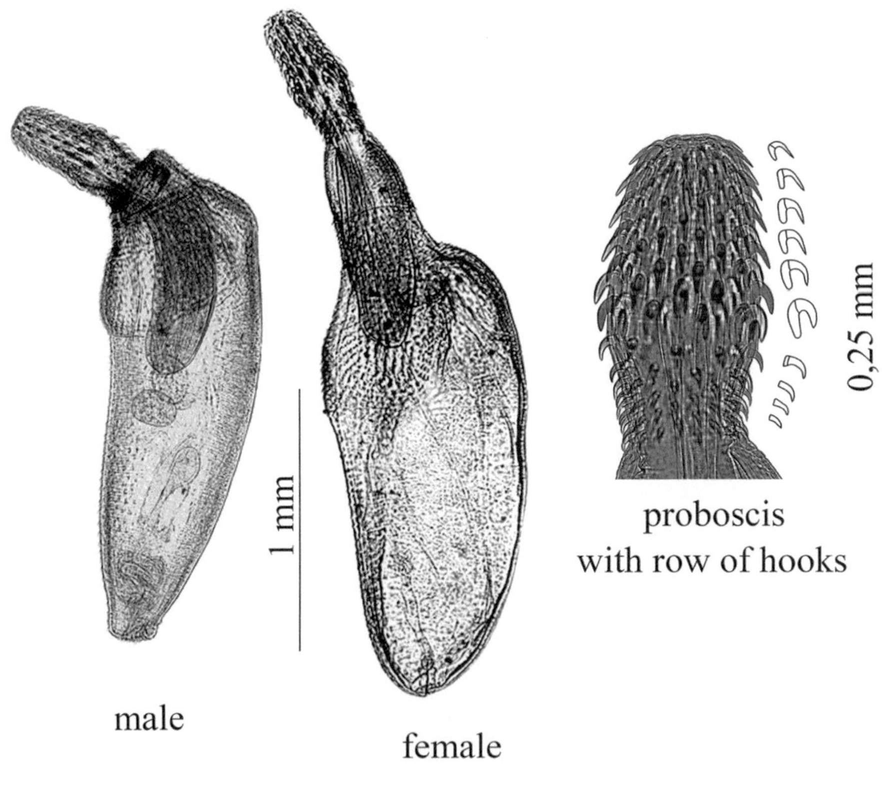

Fig. 8.12 *Corynosoma beaglense* Laskowski, Jeżewski, Zdzitowiecki, 2008: male and female cystacanth: proboscis

of 16 rows of 9/10–10/11, including 4–4/5 basal hooks; distal hooks shorter than the prebasal hooks; a fore-trunk not separated from the hind-trunk by a constriction; somatic spines contiguous with the genital spines on the ventral side of the trunk of the male and covering the entire length of the ventral side of the female trunk, and the presence of genital spines surrounding the terminal genital pore of the male.

Male. Total length approx. 2.6 mm. Proboscis 0.530×0.212 mm. Distal hook length 50 μm; prebasal hook length 56 μm. Neck retracted into trunk, c. 0.210 mm in width. Trunk 1.89×0.61 mm. Genital pore surrounded by genital spines (max. length 29 μm) contiguous with somatic spines (max. length 37 μm). Proboscis receptacle 746×239 μm. Lemnisci 0.502–0.209 and 0.458–0.206 mm. Testes oval, arranged diagonally at end of proboscis receptacle, 0.185×0.136 mm and 0.184×0.128 mm. Cement-glands elongate, pear-shaped, just posterior to testes, 6 in number. Säfftigen's pouch club-shaped.

Female. Total length 2.79 mm and 2.53 mm. Proboscis 560×210 μm and 521×209 μm. Distal hook length 48 μm and 51 μm; prebasal hook length 52 μm and 63 μm. Neck wider than long, 269×367 μm and 271×307 μm. Trunk 1.98×0.93 mm and 1.76 9 0.72 mm; whole ventral side covered with somatic spines with max. length 37 μm. Proboscis receptacle 794×298 μm and 741×202 μm. Lemnisci 382–657×210–263 μm. Length of reproductive organs (from anterior end of uterine bell to genital pore) 588 μm in one case. Vaginal sphincter double, 79×77 μm and 76×64 μm.

The definitive host of this species is unknown. *C. beaglense* is similar to two Sub-Antarctic parasites of birds, *Andracantha baylisi* and C. *clavatum* Goss, 1940, in the shape of the trunk, neck, and proboscis, as well as the proboscis armature. It differs from them in the lack of a zone of small somatic spines between two zones of large spines (a generic feature), the somatic spines on the male contiguous with the genital spines, the somatic spines on females extending to the posterior extremity, a smaller proboscis, shorter hooks, and the distal hooks shorter than the prebasal hooks.

Corynosoma evae Zdzitowiecki, 1984 (Fig. 8.13)

Diagnosis (after Zdzitowiecki 1984b): Proboscis hooks in 20–24 rows of 11/12–13, including 3–4 rootless basal ones. Prebasal hook the longest, stout. Fore-trunk constitutes 55–64% of total trunk length. Hind-trunk cylindrical. Somatic armature covers 61–69% of trunk length on ventral side. Genital spines (if present) separated from somatic ones. Lemnisci flat, shorter than proboscis receptacle.

Male. Total dimensions 3.5–4.6×1.1–1.5 mm. Proboscis 0.633–0.719×0.257–0.296 mm. Maximum hook length 57–63 μm. Genital spines arranged in 4 irregular rows, 40–60 in number. Cement glands pyriform.

Female. Total dimensions 4.3–5.2×1.1–1.9 mm. Proboscis 0.612–0.788×0.254–0.337 mm. Maximum hook length 61–73 μm. Genital spines absent. Genital pore terminal. Eggs 103–127×34–43 μm.

Suitable definitive hosts: seals (*Hydrurga leptonyx*, *Otaria flavescens*).

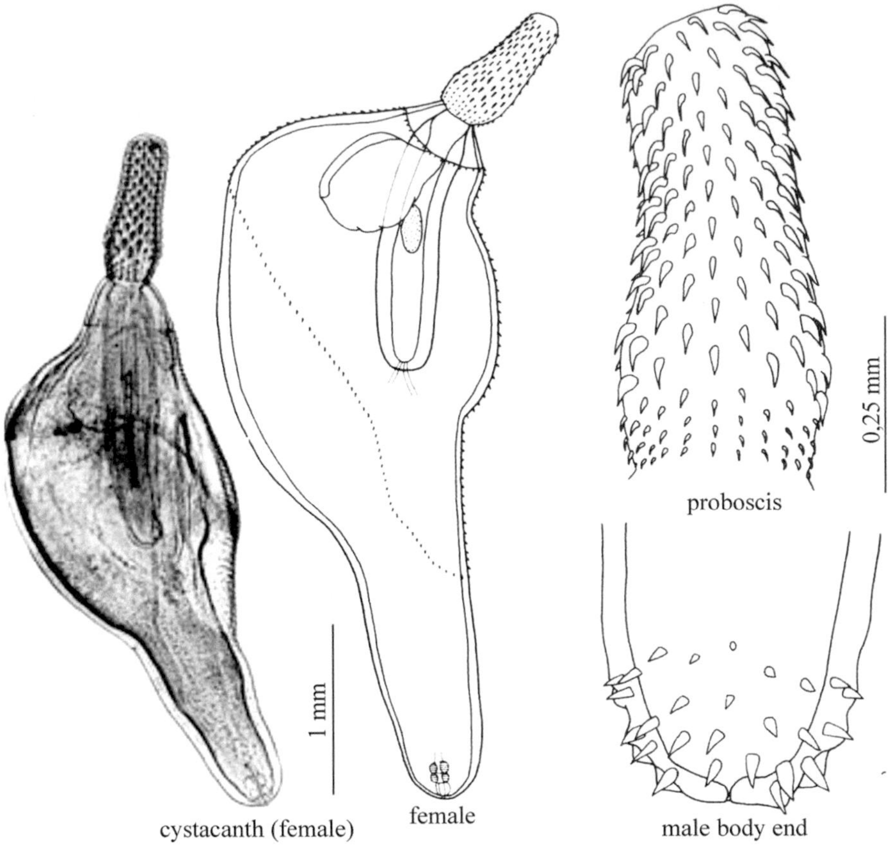

Fig. 8.13 *Corynosoma evae* Zdzitowiecki, 1984: adult female; cystacanth female; male body end; proboscis

Paratenic hosts: fishes. Bathydraconidae: *Parachaenichthys georgianus*; Nototheniidae: *Patagonotothen longipes*; Channichthyidae: *Champsocephalus esox*.

Habitat: Ileum.

Biology and ecology: Not known.

Distribution: The South Shetland Islands, South Georgia, Falkland Islands, Beagle Channel (Zdzitowiecki 1984b, 1986e; Laskowski and Zdzitowiecki 2009; Laskowski et al. 2007). It is probably rather a Sub-Antarctic than an Antarctic species. The present authors did not find cystacanths in the large sample of fishes examined at the South Shetland Islands, one cystacanth was found at South Georgia and 10 cystacanths were found in Beagle Channel. Cystacanths found by Reimer (1987) in fishes at the South Shetland Islands and South Georgia were probably wrongly determined and should be referred to *C. arctocephali*.

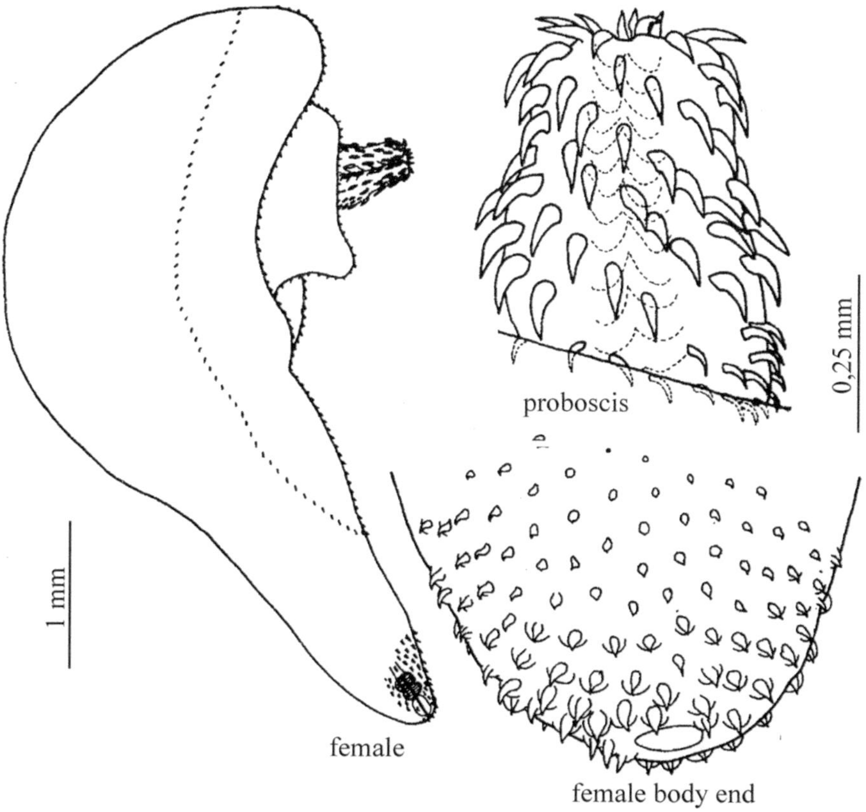

Fig. 8.14 *Corynosoma gibsoni* Zdzitowiecki, 1986: adult female; proboscis; female body end

Corynosoma gibsoni Zdzitowiecki, 1986 (Fig. 8.14)
Diagnosis (after Zdzitowiecki 1986e):

Description: All investigated specimens (five females) were partly contracted, with the proboscis, neck, and anterior part of the trunk retracted, and the proboscis partly invaginated. Total length of not contracted specimens, if attains about 6.2–6.7 mm (length of trunk about 4.6–5.2 mm). The maximum width of dilated fore-trunk 2.02–2.24 mm, width of the hind-trunk 0.66–1.01 mm. The fore-trunk is about twice as long as the hind-trunk. Approximate length of the proboscis (measured only in one specimen by adding the length of invaginated part to the length of non-invaginated part) about 1.2–1.3 mm. Width of the proboscis 0.39–0.42 mm. Hooks arranged in 19–20 rows, number of hooks per row exceeding 10 (the most probably 15), basal hooks with reduced roots 3–4 in number. The largest are the hooks situated just before the basal ones. Maximum length of the blade 100–119 μm. Neck impossible to observe. Somatic armature covers about 3/4 of the trunk at the ventral side, partly laterally. The anterior most genital spines are 40–91 μm distant from the body end. Width of the unarmed zone between somatic and genital spines 0.30–

0.81 mm. Maximum dimensions of the somatic spines 65×15 μm, of the genital spines 72×24 μm. Dimensions of the proboscis receptacle about 1.7×0.5–0.6 mm. Lemnisci not visible, screened by embryophores. The genital duct, observed only in one specimen, measures 1.4 mm. The vagina is provided with double sphincter. Genital opening terminal. Dimensions of mature embryophores, measured inside the body, through the body wall, in three specimens 155–188×43–56 μm.

Females of *C. gibsoni* sp. n. are similar to *C. hamanni* (Linstow 1892) in respect of proboscis length and embryophore dimensions (cf. Zdzitowiecki 1984b) but differ from the latter by the presence of an unarmed zone separating somatic and genital armature, as well as by the shape of the posterior part of the trunk, especially its posterior tip. All other representatives of the genus Corynosoma have smaller embryophores (cf. Golvan 1959; Zdzitowiecki 1984a, b). Of these, *C. arctocephali.* Zdzitowiecki, 1984, the most similar in embryophore dimensions, has a shorter proboscis, smaller hooks, longer hind-trunk, shorter range of somatic armature and greater distance between somatic and genital armature.

Genus ***Andracantha*** Schmidt, 1975

Diagnosis: Proboscis cylindrical. Neck conspicuous. Fore-trunk forming bulb, connected with neck by short segment similar in width as neck. Hind-trunk tapering posteriorly. Conspicuous somatic spines arranged in two circular fields separated from each other by either a bare zone or a zone covered with smaller spines. Genital spines separated from somatic ones, present at least on some specimens of both sexes. Proboscis receptacle double-walled. Testes parallel. Cement glands tubular or pyriform, six or eight in number. Vaginal sphincter in females double. Eggs with

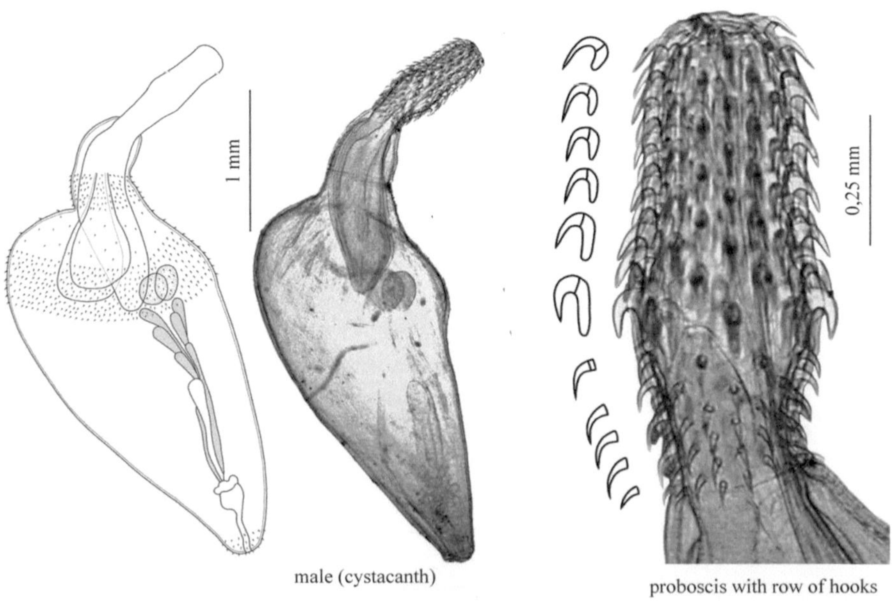

Fig. 8.15 *Andracantha baylisi* (Zdzitowiecki 1986a, b, c, d, e, f, g): male cystacanth; proboscis

or without polar prolongations of middle envelope. Parasites of birds. Paratenic hosts: fishes.

Andracantha baylisi (Zdzitowiecki 1986a, b, c, d, e, f, g) (Fig. 8.15)
Synonyms: *Corynosoma* sp. Zdzitowiecki 1985.

Diagnosis (after Zdzitowiecki 1985, 1989a; Laskowski et al. 2008): Proboscis almost cylindrical slightly dilated just beyond mid-length, with length/width ratio 2.69–2.92:1. Hooks arranged in 16 rows of 9/10–10/11, including 5–5/6 rooted ones and 4/5 basal ones with reduced roots. Anterior hooks gradually increase from apex in blade width and root length, but distal-most hook is longest by far. Blades of anterior 4–5 hooks longer than roots; blade of prebasal hook shorter than root. Area of basal hooks constitutes 33–39% of proboscis length. Neck trapezoid, may be longer or shorter than wide, curved towards ventral side. Fore-trunk not separated from hind-trunk by constriction. Short anterior part of trunk similar in width to neck, then trunk dilates greatly before tapering posteriorly. Anterior 36–40% of trunk length covered with somatic spines (max. length 48 μm), which are arranged in 2 densely spined zones separated by zone of smaller, loosely arranged spines. Anterior zone of large spines constitutes 12–22% of length of whole armature area, zone of minute spines 36–42% and posterior zone of large spines 39–46% (measured along ventral side of trunk). Approximately 20–30 genital spines (max. length 21 μm) present at posterior extremity of trunk. Genital pore terminal, surrounded by genital spines in both sexes. Proboscis receptacle double-walled, extending to level of posterior zone of larger somatic spines. Lemnisci flat, rounded to ellipsoid, shorter than proboscis receptacle.

Male. Only juvenile specimens from paratenic hosts were available. Total dimensions 3.66–5.0 × 1.12–1.74 mm. Proboscis 0.823–0.920 × 0.290–0.350 mm. Distal hook length 107–119 μm, prebasal hook length 79–95 μm. Neck length 0.404–0.407 × 0.373–0.461 mm. Trunk 2.485–2.975 × 1.133–1.351 mm. Proboscis receptacle 1.257–1.411 × 0.275–0.378 mm. Lemnisci 0.646–0.930 × 0.341–0.464 mm. Testes parallel, at end of proboscis receptacle, 0.219–0.272 × 0.120–0.192 mm. Cement glands elongate, pear-shaped, 6 in number. Säfftigen's pouch club-shaped.

Female. Total dimensions of adult specimens about 5–5.7 × 1.60–1.86 mm. Proboscis 0.820–0.970 × 0.240–0.380 mm. Distal hook length 119–136 μm. Prebasal hook length 92–119 μm. Genital spines present (1–20 in number) or absent. Genital pore terminal. Eggs with polar prolongations of middle envelope, 81–101 × 27–30 μm.

Total dimensions of juvenile specimens 4.0–5.7 × 1.31–1.86 mm. Proboscis 0.820–0.970 × 0.240–0.380 mm. Distal hook length 104–136 μm. Prebasal hook length 89–119 μm. Neck 0.421–0.461 μm. Trunk 2.714 × 1.091 mm. Genital spines present (1–20 in number) or absent. Genital pore terminal. Eggs with polar prolongations of middle envelope, 81–101 × 27–30 μm.

Suitable definitive hosts: birds (*Chionis alba, Phalacrocorax albiventer*).
Paratenic hosts: fishes. Nototheniidae: *Notothenia rossii, Patagonotothen longipes*; Bathydraconidae: *Parachaenichthys* georgianus; Channichthyidae: *Chaenocephalus aceratus, Champsocephalus esox*

Habitat: Intestine.

Biology and ecology: Cystacanths are present, though rare, in fishes at South Georgia and in Beagle Channel. Thus, the life cycle is completed in this area.

Distribution: Western Antarctic and Sub-Antarctic: the South Shetland Islands, South Georgia, Patagonia, Beagle Channel. The only specimen found in the definitive host (*Chionis alba*) on King George Island (South Shetland Islands) probably arrived from another area, as cystacanths of the species were not found in fishes at the South Shetland Islands. Six out of 290 notothenioid fishes examined at South Georgia housed few cystacanths (1–2 specimens per host) (Zdzitowiecki 1985, 1986f, 1989a, 1990b; Laskowski et al. 2007; Laskowski and Zdzitowiecki 2009).

Genus ***Bolbosoma*** Porta, 1908

Diagnosis: Proboscis cylindrical or conical. Neck short. Fore-trunk consists of short conical anterior part, large bulb, and narrow part beyond bulb. Hind-trunk cylindrical. Somatic spines present on prebulbar part of fore-trunk and usually on bulb. Genital spines absent. Proboscis receptacle double-walled. Testes in tandem. Cement glands tubular. Vaginal sphincter in females double. Eggs with polar prolongations of middle envelope. Parasites of mammals, mainly whales. Intermediate hosts – crustaceans (till now found only in euphausiids). Fishes may play a role as paratenic hosts.

Bolbosoma brevicolle (Maim 1867) (Fig. 8.16)
Synonyms: *B. paramuschiri* Skryabin, 1959.

Diagnosis (according Zdzitowiecki 1991): Proboscis hooks arranged in 20–22 rows of 7 (rarely 6 or 8), including one small basal hook, which may be rooted or not. Subdistal (second) hook the longest one. Somatic spines arranged in 20 irregular circles, covering the whole prebulbar part of fore-trunk and reaching beyond half of length of bulb. Anterior spines small, posterior spines two to three times longer. Hind-trunk constituting 74–84% of trunk length. Lemnisci very long, filiform, as long as trunk. Proboscis receptacle ends inside bulb.

Male. Total length 23–32 mm. Bulb 2.3–3.1 × 1.9–2.3 mm. Hind-trunk width 1.70–2.75 mm. Proboscis 0.51–0.57 × 0.42–0.51 mm. Anterior spines 40–60 × 15–32 μm. Posterior spines 100–160 × 60–90 μm. Testes in tandem, oblique, not separated.

Female. Total length 21–38 mm. Bulb 2.5–2.8 × 2.0–2.65 mm. Hind-trunk width 2.0–3.3 mm. Proboscis 0.54–0.60 × 0.45–0.52 mm. Maximum hook length 113 μm. Anterior spines 60–80 × 20–30 μm. Posterior spines 95–120 × 48–75 μm. Eggs 118–131 × 25–29 μm (mean 124 × 26 μm).

Suitable definitive hosts: whales (*Balaenoptera musculus*).
Unsuitable definitive hosts: whales (*Balaenoptera borealis*).
Other suitable and unsuitable definitive hosts: whales (*Balaenoptera acutorostrata*, *B. physalus*, *Eubalaena glacialis sieboldi*, *Physeter catodon*).
Habitat: Intestine.

Biology and ecology: The species is abundant in blue whales (*B. musculus*) at South Georgia.

Distribution: Cosmopolitan, including the Antarctic: environs of the South Shetland Islands and South Georgia (Baylis 1929; Petrotschenko 1958; Yamaguti 1963; Zdzitowiecki 1986a).

Key to the classes of acanthocephala
(After Amin 1987, modified)

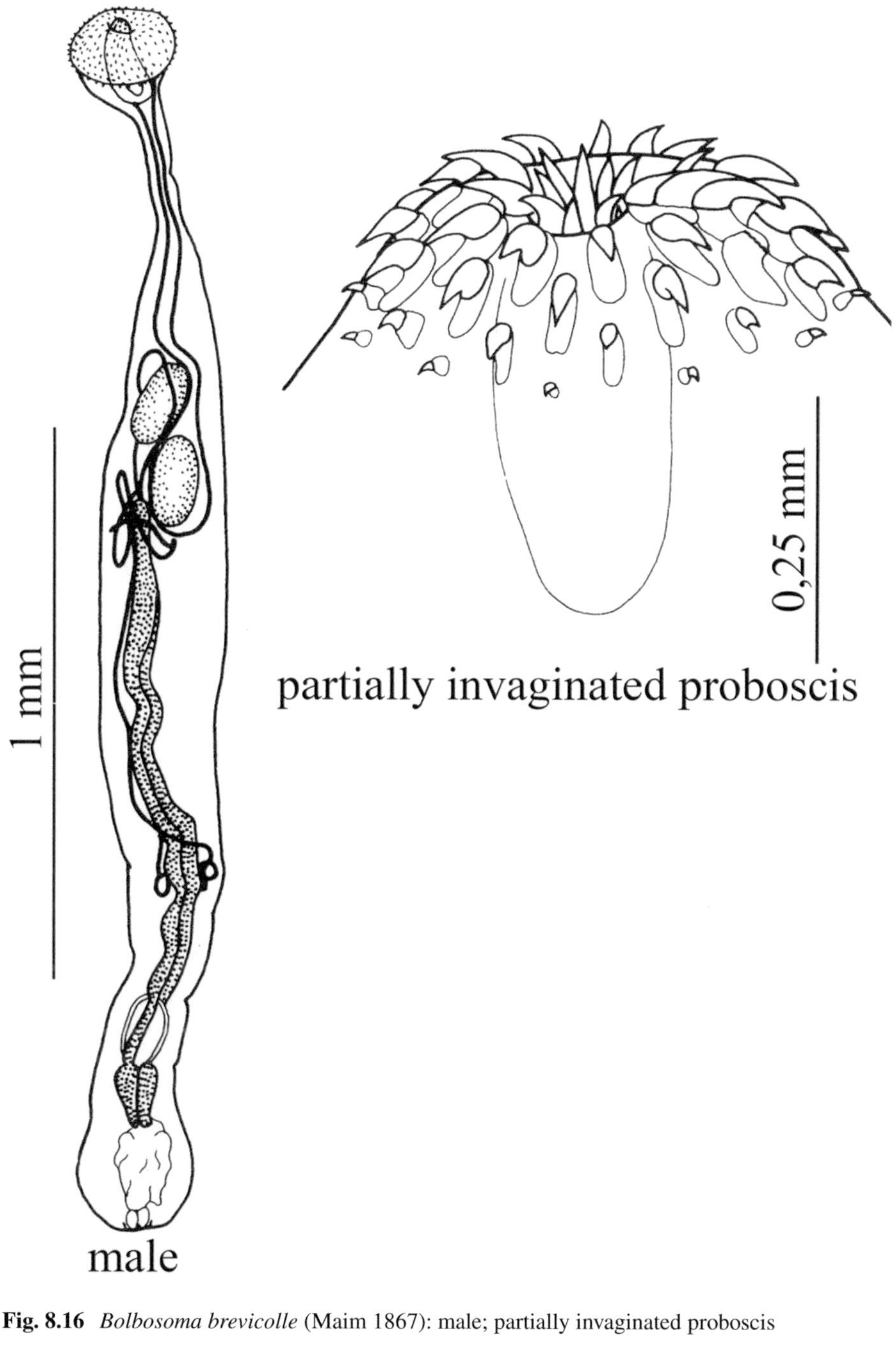

Fig. 8.16 *Bolbosoma brevicolle* (Maim 1867): male; partially invaginated proboscis

1a Main longitudinal lacular canals lateral. Nuclei of lemnisci and cement glands and hypodermal nuclei fragmented. Ligament sacs in females single, not persistent. Proboscis receptacle double-walled. Definitive hosts: fishes, amphibians, reptiles, birds and mammals. Intermediate hosts: crustaceans	Class **Palaeacanthocephala***
1b Main longitudinal lacular canals dorsal and ventral, or only dorsal. Nuclei of lemnisci and cement glands and/or hypodermal nuclei not fragmented, usually giant. Ligament sacs in females double, persistent. Proboscis receptacle single-walled, complex, or absent	2
2a(lb) Protonephridia present or absent. Trunk not spined. Proboscis receptacle absent or single-walled. Cement glands separate, pyriform. Eggs usually oval, thick-shelled. Definitive hosts: birds and mammals. Intermediate hosts: insects, rarely myriapods	Class **Archiacanthocephala**
2b Protonephridia absent. Trunk spined or not. Proboscis receptacle single-walled. Cement glands elongate to tubular, or syncytial. Eggs variable. Definitive hosts: fishes, amphibians, and reptiles. Intermediate hosts: probably crustaceans	3
3a(2b) Trunk spined. Proboscis claviform, with numerous longitudinal rows of hooks. Cement glands separate, elongate pyriform to tubular. Eggs oval, with radial sculpturings at right angles to surface. Definitive hosts: fishes and Crocodilia. Intermediate hosts unknown, probably crustaceans	Class **Polyacanthocephala**
3b Trunk spined or not. Proboscis usually small, with few radially arranged hooks. Cement gland single, syncytial. Eggs variably shaped, but not like those of Polyacanthocephala. Definitive hosts: fishes and occasionally amphibians and reptiles. Intermediate hosts: crustaceans	Class **Eoacanthocephala**

Key to the orders, families, subfamilies, genera, and species of the antarctic acanthocephala (palaeacanthocephala)

(After Zdzitowiecki 1991, modified)

la Mature stage parasite of fishes. Trunk armed or not. (Order **Echinorhynchida**)	2
lb Mature stage parasite of mammals and birds. Fishes are paratenic hosts of many species. Trunk armed. (Order Polymorphida, family **Polymorphidae**, subfamily **Polymorphinae**)	
2a(1a) Anterior part of trunk armed with spines	3
2b Trunk unarmed	5
3a(2a) Proboscis cylindrical, slightly dilated subterminally. Ventral proboscis hooks larger than dorsal. Vulvar sphincter in females double	(Family **Heteracanthocephalidae**, subfamily **Aspersentinae**)
	Aspersentis megarhynchus
3b Proboscis globular. Ventral proboscis hooks not different from dorsal. 2–3 types of hooks arranged along proboscis. Vulvar sphincter in females single	(Family **Arhythmacanthidae**, subfamily **Arhythmacanthinae**, genus **Heterosentis**) 4
3c Proboscis relatively short, cylindrical to globular, armed with ten basal spines between 10 rows with rooted hooks (one or two) and basal spines (one or two). Parasites of fishes (Family **Arhythmacanthidae**, genus ***Hypoechinorhynchus***)	***Hypoechinorhynchus magellanicus***
4a(3b) Proboscis hooks arranged in 10 longitudinal rows. One large and 2–4 small hooks in row. Blade and root of large hook similar in length	***Heterosentis heteracanthus***
4b Proboscis hooks arranged in circa 15 rows. Probably 1-2 large and 1–3 small hooks in row. Blade of large hook considerably longer than root	***Heterosentis magellanicus***
5a(2b) Parasite of *Zanclorhynchus spinifer* at Macquarie Island. Proboscis circa 1 mm long, armed with 14–16 rows of hooks, circa 10–12 in row	***Echinorhynchus zanclorhynchi***
5b Parasites of other Antarctic and Sub-Antarctic fishes	6
6a(5b) Eight pyriform cement glands arranged in compact group in males. Vulvar sphincter in females double	(Family **Rhadinorhynchidae**, subfamily **Gorgorhynchinae**, genus **Metacanthocephalus**) 7
6b Six spherical or ovoid cement glands usually arranged along trunk of males. Vulvar sphincter in females single	(Family **Echinorhynchidae**, subfamily **Echinorhynchinae**, genus **Echinorhynchus**) 10
7a(6a) Trunk cylindrical, slightly dilated anteriorly. Eggs longer than 100 μm	8
7b Trunk elongate, oval with maximum width at mid-body. Mean length of eggs smaller than 100 μm	9

8a(7a) Proboscis cylindrical, 0.54–0.68 mm long. Hooks in 13–15 rows of 8–10. Length of eggs 110–150 µm	***Metacanthocephalus campbelli***
8b Proboscis ovoid to cylindrical, 0.30–0.44 mm long. Hooks in 11–16 rows of 4–6 (usually 5). Length of eggs 100–120 µm	***Metacanthocephalus dalmori***
9a(7b) Proboscis 0.43–0.60 mm long. Hooks in 12–17 rows of 5–8 (usually 6–7). Length of eggs 80–110 µm (mean 97 µm)	***Metacanthocephalus johnstoni***
9b Proboscis 0.30–0.42 mm long. Hooks in 12–13 rows of 5–7 (usually 6). Length of eggs 80–90 µm	***Metacanthocephalus rennicki***
10a(6b) Proboscis hooks in 14–20 rows. Length of eggs 90–120 µm	***Echinorhynchus petrotschenkoi***
10b Proboscis hooks in 12 rows. Length of eggs 70–100 µm	***Echinorhynchus muraenolepisi***
11a (1b) Proboscis spherical. Neck very long and narrow. Trunk without anterior dilatation. Parasite of birds	***Profilicollis antarcticus***
11b Proboscis cylindrical or conical. Trunk dilated anteriorly. Parasites of birds and mammals	12
12a (11b) Proboscis cylindrical. Fore-trunk forming bulb not separated from hind-trunk. Parasites of seals and birds (males of some species may be found in whales)	13
12b Proboscis conical, rarely cylindrical. Fore-trunk forming bulb separated from hind-trunk by constriction. Parasites of whales	(Genus **Bolbosoma**) 23
13a(12a) Somatic armature divided into anterior and posterior fields. Genital armature present or absent in specimens of both sexes. Parasites of birds, mainly cormorant	(Genus **Andracantha**) 14
13b Somatic armature not divided. Genital spines present in all males and usually in females. Parasites of seals and penguins	(Genus **Corynosoma)** 15
14a(13a) Length of proboscis 0.82–0.97 mm. Distal proboscis hooks longer than prebasal. Length of eggs 90–100 µm. The species occurs in western Antarctic and Sub-Antarctic	***Andracantha baylisi***
14c Length of proboscis 0.63–0.75 mm. Distal proboscis hooks shorter than prebasal. Length of eggs 70–80 µm. The species occurs in environs of South Australia, New Zealand, and Kerguelen	***Andracantha clavata***
15a(13b) Somatic and genital armature connected on ventral side of trunk	16
15b Genital armature separated from somatic or absent in females	20

16a(15a) Lemnisci flat	17
16b Lemnisci consist of many irregular folds	19
17a(16a) Proboscis ovoid to cylindrical, 0.88–1.12 mm long. Length of largest hooks 130–160 μm. Parasite of penguins	***Corynosoma shackletoni***
17b(16a) Proboscis almost cylindrical, dilated just posterior to mid-length, 0.52–0.56 mm long, shorter than proboscis receptacle	***Corynosoma beaglense***
17c Proboscis cylindrical, dilated before base, shorter than 0.75 mm. Largest hooks shorter than 90 μm. Parasites of seals	18
18a(17b) Proboscis hooks in 16–18 rows of 11–15, including 2–4 rootless basal hooks. Genital pore in females subterminal	***Corynosoma australe***
18b Proboscis hooks in 22 rows of 12–13, including 4–6 rootless basal hooks. Genital pore in females terminal	***Corynosoma hannae***
19a(16b) Proboscis longer than 1 mm. Number of proboscis hooks in row 12–16 (usually 14–15). Genital pore in females on the bottom of the hollow between two lateral folds. Length of eggs 160–200 μm	***Corynosoma hamanni***
19b Proboscis shorter than 1 mm. Number of proboscis hooks in row 10–14 (usually 12–13). Genital pore in females terminal. Length of eggs 90–120 μm	***Corynosoma pseudohamanni***
20a(15b) Number of rows of proboscis hooks 15–18	21
20b Number of rows of proboscis hooks 19–24	22
21a(20a) Hind-trunk cylindrical, considerably longer than dilated fore-trunk. Proboscis longer than 0.9 mm. Cement glands in males tubular. Length of eggs 110–130 μm. Parasite of elephant seals	***Corynosoma bullosum***
21b Hind-trunk cylindrical, a little shorter than dilated fore-trunk. Proboscis shorter than 0.75 mm. Cement glands in males pyriform. Length of eggs 70–80 μm. Parasite of fur seals and leopard seals	***Corynosoma australe***
22a(20b) Length of proboscis 0.7–1.0 mm. Genital spines in males arranged in 8–9 circles, circa 150 in number. Genital spines in females present or absent. Length of eggs 130–160 μm	***Corynosoma arctocephali***
22b Length of proboscis 0.6–0.8 mm. Genital spines in males arranged in 4 circles, circa 40–60 in number. Genital spines in females absent. Length of eggs 100–130 μm	***Corynosoma evae***

23a(12b) Total length circa 20 mm. Fore-trunk spines arranged in 6–10 circles before bulb. Lemnisci short, flat	***Bolbosoma balaenae***
23b Total length less than 7 mm. Somatic spines cover anterior part of fore-trunk, including bulb, arranged in at least 15 circles. Lemnisci very long, filiform	24
24a(23b) Proboscis hooks usually in 19–22 (rarely 23 or 24) rows of usually 6–7 (rarely 5 or 8)	25
24b Proboscis hooks in 24–27 rows of 7–8	26
25a(24a) Total length 11–25 mm. Fore-trunk spines arranged in circa 15 circles. Length of eggs 130–170 μm. Parasite of sei whales of southern hemisphere	***Bolbosoma turbinella australis***
25b Total length 21–38 mm. Fore-trunk spines arranged in circa 20 circles. Length of eggs 120–130 μm. Parasite of blue whales and fin whales	***Bolbosoma brevicolle***
26a(24b) Total length 60–64 mm. Length of eggs 110–140 μm	***Bolbosoma hamiltoni***
26b Total length 16–39 mm. Length of eggs 90–120 μm	***Bolbosoma tuberculata***

References

Amin OM (1987) Key to the families and subfamilies of Acanthocephala, with the erection of a new class (Polyacanthocephala) and a new order (Polyacanthorhynchida). J Parasitol 73:1216–1219

Baylis HA (1929) Parasitic Nematoda and Acanthocephala collected in 1925–1927. Discov Rep 1:541–560

Dollfus RP (1965) Acanthocephale d'un téléostéen du genre Notothenia Richardson des Kerguelen (Mission Jean-Claude Hureau, 1963–1964). Bull Mus Natn Hist Nat Paris Ser 2(36):641–646

Edmonds SJ (1955) Acanthocephala collected by the Australian national Antarctic research expedition on heard island and Macquarie island during 1948–50. Trans R Soc S Aust 78:141–144

Edmonds SJ (1957) Acanthocephala. Rep BANZ Antarct Res Expect Ser B 6:91–98

Golvan YJ (1959) Acanthocephales du genre Corynosoma Ltihe, 1904, parasites de mammiferes d'Alaska et de Midway. Ann Parasit Hum Comp 34:288–321

Golvan YJ (1969) Systematique des acanthocephales (Acanthocephala Rudolphi 1801). Premiere partie: l'ordre des Palaeacanthocephala Meyer 1931, premier fascicule: la super-famille des Echinorhynchoidea (Cobbold 1876) Golvan et Houin 1963. Mem Mus Natn Hist Nat Paris Ser A Zool 57:1–373

Hoberg EP (1986) Aspects of ecology and biogeography of Acanthocephala in Antarctic seabirds. Ann Parasitol Hum Comp 61:199–214

Holloway HL Jr, Nickol BB (1970) Morphology of the trunk of *Corynosoma hamanni* (Acanthocephala: Polymorphidae). 1. Morph 130:151–161

Holloway HL Jr, Spence JA (1980) Ecology of animal parasites in McMurdo Sound. Antarctica Comp Phys Ecol 5:262–284

Hoogesteger JN, White MG (1981) Notes on parasite infestation of inshore fish at Signy Island, South Orkney Islands. Br Antarct Surv Bull 54:23–31

Johnston TH, Best EW (1937) Acanthocephala. Aust. Antarct. Exped., 1911–14. Sci Rep Ser C 10(2):1–20

Joyeux C, Baer JG (1954) Cestodes et Acanthocephales recoltes par M. Patrice PAULIAN aux iles Kerguelen et Amsterdam. Mem Lnst Scient Madagascar Ser A 9:23–40

Kagei N, Watanuki T (1975) On the parasites of fishes from the Antarctic ocean. Antarctic Res 54:84–93

Kamegai S, Ichihara A (1973) Parasitic helminths of Antarctic animals (2). *Corynosoma hamanni* (Linstow, 1892) from Weddell seal, *Leptonychotes weddelli*, caught on the Ongul Isl. Antarctic Res Bull Meguro Parasit 7:26–27

Kumar P. (1992) Arhythmacanthus zdzitowieckii, new species (Acanthocephala: Arhythmacanthidae) from estuarine fish, Clariasbatrachus of Chilka lake, Orissa, India. Pak. J. Zool. 24: 143–144

Laskowski Z, Zdzitowiecki K (2004) New morphological data on a sub-Antarctic acanthocephalan, *Aspersentis johni* (Baylis, 1929) (Palaeacanthocephala: Heteracanthocephalidae). Syst Parasitol 59:39–44

Laskowski Z, Zdzitowiecki K (2005) The helminth fauna of some notothenioid fishes collected from the shelf of Argentine Islands, West Antarctica. Pol Polar Res 26:315–324

Laskowski Z, Rocka A, Zdzitowiecki K, Ghigliotti L, Pisano E (2005) New data on the occurrence of internal parasitic worms in the *Gymnodraco acuticeps* and *Cygnodraco mawsoni* (Bathydraconidae) fish in the Ross Sea, Antarctica. Pol Polar Res 26:37–40

Laskowski Z, Rocka A, Zdzitowiecki K, Ozouf-Costaz C (2007) Occurrence of endoparasitic worms in dusky notothen, *Trematomus newnesi* (Actinopterygii Nototheniidae), at Adelie Land, Antarctica. Pol Polar Res 28:37–42

Laskowski Z, Zdzitowiecki K (2008) New morphological data on the acanthocephalan *Hypoechinorhynchus magellanicus* Szidat, 1950 (Palaeacanthocephala: Arhythmacanthidae). Syst Parasitol 69:179–183

Laskowski Z, Jeżewski W, Zdzitowiecki K (2008) Cystacanths of Acanthocephala in notothenioid fish from the Beagle Channel (sub-Antarctica). Syst Parasitol 70:107–117

Laskowski Z, Zdzitowiecki K (2009) Occurrence of acanthocephalans in notothenioid fishes in the Beagle Channel (Magellanic sub-region, sub-Antarctic). Pol Polar Res 30:179–186

Laskowski Z, Zdzitowiecki K (2010) Contribution to the knowledge of the infection with Acanthocephala of a predatory Antarctic ice-fish *Chaenocephalus aceratus*. Pol Polar Res 31:303–308

Laskowski Z, Jeżewski W, Zdzitowiecki K (2010) New data on the occurrence of Acanthocephala in Antarctic Amphipoda. Acta Parasitologica 55:161–166

Laskowski Z, Korczak-Abshire M, Zdzitowiecki K (2012) Changes in acanthocephalan infection of the Antarctic fish *Notothenia coriiceps* in Admiralty Bay, King George Island, over 29 years. Pol Polar Res 33:99–108

Leiper RT, Atkinson EL (1914) Helminthes of the British Antarctic Expedition, 1910–1913. Proc Zool Soc Lond 1:222–226

Leiper RT, Atkinson EL (1915) Parasitic worms with a note on a freeliving nematode. Nat Hist Rep Br Antarct Terra Nova Exped Zool 2:19–60

Linstow O (1892) Helminthen von Stidgeorgien. Nach der Ausbeute der deutschen Station von 1882–1883. Jb Hamb Wiss Anst 9:59–77

Linstow O (1896) Nemathelminthen. Hamburger Magalhaensische Sammelreise. L. Friedrichsen u. Co. Hamburg, p 22

Markowski S (1971) On some species of parasitic worms in the "Discovery" collections obtained in the years 1925–36. Bull Brit Mus (Nat Hist) Zool 21:53–65

Meyer A (1931) Die Stellung des Genus Heterosentis van Cleave 1931 in Acanthocephalensystem. Zool Anz 94:258–265

Nickol BB, Holloway HL Jr (1968) Morphology of the presoma of *Corynosoma hamanni* (Acanthocephala: Polymorphidae). J Morph 124:217–226

Nikolsky OR (1974) [Acanthocephalan fauna of 105 pinnipeds of the Pacific Ocean sector of the marine Antarctic]. lzv. tikhookean nauchnoissled. Inst Ryb Khoz Okeanogr 88:101–106 [In Russian]

Parukhin AM, Lyadov VN (1982) Helminth fauna of food Nototheniidae fishes from Kerguelen subregion. Ekol Morya Kiev 10:49–57 [In Russian]
Parukhin AM, Sysa VN (1975) [The incidence of infection in Notothenioidei fish in Sub-Antarctic waters]. Problemy Parazitologii. Materialy nauchnojj konferencii parazitologov USSR, 2, Kiev, "Naukova Dumka", pp 97–99 [In Russian]
Petrotschenko VI (1956) [Acanthocephala of domestic and wild animals]. Tom I. lzdatelstvo Akademii Nauk SSSR, Moskva, p 435 [In Russian]
Petrotschenko VI (1958) [Acanthocephala of domestic and wild animals]. Tom II. Izdatelstvo Akademii Nauk SSSR, Moskva, p 458 [In Russian]
Pichelin S, Cribb TH (1999) A review of the Arhythmacanthidae (Acanthocephala) with a description of *Heterosentis hirsutus* n.sp. from Cnidoglanis macrocephala (Plotosidae) in Australia. Parasite 6:293–302
Railliet A, Henry A (1907) Nemathelminthes parasites. Exped. antarctique française, 1903–1905. Masson et Cie, Paris, p 15
Reimer LW (1987) Helminthen von Fischen der Antarktis. Fischerei-Forschung Rostock 25:36–40
Rennie J (1906) "Scotia" Collections. On *Echinorhynchus antarcticus* n. sp., and its Allies. Proc R Soc Edinburgh 26:437–446
Rodjuk GN (1984) New representatives of the genus Metechinorhynchus (Acanthocephala), parasites of fishes of the western Antarctic. Zool Zh 63:1893–1896 [In Russian]
Rodjuk GN (1985) Parasitic Fauna of the Fishes of the Atlantic Part of the Antarctic (South Georgia Island and South Shetland Isles). In: Hargis Jr WJ (ed) Parasitology and pathology of marine organisms of the World Ocean. NOAA Tech. Rep. NMFS 25, pp 31–32
Rodjuk GN (1986) New species of Acanthocephala of the genus Echinorhynchus (Echinorhynchidae) from the southwestern Atlantic. Parazitologiya 20:224–227 [In Russian]
Zdzitowiecki, Rokosz B (1986) Prevalence of acanthocephalans in fishes of South Shetlands (Antarctic). II. *Aspersentis austrinus* Van Cleave, 1929 and remarks on the validity of *Heteracanthocephalus hureaui* Dollfus, 1965. Acta Parasit Pol 30:161–171
Skryabin AS (1966) [New corinosome *Corynosoma mirabilis* n. sp. – a parasite of the sperm whale]. In: Delyamure SL (ed) [Helminth fauna of animals of southern seas]. Izd. "Naukova Dumka", Kiev, pp 10–12 [In Russian]
Skryabin AS (1970) A new species, *Bolbosoma tuberculata* sp.n. (Polymorphydae Meyer, 1931), a parasite of whales. Parazitologiya 4:334–337 [In Russian]
Skryabin AS (1972) On morphological differences between acanthocephals *Bolbosoma turbinella* (Diesing, 1851) (fam. Polymorphidae) from the northern and southern hemispheres. Parazitologiya 6:57–64 [In Russian]
Skryabin AS, Nikolsky OR (l971) [*Corynosoma singularis* sp. nov. (family Polymorphidae) – a parasite of marine mammals of the Antarctic]. Nauch Dokl Wyss Skholy Biol Nauki (1971):7–9 [In Russian]
Smales LR (1996) A redescription of Aspersentis zanclorhynchi (Johnston and Best, 1937) comb. nov. (Heteracanthocephalidae: Acanthocephala). Trans R Soc S Aust 120:167–171
Szidat L (1950) Los Parasitos del Robalo (Eleginops maclovinus (Cuv. & Val.)). Prim. Congr. nac. de pesquerias Marit. e Indust., Mar del Plata, 1949, vol 2, Buenos Aires, pp 235–270
Szidat L (1965) Estudios sobre la fauna de panisitos de peces antarcticos. I – Los panisitos de Notothenia neglecta Nybelin. Servicio de Hidrografia Naval, Secretaria de Marina, Republica Argentina, Publico H 910, Buenos Aires, p 84
Szidat L, Graefe G (1967) Estudios sobre la fauna de panisitos de peces antarcticos. II – Los parasitos de Parachaenichthys charcoti. Servicio de Hidrografia Naval, Armada Argentina, Republica Argentina, Publico H 911, Buenos Aires, p 27
Van Cleave HJ (1929) New Genus and new Species of Acanthocephala from the Antarctic. Ann Mag Nat Hist 10(4):229–232
Van Cleave HJ (1931) Heterosentis, a new Genus of Acanthocephala. Zool Anz 93:144–146
Yamaguti S (1963) Systema Helminthum. Volume V. Acanthocephala. Interscience Publishers, New York/London, p 423

Zdzitowiecki K (1978) On the occurrence of juvenile acanthocephalans of the genus Corynosoma Lühe, 1904 in fishes off South Georgia and South Shetland Islands (the Antarctic). Acta Ichthyologica et Piscatoria 8:111–127

Zdzitowiecki K (1981) Redescription of *Aspersentis austrinus* Van Cleave, 1929 (Acanthocephala). Acta Parasit Pol 28:73–83

Zdzitowiecki K (1983) Antarctic acanthocephalans of the genus Metacanthocephalus. Acta Parasit Pol 28:417–437

Zdzitowiecki K (1984a) Description of *Heterosentis heteracanthus* (Linstow, 1896) (Acanthocephala, Arythmacanthidae) from Antarctic fishes, and remarks on the taxonomic status of Heterosentis Van Cleave, 1931. Acta Parasit Pol 29:111–115

Zdzitowiecki K (1984b) Some Antarctic acanthocephalans of the genus Corynosoma parasitizing Pinnipedia, with descriptions of three new species. Acta Parasit Pol 29:359–377

Zdzitowiecki K (1984c) Redescription of *Corynosoma hamanni* (Linstow, 1892) and description of *C. pseudohamanni* sp. n. (Acanthocephala) from the environs of South Shetlands (Antarctic). Acta Parasit Pol 29:379–393

Zdzitowiecki K (1985) Acanthocephalans of birds from South Shetlands (Antarctic). Acta Parasit Pol 30:11–24.*

Zdzitowiecki K (1986a) Acanthocephala of the Antarctic. Pol Polar Res 7:79–117.*

Zdzitowiecki K (1986b) Prevalence of acanthocephalans in fishes of South Shetlands (Antarctic). I. Juvenile Corynosoma spp. Acta Parasit pol 30:143–160

Zdzitowiecki K (1986c) A contribution to the knowledge of morphology of *Corynosoma bullosum* (Linstow, 1892) (Acanthocephala). Acta parasit Pol 30:225–232

Zdzitowiecki K (1986d) *Echinorhynchus nototheniae* sp. n. (Acanthocephala) from nototheniid fishes from the environs of South Shetlands (Antarctic). Acta Parasit Pol 31:23–27

Zdzitowiecki K (1986e) *Corynosoma gibsoni* sp. n., a parasite of *Otaria flavescens* (Shaw, 1800) from the Falkland Islands and a note on the occurrence of *C. evae* Zdzitowiecki, 1984. Acta Parasit Pol 31:29–32.*

Zdzitowiecki K (1986f) Redescription of *Corynosoma tunitae* (Weiss, 1914) and description of *C. baylisi* sp. n. (Acanthocephala, Polymorphidae), parasites of piscivorous birds. Acta Parasit Pol 31:117–123

Zdzitowiecki K (1986g) Prevalence of acanthocephalans in fishes of South Shetlands (Antarctic). III. *Metacanthocephalus johnstoni* Zdzitowiecki, 1983, *M. dalmori* Zdzitowiecki, 1983 and notes on other species; general conclusions. Acta Parasit Pol 31:125–141

Zdzitowiecki K (1987) Acanthocephalans of marine fishes in the regions of South Georgia and South Orkneys (Antarctic). Acta Parasit Pol 31:211–217

Zdzitowiecki K (1989a) New data on the morphology and distribution of two acanthocephalans, *Andracantha baylisi* (Zdzitowiecki, 1986) comb. n. and *Corynosoma australe* Johnston, 1937. Acta Parasit Pol 34:167–172

Zdzitowiecki K (1989b) A redescription of *Echinorhynchus petrotschenkoi* (Rodjuk, 1984) comb. n. (Acanthocephala). Acta Parasit Pol 34:173–180

Zdzitowiecki K (1990a) Reexamination of five Antarctic and Subantarctic digenean and acanthocephalan species from Professor Szidat's collection. Acta Parasit Pol 35:31–36

Zdzitowiecki K (1990b) Occurrence of acanthocephalans in the open sea fishes off South Shetlands and South Georgia (Antarctic). Acta Parasit Pol 35:131–142

Zdzitowiecki K, White MG (1996) Acanthocephalan infection of inshore fishes at the South Orkney Islands. Antarct Sci 8:273–276

Zdzitowiecki K (1991) Antarctic Acanthocephala. In: Wägele JW, Sieg J (eds) Synopses of the Antarctic benthos, vol 3, Koeltz Scientific Books, Koenigstein pp 1–116

Zdzitowiecki K (2001) Acanthocephala occurring in intermediate hosts, amphipods, in Admiralty Bay (South Shetland Islands, Antarctica). Acta Parasit 46:202–207

Zdzitowiecki K, Presler P (2001) Occurrence of Acanthocephala in intermediate hosts, Amphipoda, in Admiralty Bay, South Shetland Islands, Antarctica. Pol Polar Res 22:205–212

Zdzitowiecki K, Laskowski Z (2004) Helminths of an antarctic fish, *Notothenia coriiceps*, from the Vernadsky Station (Western Antarctica) in comparison with Admiralty bay (South Shetland Islands). Helminthologia 41:201–207

Zdzitowiecki K, Laskowski Z (2013) New data on the occurrence of Acanthocephala in some fish in Admiralty Bay (South Shetland Islands). Acta Parasitol 58:547–550. doi:10.2478/s11686-013-0175-1

Chapter 9
Macroparasites in Antarctic Penguins

Julia I. Diaz, Bruno Fusaro, Virginia Vidal, Daniel González-Acuña, Erli Schneider Costa, Meagan Dewar, Rachael Gray, Michelle Power, Gary Miller, Michaela Blyton, Ralph Vanstreels, and Andrés Barbosa

9.1 Introduction

Parasites are the majority of species on Earth (Windsor 1998). The total number of parasite species is likely to be huge, because practically all free-living metazoan species harbor at least one parasite species and almost every individual of every species is parasitized by at least one parasite during its life cycle (Poulin and Morand 2004). The number of parasite species has been estimated as a range from 30 to 71 %

Julia I. Diaz, Daniel González-Acuña, Erli Schneider Costa, Meagan Dewar, Rachael Gray, Michelle Power, Gary Miller, Ralph Vanstreels, and Andrés Barbosa are Working Group of Health Monitoring of Birds and Marine Mammals of the SCAR.

J.I. Diaz (✉)
Centro de Estudios Parasitológicos y de Vectores (CEPAVE), FCNyM, UNLP, CONICET, La Plata, Argentina
e-mail: jidiaz@cepave.edu.ar

B. Fusaro
Instituto Antártico Argentino (IIA), Plan Anual, Argentina

V. Vidal • A. Barbosa (✉)
Departamento de Ecología Evolutiva, Museo Nacional de Ciencias Naturales – CSIC, Madrid, Spain
e-mail: barbosa@mncn.csic.es

D. González-Acuña
Facultad de Ciencias Veterinarias, Universidad de Concepción, Concepción, Chile

E.S. Costa
Universidade Estadual do Rio Grande do Sul and APECS-Brazil, Porto Alegre, Brazil

M. Dewar
Deakin University, Waurn Ponds, Australia

S. Klimpel et al. (eds.), *Biodiversity and Evolution of Parasitic Life in the Southern Ocean*, Parasitology Research Monographs, Vol. 9,
DOI 10.1007/978-3-319-46343-8_9

R. Gray
Faculty of Veterinary Science, School of Life and Environmental Sciences,
The University of Sydney, Sydney, Australia

M. Power
Department of Biological Sciences, Macquarie University, Sydney, Australia

G. Miller
Institute for Marine and Antarctic Studies, University of Tasmania, Sandy Bay, TAS,
Hobart, Australia

M. Blyton
Western Sydney University, Richmond, Australia

R. Vanstreels
Department of Pathology, Faculty of Veterinary Medicine and Animal Science,
University of São Paulo, São Paulo, Brazil

of the living species (Price 1980; de Meeus and Renaud 2002). Therefore, parasites can be considered a selective pressure affecting different aspects of the host life which can modulate host populations (Morand and Deter 2009). Moreover, parasite diversity provides insights into the history and biogeography of other organisms, into the structure of ecosystems, and into the processes behind the diversification of life (Poulin and Morand 2004).

Helminths and ectoparasites are the main macroparasites of birds. Helminth is a Greek word that means "worm" and is a conventional name, but not a taxon of animal classification (Miyazaki 1991). Among helminth parasites are included those metazoan "worms" that in any stage of their life cycle live in or on other metazoan species (host). Helminths living inside bird hosts are represented by the major groups, Digenea, Cestoda, Nematoda, and Acanthocephala.

Helminths occupy diverse sites within the host including the gastrointestinal, respiratory, and urinary systems and in organs and tissue spaces of their host. Depending on the parasitic species, their intensity of infection, the host immune status, and the environmental conditions, their presence might not lead to obvious clinical manifestations, or it may manifest itself in terms of individual morbidity and mortality or produce more subtle negative effects on host fitness (Hoberg 2005).

Ectoparasites include arthropod parasites such as ticks, mites (Acari), lice, bugs, fleas, and flies (Insecta). The effects of ectoparasites may include anemia (Gauthier-Clerc et al. 1998; Mangin et al. 2003), feather damage (Barbosa et al. 2002), transmission of pathogens (Allison et al. 1978; Morgan et al. 1981; Siers et al. 2010; Yabsley et al. 2012), and, in the case of some ticks, injection of neurotoxins (Gothe et al. 1979). The consequences of these infestations can vary greatly depending on their intensity and on host health and immune status, and can include negative impacts in terms of mortality, breeding success, and behavior (Gauthier-Clerc et al. 1998; Mangin et al. 2003).

Animals living in Antarctica including penguins are also affected by parasites; however, there is limited knowledge available on their presence, their distribution, epidemiology, life cycles, and health effects on the host in Antarctic fauna (Barbosa and Palacios 2009; Kerry and Riddle 2009). Among Antarctic vertebrates, penguins represent more than 90 % of the terrestrial biomass and are the most studied group on this matter. However, available information is sparse and fragmented. In this chapter, we examine the published information on macroparasites of Antarctic penguins, using these species as a model to understand the broader picture on the parasitology of Antarctic birds.

9.2 Diversity and Richness of Helminth Parasites

Former Antarctic expeditions, such as the ones led by James Clark Ross (1839–1843), Jean-Baptiste Charcot (1903–1905), and Robert Falcon Scott (1910–1913), among others, already collected parasites and left us a valuable source of information. One of the most thorough publications on helminths from that time was the one written by Johnston in 1937–1938, dealing with parasites collected during the 1911–1914 Australian Antarctic expedition. He not only supplied descriptions and drawings of helminths, but also included the review and history of each one of them. In general, former published surveys on helminths parasitizing Antarctic penguins often provided only a list of hosts and the parasites collected from them and with few cases reporting on the proportion of infected hosts (Johnston and Mawson 1945; Mawson 1953). Nevertheless, data about their prevalence, intensity, or abundance are scarce, and have only started to be provided in the last decades (Fonteneau et al. 2011; Vidal et al. 2012; Diaz et al. 2013, 2016). Despite this apparent gap, there is a sufficient number of publications that, when compiled and compared, allow as a fairly comprehensive assessment of the richness of helminths present in Antarctic penguins.

Antarctic and Sub-Antarctic penguins act as definitive host of only 13 recognized helminth species (Table 9.1). The core component of the helminth fauna of Antarctic penguins are cestodes, mainly *Parorchites zederi* (Dilepididae). This species is the only cyclophyllidean present in pelagic birds and is widely distributed among Antarctic penguins, including the three pygoscelid species and the Emperor penguin (Cielecka et al. 1992; Vidal et al. 2012; Diaz et al. 2013, 2016; Kleinertz et al. 2014). The presence of Cyclophyllidea eggs has also been demonstrated in the feces of Adélie penguins (Fredes et al. 2008), and it is reasonable to presume these were *P. zederi*.

Members of the Tetrabothriidea are also important components of the helminth communities of Antarctic penguins (Baer 1954). *Tetrabothrius pauliani* Joyeux and Baer 1954 was registered parasitizing all pygoscelid species and also the King penguin, *Tetrabothrius joubini* Railliet and Henry, 1912 was only reported in the Chinstrap penguin (Prudhoe 1969; Cielecka et al. 1992; Georgiev et al. 1996), and *Tetrabothrius wrighti* Leiper and Atkinson 1914 was registered in Adélie, King, and Emperor penguins (Leiper and Atkinson 1914; Johnston 1937; Prudhoe 1969; Fonteneau et al. 2011).

Table 9.1 Summary of macroparasites recorded in Antarctic penguins

	Emperor *Aptenodytes forsteri*	King *Aptenodytes patagonicus*	Adélie *Pygoscelis adeliae*	Chinstrap *Pygoscelis antarctica*	Gentoo *Pygoscelis papua*	Macaroni *Eudyptes chrysolophus*	Southern Rockhopper *Eudyptes chrysocome*	Royal *Eudyptes schlegeli*
Breeding distribution	Antarctic	***Sub-Antarctic	Antarctic	Antarctic/Sub-Antarctic	Antarctic/Sub-Antarctic	Sub-Antarctic	Sub-Antarctic	Sub-Antarctic
Diet (% by weight)								
Fishes	96.8	68.7–99.8	67.3–100	0.1–38.8	0.2–70.0	2.0–25.0	1.9–28.3	45.8
Crustaceans	0.4	0.1	54.6–99.6	55.1–99.9	30.0–99.8	75.0–98.3	45.1–99.7	51.4
Cephalopods	2.7	0.2–31.3	0–0.4	0	0.5–2.1	0.1–2.2	0.3–53.0	2.8
Cestoda								
Parorchites zederi	X (1, 2)		X (1, 18, 19)	X (18, 24, 25)	X (18, 26, 29, 34)			
Tetrabothrius joubini				X (18, 23, 24)				
Tetrabothrius pauliani		X (1)	X (18)	X (18, 25, 26, 27)	X (1)			
Tetrabothrius wrightii	X (1)	X (8)						
Tetrabothrius sp.	X (2)		X (19, 21)	X (31, 32)		X (27)		
Diphyllobotrhium sp.	X (2)				X (34)			
Nematodes								
Stegophorus macronectes			X (19, 20)	X (25)	X (7, 29)	X (7)	X (7)	
Stegophorus adeliae					X (34)			
Tetrameres wetzeli		X (8)			X (29)			
Tetrameres sp.			X (19)					
Contracaecum heardi		X (8)			X (7)	X (7)		
Contracaecum sp.		X (7)			X (31, 32)		X (7)	
Terranova piscium								X (7)[a]
Streptocara sp.			X (21)		X (31)			

Stomachus = *Anisakis* sp.					X (7)[a]			X (7)[a]
Acanthocephalans								
Corynosoma shackletoni					X (29, 33, 34)			
Corynosoma hamanni					X (29)[a]			
Corynosoma bullosum					X (29)[a]			
Corynosoma pseudohamanni				X (28)[a]				
Corynosoma sp.				X (25)				
Chewing lice								
Austragonioides antarcticus			X (3, 4, 5, 10)					
A. brevipes		X (4, 5, 9, 10)						
A. bicornutus						X (4, 5, 10)		
A. bifasciatus			X (4, 5)					
A. concii						X (4, 5)	X (5, 10)	
A. cristati						X (3, 4, 5, 10, 11)	X (3, 4, 5, 10, 11)	X (4, 5, 10)
A. demersus						X (4)		
A. gressitti				X (3, 4, 5, 10)	X (3, 4, 5, 10, 34)	X (4)		
A. keleri					X (4, 5)	X (3, 4, 5, 10	X (3, 4, 5)	
A. macquariensis				X (4, 5)	X (4, 5)	X (3, 4, 5, 10, 11)	X (3, 4, 5, 9, 10, 11)	X (10)
A. strutheus								X (4, 5)
A. mawsoni	X (3, 4, 5)							

(continued)

Table 9.1 (continued)

	Emperor *Aptenodytes forsteri*	King *Aptenodytes patagonicus*	Adélie *Pygoscelis adeliae*	Chinstrap *Pygoscelis antarctica*	Gentoo *Pygoscelis papua*	Macaroni *Eudyptes chrysolophus*	Southern Rockhopper *Eudyptes chrysocome*	Royal *Eudyptes schlegeli*
A. hamiltoni						X (11)	X (4, 5, 10, 11)	X (4, 5, 9, 10)
Nesiotinus demersus		X (11)						
Naubates prioni					X (11)			
Ticks								
Ixodes uriae		X (10, 12, 13, 14, 15, 16, 17)	X (10, 23)	X (23)	X (10, 23, 34)	X (10, 12, 15, 38, 40)	X (10, 12, 38, 39)	X (10)
Mites *Rhinonyssus scheili*			X (22)					
Ayersacarus tilbrooki					X (36)			
Gamasellus antarcticus					X (37)			
Gamasolaelaps arenosus					X (36)			
Fleas *Parapsyllus heardi*						X (10)	X (10)	
Parapsyllus longicornis					X (35)		X (10)	
Parapsyllus magellanicus							X (10, 12)	

Diet from Williams (Williams 1995)

Numbers in brackets refers to references in the table notes

[a]Denotes immature individuals

References: (1) Prudhoe (1969), (2) Kleinertz et al. (2014), (3) Clay and Moreby (1967), (4) Clay and Moreby (1970), (5) Clay (1967), (7) Mawson (1953), (8) Fonteneau et al. (2011), (9) Banks et al. (2006), (10) Murray et al. (1991), (11) Palma and Horning (2002), (12) Murray and Vestjens (1967), (13) Gauthier-Clerc et al. (1998), (14), Gauthier-Clerc et al. (1999), (15) Frenot et al. (2001), (16) Gauthier-Clerc et al. (2003), (17) Mangin et al. (2003), (18) Cielecka et al. (1992), (19) Diaz et al. (2016), (20) Zdzitowiecki and Drózdz (1980), (21) Fredes et al. (2008), (22) Wilson (1967), (23) Barbosa et al. (2011), (24) Ippen et al. (1981), (25) Vidal et al. (2012), (26) Georgiev et al. (1996), (27) Andersen and Lysfjord (1982), (28) Dimitrova et al. (1996), (29) Diaz et al. (2013), (31) Fredes et al. (2006), (32) Fredes et al. (2007), (33) Hoberg (1986), (34) Gonzalez-Acuña et al. (2013), (35) De Meillon (1952), (36) Hunter (1970), (37) Tragardh (1908), (38) Brooke (1985), (39) Schultz and Petersen (2003), (40) Bergstrom et al. (1999)

Undetermined species of *Tetrabothrius* were also mentioned in Antarctic and Sub-Antarctic regions (Barbosa and Palacios 2009; Kleinertz et al. 2014).

Eggs of *Diphyllobothrium* sp. have been documented in fecal samples of Emperor (Kleinertz et al. 2014) and only in one Gentoo penguin specimen (Gonzalez-Acuña et al. 2013). Recently, some mature and gravid specimens identified as *Diphyllobothrium* sp. were recovered from different colonies of the three pygocelid species (Fusaro and Diaz unpublished data), and in some instances these parasites can be found on the penguin nests (Barbosa unpublished data). Diphyllobothriidae is a very common group in Antarctic marine mammals but does not seem as common in seabirds. It is worth noting that even though *Diphillobotrium scoticum* (see Meggitt 1924; Markowski 1952) has been registered as parasites of pygoscelid penguins (Adélie and Chinstrap), this finding was later denied by Johnston (1937).

Spirurid nematodes occur in the esophagus and stomach of seabirds and are one of the more abundant components in the helminth communities of penguins. *Stegophorus macronectes* (Johnston and Mawson 1942) (Acuariidae) is the best represented species. This acuarid nematode has a wide host and geographical distribution, having been reported in all pygoscelid species (Vidal et al. 2012; Diaz et al. 2013, 2016) and in the Rockhopper and Macaroni penguins in Sub-Antarctic regions (Johnston and Mawson 1945; Mawson 1953; Zdzitowiecki and Drózdz 1980). The taxonomical and nomenclatural history of this species is complex, and different synonyms were employed in the past including *Stegophorus adeliae* Johnston and Mawson 1945 and *Stegophorus paradelia* Johnston, 1938 sensu Petter, 1959 (see Vidal et al. 2016).

In addition to acuarids, nematodes of the genus *Tetrameres* (Spirurida, Tetrameriidae) parasitized the proventricular glands in Antarctic penguins (Schmidt 1965). *Tetrameres wetzeli* (Schmidt 1965) is the only species on the genus described parasitizing penguin hosts, Rockhopper, King, and Gentoo penguins (Schmidt 1965; Fontaneau et al. 2011; Diaz et al. 2013). Undetermined species of *Tetrameres* were also found in Adélie penguins (Diaz et al. 2016).

Contracaecum ascaridoid nematodes are commonly found in the stomach of piscivorous birds (Garbin et al. 2007, 2008; Diaz et al. 2010). *Contracaecum heardi* Johnston and Mawson 1942 is the species best documented among Sub-Antarctic penguins infecting King, Macaroni, and Gentoo penguins (Mawson 1953; Fonteneau et al. 2011).

Other nematode species have been found in Antarctic and Sub-Antarctic penguins. However, most of these reports were based on eggs, few, immature, or fragmented specimens, or corresponded to fish or mammal parasites, so their identification was not possible or is doubtful (e.g., *Contracaecum* spp., *Stomachus* = *Anisakis* sp., *Streptocara* sp., *Terranova* sp., *Capillaria* sp., among others (Mawson 1953; Fredes et al. 2006, 2007, 2008).

Acanthocephalans are not common in pelagic birds. Only *Corynosoma shackletoni* Zdzitowiecki 1978 has been found at the adult stage in Gentoo penguins (Hoberg 1986; Diaz et al. 2013). Other *Corynosoma* species were registered in pygoscelid penguins (e.g., *Corynosoma bullosum*, *Corynosoma hamanni*, and *Corynosoma pseudohamanni*). However, all those reports correspond to immature specimens (see Zdzitowiecki 1991; Dimitrova et al. 1996; Vidal et al. 2012; Diaz et al. 2013), and it is thought that these parasites only reach maturity in cetaceans or pinnipeds with penguin infections being accidental (Holloway and Bier 1967; Hoberg 2005).

Digenea parasites have not been recorded in Antarctic or Sub-Antarctic penguins. This likely occurs due to the limitation of their life cycle, the focal nature of transmission near island systems, and the dilution effect of the marine costal environment, which diminishes their ability to thrive in this kind of hosts (Hoberg 2005).

It is well established that pelagic birds generally support a depauperate parasite fauna, with a much lower diversity than that of birds inhabiting in neritic and littoral waters (Hoberg 2005). A noticeable pattern that emerges by comparing the community of helminths present in Antarctic penguins to that of seabirds from other continents is that the helminth community of penguins is remarkably less diverse. For instance, seabirds of the Alcidae family there are reported in more than 40 helminth species (Muzaffar and Jones 2004), while Antarctic penguin species are parasitized by a total of 10 species (Barbosa and Palacios 2009). Nevertheless, such comparison should be taken with caution as the different number host species might allow more parasite species; in addition, differences in research effort could also affect the comparison. Within penguins, differences in helminths richness between Antarctic and non-Antarctic penguins are similar. Non-Antarctic penguins harbor 12 helminth species, while Antarctic penguins present eight recognized species and seven species parasitize penguin species distributed in the Sub-Antarctic region (Clarke and Kerry 2000; Barbosa and Palacios 2009). Moreover, penguins included in the genus *Spheniscus* (non-Antarctic) have helminth communities richer than those of *Pygoscelis* genus (Clarke and Kerry 2000; Barbosa and Palacios 2009; Brandão et al. 2014). Infracommunities of three pygoscelid species present in the Antarctic Peninsula harbor between one or three helminth species, while those of the Magellanic penguins in Patagonia harbor up to five species (Diaz et al. 2010, 2013, 2016; Vidal et al. 2012). In general, the low number of helminths found in pygoscelid penguins can be explained by the narrow range of variety of prey present in their diet which is form mainly by krill and some few species of squid and fishes (Williams 1995). A wider diet and/or foraging plasticity facilitate the exposure to a high number of parasite species through the ingestion of a high number of intermediate hosts (Hoberg 1996).

9.3 Life Cycles and Source of Infection of Helminths

Most helminths that infect seabirds have indirect life cycles, involving a definitive host, the bird in which adults develop and sexual reproduction occurs, and one or more intermediate/paratenic hosts (invertebrates, fishes) carrying the larval stages. As a result, infestations by helminths are strongly influenced by the trophic relationships of the hosts (Hoberg 1996). Specialized foragers, such as some Antarctic penguins, can therefore be expected to be infested by fewer parasites than more generalist species.

The trophic webs of the Southern Ocean have macrozooplankton such as euphausiids (krill) playing a key role as an intermediate between primary producers and top predators. Krill (especially *Euphausia* spp.) are the main prey item for most Antarctic penguins (Cherel and Kooyman 1998) and are therefore plausible intermediate hosts for their helminths (Hoberg 2005; Bush et al. 2012).

Larval stages of penguin cestodes use a variety of prey crustaceans/fishes as intermediate hosts (Hoberg 2005). *Parorchites zederi* is probably widely distributed among Antarctic penguins due to a broad oceanic distribution of euphausiids (Hoberg 2005; Vidal et al. 2012; Diaz et al. 2013).

The complete life cycle of *Tetrabothrius* species remains unclear and further investigations are needed. It has been suggested that the first intermediate host of tetrabothriidean cestodes are marine crustaceans and second intermediate or paratenic host could be cephalopods or fishes (Baer 1954; Hoberg 1987). Larval stages identified as Tetrabothriidae were found in nototheniid fishes in Sub-Antarctic waters (Rocka 2003). Presence of tetrabothrids could therefore be higher in penguin species that include cephalopods or fishes in their diets (Diaz et al. 2016).

Acuarid and tetramerid nematodes that parasitize aquatic vertebrates are known to develop to the third infective stage in the hemocoel of crustaceans (Anderson 2000). The high prevalence of *S. macronectes* in Antarctic penguins could thus be a consequence of the broad oceanic distribution of euphausiids and their key role in the Southern Ocean trophic web, since they likely serve as suitable intermediate/paratenic hosts. This is corroborated by the observation of a third stage nematode larva in a krill specimen during a survey from Punta Stranger (Diaz pers. obs.). Morphological features observed in that case (Fig. 9.1) are consistent with those of an Acuariidae third stage larva (see Anderson 2000).

However, considering that this larva was the only parasite specimen found after having dissected hundreds of krill individuals (Vidal and Barbosa unpublished data) prevalence of helminth larvae in krill is likely very low. In fact, it is striking that Kagei et al. (1978) found no helminth stages in two large samples of more than 35000 and 55000 Antarctic krill (*E. superba*) each one.

Fishes serve as paratenic hosts for the infective third stage larvae of Anisakidae nematods, which mature after being ingested by the definitive hosts. Species of Nototheniidae have been registered as intermediate hosts of *Contracaecum* larvae in the Antarctic region (Kloser et al. 1992; Rocka 2004). The diet of Antarctic penguins includes varying proportion of nototheniid fish, particularly like *Pleuragramma antarcticum* in different proportions (Adams and Klages 1989; Pütz 1995; Ainley et al. 1998; Lescröel et al. 2004), and it is reasonable to speculate that these species may be involved in the transmission of *Contracaecum* to penguins. However, considering that Antarctic penguins generally do not have a strictly piscivorous diet, reports of Anisakidae are very scarce.

Acanthocephalans appear to be almost absent from pelagic birds (Anderson 2000). *Corynosoma* matures in the gut of mammals and birds, whereas fishes and

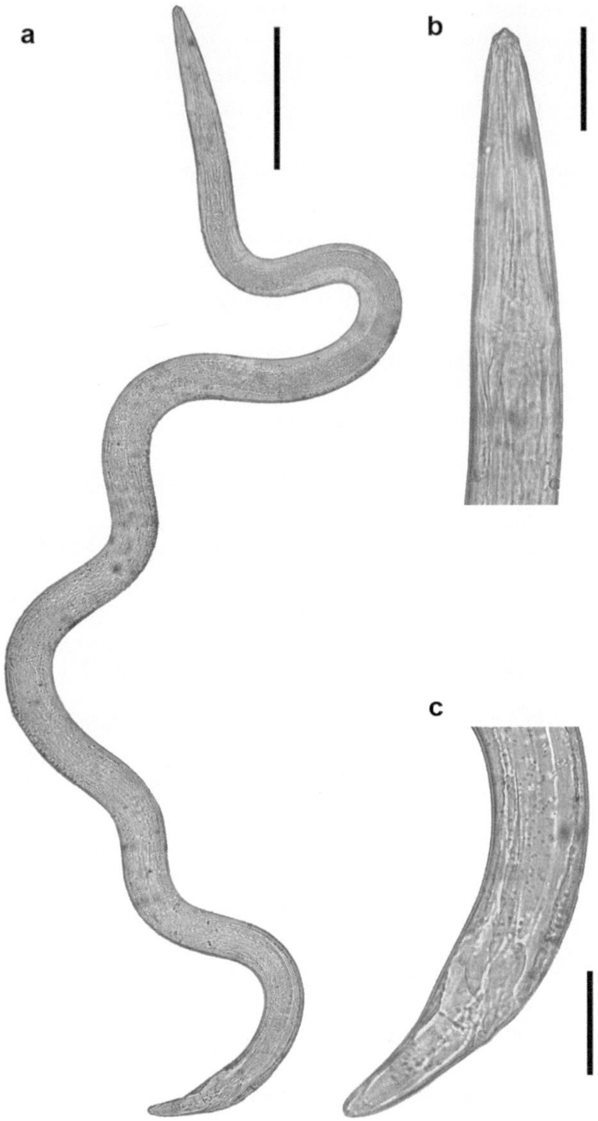

Fig. 9.1 Acuariidae third stage larvae found in the hemocoele of *E. superba*. (**a**) Complete specimen. (**b**) Detail of anterior end. (**c**) Detail of posterior end. Scale bards: (a) 500 μ; (b, c) 100 μ

aquatic invertebrates serve as intermediate hosts. However, since euphausiids are not part of the life cycle of *Corynosoma*, infestation rates are low in krill-dependent species like penguins (Muzaffar and Jones 2004). Notothenid fishes such as *Notothenia coriiceps* have been reported harboring cystacanths of *C. shackletoni* in the studied area (Laskowski and Zdzitowiecki 2005; Laskowski et al. 2012), and therefore are likely to play a role in the transmission of *Corynosoma* spp. to penguins in the Antarctic.

Finally, it should be noted that many helminth species that were reported parasitizing Antarctic penguins only develop to maturity on mammal definitive hosts.

However, taking account that some marine mammals (i.e., pinnipeds, cetaceans) and penguins feed on the same prey items, several larvae or immature stages could appear in the intestinal tract of the birds (e.g., *C. bullosum*, *C. hammani*, and *C. pseudohammani* (Mawson 1953; Zdzitowiecki 1991).

9.4 Ectoparasites

Due to the harsh conditions in Antarctica, the number of species of ectoparasites present in Antarctic penguins is relatively small and limited to ticks, fleas, and chewing lice (Barbosa and Palacios 2009) (Table 9.1). There is only one tick species (*Ixodes uriae*) which is distributed in both Sub-Antarctic (Gauthier-Clerc et al. 1998) and Antarctic regions (Barbosa et al. 2011). Flea species of Antarctic penguins (*Parapsyllus heardi*, *P. longicornis*, *P. magellanicus*) are only present in Sub-Antarctic islands (De Meillon 1952; Murray and Vestjens 1967; Murray et al. 1991). Finally, chewing lice species are the more diverse group of ectoparasites with 17 species (*Austrogoniodes antarcticus, A. bicornutus, A. bisfasciatus, A. brevipes, A. chrysolophus, A. concii, A. cristati, A. gressitti, A. hamiltoni, A. keleri, A. mawsoni, A. macquiariensis, A. strutheus, A. vanalphenae, A. watersoni, Naubates prioni, Nesiotinus demersus*) only five of which occur in the Antarctic continent and adjacent islands (*Austrogoniodes antarcticus, A. bifasciatus, A. chrysolophus, A. gressitti, A. mawsoni*) (Clay 1967; Clay and Moreby 1967, 1970; Murray et al. 1991; Palma and Horning 2002; Banks et al. 2006).

9.5 Prevalence and Parasitism Intensity

Information on the prevalence or infection intensity of helminths and ectoparasites of Antarctic penguins is scarce, with only 12 out of 33 published studies examined in this chapter providing information on prevalence (Table 9.2). Prevalence of metazoan parasites can differ considerably among parasites species, host species, regions, years and season. As a result, the interpretation of the prevalence data herein compiled should be cautious, especially because most of the information is based on relatively small simple sizes.

A remarkable trend is that penguin helminths tend to occur at higher prevalence than ectoparasites, with a maximum prevalence in several worm species (*P. zederi, T. pauliani, S. macronectes*). Current data indicate that *P. zederi* has the widest distribution of prevalence information, from East Antarctica showing the lowest prevalence in the Emperor penguin to Avian Island and Deception Island with the highest prevalence in both Adélie and Chinstrap penguins. Among penguin species, *P. zederi* parasitizing Gentoo penguin seems to be more prevalent in the South Shetlands than in more Southern locations although the opposite is shown in Adélie penguin with the higher prevalence in the more Southern location in Avian Island than in the Northern populations. *Stegophorus macronectes* does not show any clear geographical pattern

Table 9.2 Parasite prevalences in Antarctic penguins

Host species	Location	*N*	Parasite	*P* %	Reference
Eudyptes chrysolophus	Sub-Antarctic ±		*Ixodes uriae*	6	Bergstrom et al. (1999)
	Antarctic Peninsula	13	*Tetrabothrius* sp.	23	Andersen and Lysfjord (1982)
Aptenodytes patagonicus	Crozet ± Archipelago	41	*Tetrabothrius wrighti*	100	Fonteneau et al. (2011)
			Tetrameres wetzeli	41.5	
			Contracaecum heardi	14.6	
			Ixodes uriae	15	Gauthier-Clerc et al. (1999)
Aptenodytes forsteri	East Antarctica	50f	*Parorchites zederi*	2	Kleinertz et al. (2014)
			Tetrabothrius sp.	24	
			Diphyllobothrium sp.	2	
Pygoscelis adeliae	Hope Bay	7C	*Stegophorus macronectes*	50	Diaz et al. (2016)
			Tetrameres sp.	33	
	Avian Is.	2	*Parorchites zederi*	100	
			Tetrabothrius sp.	50	
	25 de Mayo/ King George Is.	7/19C	*Parochites zederi*	29/16C	
			Stegophorus macronectes	14/21C	
			Tetrameres sp.	14	
		3	*Parochites zederi*	33	Cielecka et al. (1992)
			Tetrabothrius pauliani	33	
			Ixodes uriae	9*	Barbosa et al. (2011)
Pygoscelis antarctica	Bouvet Is.	9	*Tetrabothrius pauliani*	88	Andersen and Lysfjord (1982)
	Deception Is.	4/61C	*Parorchites zederi*	100/26C	Vidal et al. (2012)
			Tetrabothrius pauliani	100/13C	
			Stegophorus macronectes	67/72C	
			Ixodes uriae	26*	Barbosa et al. (2011)
	Ronge Is.		*Ixodes uriae*	2*	
	Livingston Is.		*Ixodes uriae*	10*	
		3	*Tetrabothrius joubini*	66	Georgiev et al. (1996)
			Tetrabothrius pauliani	33	
	25 de Mayo/ King George Is.	3	*Parorchites zederi*	100	Cielecka et al. (1992)
			Tetrabothrius pauliani	100	
			Tetrabothrius joubini	100	

Table 9.2 (continued)

Host species	Location	*N*	Parasite	*P* %	Reference
Pygoscelis papua	Paradise Bay	5/100e	*Parorchites zederi*	20	Gonzalez-Acuña et al. (2013)
			Stegophorus adeliae	40	
			Corynosoma shackletoni	40	
			Ixodes uriae	5	
			Austrogonioides gressitti	4	
	Antarctic Peninsula	6/100e	*Parorchites zederi*	33	
			Stegophorus adeliae	16	
			Corynosoma shackletoni	33	
			Austrogonioides gressitti	1	
	Ardley Is.	3/100e	*Diphyllobothrium* sp.	100	
			Stegophorus adeliae	33	
			Austrogonioides gressitti	1	
	25 de Mayo/	3/8C	*Parorchites zederi*	100/0C	Cielecka et al. (1992)
	King George Is.	1	*Parorchites zederi*	100	Georgiev et al. (1996)
		37	*Parorchites zederi*	54	Diaz et al. (2013)
			Stegophorus macronectes	48.6	
			Tetrameres wetzeli	5.4	
			Corynosoma shackletoni	13.5	
			Ixodes uriae	9*	Barbosa et al. (2011)
	Livingston Is.		*Ixodes uriae*	8–10*	
	Ronge Is.		*Ixodes uriae*	2*	

Only were considered those papers in which prevalences were provided or they were possible to be calculated, and those parasites that only reach maturity in birds
N number of birds examined, *P* prevalence, ± Sub-Antarctic Regions, *f* fecal samples, *C* chicks, * collected under stones, *e* external examination

in prevalence although seems to be more prevalent in chicks in Deception Island, while the remaining locations show prevalences around 50 %. As was mentioned above, *P. zederi* and *S. macronectes* are the most prevalent and frequent helminth species among Antarctic and Sub-Antarctic penguins, which could be due to the potential role played by euphausiids, the mean prey item in this system, as intermediate hosts.

The prevalence of *Tetrabothrius* infections in Antarctic penguins varies greatly even at the species level, with higher prevalence being recorded in the Sub-Antarctic region and South Shetlands islands whereas more austral populations have less prevalence. Data from *Tetrameres* indicate that *T. wetzeli* is more prevalent in the Sub-Antarctic region (King penguins at Crozet Island) than in the South Shetlands

(Gentoo penguins at 25 de Mayo/King George Island). Finally, *Corynosoma* species show higher prevalence in the Southern locations than in the North.

Information on the prevalence of ectoparasites is even scarcer than for helminths. Ticks are present in both Sub-Antarctic and Antarctica regions, but they present different behavior that precludes any comparisons. In Sub-Antarctic islands, ticks are found on the penguins (Gauthier-Clerc et al. 1999), while in the Antarctic Peninsula they are much less common and are usually found under the stones close to the penguin colonies (Barbosa et al. 2011). Nevertheless, data from the Antarctic Peninsula indicates a North-South decrease in the abundance and prevalence of ticks present under the stones at the penguin rookeries (Barbosa et al. 2011). However, such pattern is not coherent with a hypothesis of tick colonization from North to South because genetic studies showed that there is no latitudinal genetic cline; on the contrary, results have shown two different genetic populations of ticks in these regions (McCoy et al. 2013).

In general, the data seem to indicate a broader trend of decreased macroparasite prevalence towards more southerly localities; however, this conclusion should be considered judiciously due to the small number of studies and in some cases their small sample size. With regard to age, prevalence appears to be generally higher in adults than in chicks that could be explained due to the longer time of exposure to the parasites in adult individuals and the shorter period of time for parasite development in chicks, but again caution should be taken with this conclusion due to the small sample size in the case of adults. In fact, the opposite patterns can also be found which is explained by the less development of the immune system in the case of chicks.

Information on parasite intensity is even scarcer than prevalence information. There are only four studies giving such information from Crozet archipelago in King Penguin (mean intensity (MI) = 178.6) (Fonteneau et al. 2011), 25 de Mayo/King George Island in Gentoo penguin (MI = 22.02) (Diaz et al. 2013), Deception Island in Chinstrap penguin (MI = 23.21) (Vidal et al. 2012), and 25 de Mayo/King Gorge Island, Bahia Esperanza/Hope Bay, and Avian Island in Adélie penguin (MI = 26) (Diaz et al. 2016). These studies are generally consistent with the interpretation that the mean intensity of infection is higher in penguins inhabiting the Sub-Antarctic region than those on the South Shetland Islands or at the Antarctic Peninsula. A similar result was found comparing the mean intensity between Antarctic and non-Antarctic penguin species with higher values for the latter (D'Amico et al. 2014).

9.6 Parasite Effects on Antarctic Penguins

The effect of macroparasites on the health and fitness of Antarctic penguins is a topic that barely has been addressed, with only a few studies dealing with ticks infecting penguins living in Sub-Antarctic islands and others investigating the potential effects of helminths in the South Shetlands Islands. Reported effects of ticks on penguins include mortality due to hyperinfestation (Gauthier-Clerc et al. 1998), reduced breeding success (Mangin et al. 2003), and transmission of tick-borne diseases such as borreliosis (Olsen et al. 1995; Schramm et al. 2014; Barbosa et al. unpublished data) and babesiosis (Earle et al. 1993; Montero et al. 2016).

Helminth effects on Antarctic penguin have been reported at the level of the tissue damage, specifically, Martin et al. (2016) described lesions companied by hemorrhage, edema, degeneration, and necrosis of the intestine. More generally, using an experimental approach by means of the administration of anti-helminthic drugs, Palacios et al. (2012) estimated the effect of helminth parasites as a loss of 6 % of the body mass in infected chicks of Chinstrap penguins. Body mass loss has been also reported in Gentoo penguin chicks in a similar experiment (Palacios et al. unpublished data). Effects on the immune system of Antarctic penguins have also been demonstrated in terms of an increased foot-web swelling response to phytohemagglutinin and a decreased concentration of eosinophils and monocytes in the blood of individuals treated with anti-helminthic drugs (Bertellotti et al. 2016).

9.7 Potential Effects of Climate Change

Climate change can affect the distribution, abundance, and/or virulence of parasites (Sutherst 2001). Antarctica, however, is a region where the effects of climate change are complex and sometimes even contradictory. While the Antarctic Peninsula is one of the parts of the Earth where the temperatures have increased more rapidly in recent decades (Meredith and King 2005) and as a consequence a substantial reduction in sea ice extent has been detected (Stammerjohn et al. 2008; Fan et al. 2014), the Eastern continental region has shown an opposite trend of gradual decrease in land air temperatures and increase in sea ice extent (Fan et al. 2014). As a result, the expected effects of climate on the Antarctic fauna, including penguins and their parasites, will certainly differ between these regions.

Climate change in the Antarctic Peninsula is producing profound environmental changes affecting the trophic web from the bottom to the top through a significant reduction in the primary production (Montes-Hugo et al. 2009). With the consequent reduction in krill abundance (Atkinson et al. 2004; Flores et al. 2012), top predators such as penguins are changing their population trends (Carlini et al. 2009; Trivelpiece et al. 2011; Barbosa et al. 2012). However, not all species inhabiting the same areas have responded similarly, as is notoriously the case of the ice-intolerant Gentoo penguins, which have often benefitted from climate change, whereas the ice-dependent Adélie penguins in the same areas have experienced sharp population decreases (Forcada et al. 2006; Forcada and Trathan 2009). Dietary changes as a response to climate change could be predicted based on changes occurred during past climate changes in which penguins change their diet from krill to squid during warm periods (Emslie et al. 1998). Such changes would certainly affect not only the overall nutritional and health status of these seabirds, but it would also affect the rate of ingestion of parasite cysts/larvae and of exposure to new parasites. Similarly, because the life cycles of ectoparasites are greatly influenced by ambient temperature, it is expected that the increase of temperatures affect these parasites. For instance, there are already data to suggest that warmer years produce an increase in the abundance of ticks in the Antarctic Peninsula (Benoit et al. 2009).

9.8 Conclusions and Future Prospective

Although Antarctic penguins have been far more studied than other Antarctic seabirds, the scarce and fragmented nature of the available information has limited our broader understanding on the pathogens and disease that affect them and how they may impact their ecology, conservation, and evolution (Barbosa and Palacios 2009).

Published information is based on a geographically uneven sampling area, with few areas (e.g., South Shetland Islands) having been the subject of extensive research whereas virtually no information is available for the most of the continent (e.g., Ross Sea). As a consequence, there is not enough information yet to allow us to establish biogeographical patterns of presence and abundance of parasites. An additional complicating factor is that the information has often been collected during relatively short and discontinuous periods of time and long-term studies or surveillance of the temporal variation of prevalence or parasitism intensity is nonexistent. Such information is crucial to evaluate how environmental changes affect the ecology of these parasites and their impacts to the health of penguins.

Another challenge faced in health studies of Antarctic penguins is the difficulty of obtaining high quality data that faithfully reflect the occurrence of pathogens and disease, often due to the logistical limitations that are inherent to the continent or to application of diagnostic methods that were not specifically designed or validated to be used for these species. For instance, an important limitation that may influence data quality is the difficulty to obtain information of helminth parasites from live penguins through coprological studies because of the high probability of false negative results (Vidal et al. 2012). This, along with the ethical and legal restrictions and the endangered status of many species, restricts the study of endoparasites to the postmortem examination of naturally deceased individuals. As a result, quantitative information on the epidemiology of these parasites (prevalence, intensity of infection, etc.) are likely to be heavily biased and might allow for an adequate interpretation of their ecology and health effects. To solve this problem, the application of molecular techniques could help in improving the applicability and reliability of helminthological studies to living animals (Vidal et al. 2016).

Another important gap in our knowledge on the parasites of Antarctic penguins is the generalized insufficiency of information about their life cycles. This implies that we do not know which could be the intermediate hosts and, as a result, it is not possible to evaluate the risk of infection or how environmental factors affect the epidemiological dynamics.

Finally, from an ecological standpoint, the mechanisms and extent to which parasites affect their hosts is a critical gap in our understanding of Antarctic penguin parasites. Parasites can play a key role in the population dynamics of their hosts by affecting fitness traits such as survival, breeding success, or behavioral performance (Morand and Deter 2009). This can produce decline in host populations or affect host in different subtle ways through resources consumption and affecting metabolic rate, territorial behavior, phenology, intra- and interspecific interactions, mating and foraging success, etc. (Moller 1997). In addition, hosts can also adjust their behavior in order to avoid or reduce the effects of parasites (Perrot-Minnot and Cézilly 2009). The study of all these aspects has been virtually absent in Antarctica for decades, and only recently some studies have been published on this topic (see above).

It is therefore clear that an urgent effort is needed to obtain high quality data through long-term and geographically representative sampling effort, investigating not only the occurrence of parasites and pathogens but also deeper aspects of their ecology, life cycle, epidemiology, and health impacts. This will be a challenge not only for Antarctic researchers individually, but also reflects the need for broader instruments and policies by international and national Antarctic research programs to incorporate fauna health and pathogen studies as core components of scientific research in the Antarctic.

Acknowledgments This work is a contribution from the Genes to Geoscience funded workshop "Microbial and Parasitic impacts on Antarctic wildlife" held in August 2015 at Macquarie University, Sydney, Australia, and organized by the Working Group of Health Monitoring of Birds and Marine Mammals of the SCAR Expert Group of Birds and Marine Mammals. Macquarie University and the Standing Scientific group of Life Sciences of SCAR funded the workshop. JID is partially supported by PIP 0698 CONICET and N758 UNLP. AB is supported by the PINGUCLIM and CTM2011-24427 project funded by the Spanish Ministry of Economy and Competitiveness. RETV is supported by CAPES through the Department of Pathology (FMVZ-USP). DGA is supported by INACH T-12-13.

References

Adams NJ, Klages NT (1989) Temporal variation of the diet of the gentoo penguin *Pygoscelis papua* at sub-antarctic Marion Island. Col Waterbirds 12:30–36

Ainley DG, Wilson PR, Barton KJ, Ballard G, Nur N, Karl B (1998) Diet and foraging effort of Adélie penguin in relation to pack-ice conditions in the southern Ross Sea. Polar Biol 20:311–319

Allison FR, Desser SS, Whitten LK (1978) Further observations on the life cycle and vectors of the haemosporidian *Leucocytozoon tawaki* and its transmission to the Fiordland crested penguin. N Zeal J Zool 5:371–374

Andersen KI, Lysfjord S (1982) The functional morphology of the scolex of two Tetrabothrius Rudolphi 1819 species (Cestoda: Tetrabothriidae) from penguins. Parasitol Res 67:299–307

Anderson RC (2000) Nematode Parasite of Vertebrates. Their development and Transmission, CAB International (ed) 2nd edn. Oxon, Wallingford, UK, 650 p

Atkinson A, Siegel V, Pakhomov E, Rothery P (2004) Long-term decline in krill stock and increase in salps within the Southern Ocean. Nature 432:100–103

Baer JG (1954) Revision taxonomique et étude biologique des cestodes de la famille des Tetrabothriidae parasites d´oiseaux de haute mer et de mammiferes marins, vol 1. Mémoires de l´Université de Neuchatel, Neuchatel, pp 4–122

Banks JC, Palma RL, Paterson AM (2006) Cophylogenetic relationships between penguins and their chewing lice. J Evol Biol 19:156–166

Barbosa A, Merino S, de Lope F, Moller AP (2002) Effects on feather lice on flight behavior of male barn swallow (*Hirundo rustica*). Auk 119:213–216

Barbosa A, Palacios MJ (2009) Health of Antarctic birds: a revision of their parasites, pathogens and diseases. Polar Biol 32:1095–1115

Barbosa A, Benzal J, Vidal V, D'Amico V, Coria NR, Diaz JI, Motas M, Palacios MJ, Cuervo JJ, Ortiz J, Chitimia L (2011) Seabird ticks (Ixodes uriae) distribution along the Antarctic Peninsula. Polar Biol 34:1621–1624

Barbosa A, Benzal J, De León A, Moreno J (2012) Population decline of chinstrap penguins (Pygoscelis antarctica) on deception Island, South Shetlands, Antarctica. Polar Biol 35:1453–1457

Benoit JB, Lopez-Martinez G, Elnitsky MA, Lee RE, Denlinger DL (2009) Increase in feeding by the tick Ixodes uriae on Adélie penguins during a prolonged summer. Antarct Sci 21:151–152

Bergstrom S, Haemig PD, Olsen B (1999) Distribution and abundance of the tick *Ixodes uriae* in a diverse subantarctic community. J Parasitol 85:25–27

Bertellotti M, D'Amico V, Palacios MG, Barbosa A, Coria NR (2016) Effects of antihelminthic treatment on cell-mediated immunity in Gentoo penguin chicks. Polar Biol 39:1207–1212

Brandão ML et al (2014) Checklist of Platyhelminthes, Acanthocephala, Nematoda and Arthropoda parasitizing penguins of the world. Check List 10(3):562–573. doi:10.15560/10.3.562

Brooke ML (1985) The effect of allopreening on tick burdens of molting Eudyptid penguins. Auk 102:893–895

Bush MW, Kuhn T, Münster J, Klimple S (2012) Marine crustaceans as potential hosts and vectors for metazoan parasites. Parasitol Res Monogr 3:329–360

Carlini R, Coria NR, Santos MM, Negrete J, Juares MA, Daneri GA (2009) Responses of Pygoscelis adeliae and P. papua populations to environmental changes at Isla 25 de Mayo (King George Island). Polar Biol 32:1427–1433

Cherel Y, Kooyman GL (1998) Food of emperor penguins (*Aptenodytes forsteri*) in the western Ross Sea, Antarctica. Mar Biol 130:335–344

Cielecka D, Wojciechowska A, Zdzitowiecki K (1992) Cestodes from penguins on King George Island (South Shetlands, Antarctic). Acta Parasitol 37:65–72

Clarke J, Kerry K (2000) Diseases and parasites of penguin. Penguin Conserv 13:5–24

Clay T (1967) Mallophaga (biting lice) and Anoplura (sucking lice). Part I: austrogoniodes (Mallophaga) parasitic on Penguins (Sphenisciformes). Ant Res Ser 10:149–155

Clay T, Moreby C (1967) Mallophaga (biting lice) and Anoplura (sucking lice). Part II: keys and locality lists of Mallophaga and Anoplura. Ant Res Ser 10:157–196

Clay T, Moreby C (1970) Mallophaga and Anoplura of Subantarctic islands. Pac Insect Monogr 23:216–220

D'Amico VL, Bertelotti M, Diaz JI, Coria NR, Vidal V, Barbosa A (2014) Leucocyte levels in some Antarctic and non-Antarctic penguins. Ardeola 61:145–162

De Meillon B (1952) The fleas of the seabirds in the Southern Ocean. ANARE Reports Series B Vol 1 Zoology

Diaz JI, Cremonte F, Navone GT (2010) Helminths of the Magellanic penguin, *Spheniscus magellanicus* (Sphenisciformes), during the breeding season in Patagonian Coast, Chubut, Argentina. Comp Parasitol 77:172–177

Diaz JI, Fusaro B, Longarzo L, Coria NR, Vidal V, Jerez S, Ortiz J, Barbosa A (2013) Gastrointestinal helminths of Gentoo Penguins (*Pygoscelis papua*) from Stranger Point, 25 de Mayo/King George Island, Antartica. Parasitol Res 112:1877–1881. doi:10.1007/s00436-013-3341-3

Diaz JI, Fusaro B, Longarzo L, Coria NR, Vidal V, D'amico VL, Barbosa A (2016) Gastrointestinal helminths of Adelie Penguins (*Pygoscelis adeliae*) from Antarctica. Polar Res 35:28516

Dimitrova ZM, Chipev NH, Georgiev BB (1996) Record of Corynosoma pseudohamanni Zdzitowiecki, 1984 (Acanthocephala, Polymorphidae) in birds at Livingston, and South Shetlands, with a Review of Antarctic Avian Acanthocephalans. Bulg Antarct Res Life Sci 1: 102–110

Earle RA, Huchzermeyer FW, Brossy JJ (1993) *Babesia peircei* sp. nov. from the Jackass penguin. S Afr J Zool 28:88–90

Emslie SD, Fraser W, Smith RC, Walker W (1998) Abandoned penguin colonies and environmental change in the Palmer Station area, Anvers island, Antarctic Peninsula. Ant Sci 10:257–268

Fan T, Deser C, Schneider DP (2014) Recent Antarctic sea ice trends in the context of the Southern Ocean surface climate variation since 1950. Geophys Res Lett 41:2419–2426

Flores H, Atkinson A, Kawagushi S, Krafft B, Milinevsky G, Nicol S, Reiss C, Tarling GA, Werner R, Bravo Rebolledo E, Cirelli V, Cuzin-Roudy J, Fielding S, Groeneveld J, Haraldsson M, Lombana A, Marschoff E, Meyer B, Pakhomov EA, Rombola E, Schmidt K, Siegel V, Teschke M, Tonkes H, Toullec J, Trathan P, Tremblay N, Van de Putte A, van Franeker JA, Werner T (2012) Impact of climate change on Antarctic krill. Mar Ecol Prog Ser 458:1–19

Fonteneau F, Geiger S, Marion L, Le Maho Y, Robin JP, Kinsella JM (2011) Gastrointestinal helminths of King penguins (*Aptenodytes patagonicus*) at Crozet Archipelago. Polar Biol 34:1249–1252. doi:10.1007/s00300-011-0970-9

Forcada J, Trathan PN, Reid K, Murphy EJ, Croxall JP (2006) Contrasting population changes in sympatric penguin species in association with climate warming. Glob Chang Biol 12:411–423. doi:10.1111/j.1365-2486.2006.01108.x

Forcada J, Trathan PN (2009) Penguin responses to climate change in the Southern Ocean. Glob Chang Biol 15:1618–1630

Fredes F, Raffo E, Muñoz P, Herrera M (2006) Fauna parasitaria gastrointestinal en polluelos de Pinguino Papua (*Pygoscelis papua*) encontrados muertos en zona antártica especialmente protegida (ZAEP N°150). Parasitol Latinoam 61:179–182

Fredes F, Madariaga M, Ravo E, Valencia J, Herrera M, Godoy C, Alcaíno H (2007) Gastrointestinal parasite fauna of Gentoo penguins (*Pygoscelis papua*) from the Península Munita, Bahía Paraíso, Antarctica. Antarct Sci 19:93–94

Fredes F, Raffo E, Muñoz P, Herrera M, Godoy C (2008) Fauna parasitaria gastrointestinal en el pingüino Adelia (*Pygoscelis adeliae*) de zona antártica especialmente protegida (ZAEPN 150). Parasitología latinoamericana 63(1–2–3–4):64–68. doi:10.4067/S0717-77122008000100011

Frenot Y, de Oliveira E, Gauthier-Clerc M, DeunV J, Bellido A, Vernon P (2001) Life cycle of the tick Ixodes uriae in penguin colonies: relationship with host breeding activity. Int J Parasitol 31:1040–1047

Garbin L, Navone GT, Diaz JI, Cremonte F (2007) Further study of *Contracaecum pelagicum* (Nematoda: Anisakidae) in *Spheniscus magellanicus* (Aves: Spheniscidae) from two Argentine coast sites. J Parasitol 93:143–150

Garbin L, Diaz JI, Cremonte F, Navone GT (2008) *Contracaecum chubutensis* n. sp. New anisakid species parasitizing the imperial cormorant *Phalacrocorax atriceps* from the North Patagonian coast, Argentina. J Parasitol 94:852–859

Gauthier-Clerc M, Clerquin Y, Handrich Y (1998) Hyperinfestation by ticks *Ixodes uriae*: a possible cause of death in adult king penguins, a long-lived seabird. Colonial Waterbird 21: 229–233

Gauthier-Clerc M, Jaulhac B, Frenot Y, Bachelard C, Monteil H, Le Maho Y, Handrich Y (1999) Prevalence of *Borrelia burgdorferi* (the Lyme disease angent) antibodies in king penguin *Aptenodytes patagonicus* in Crozet Archipielago. Polar Biol 22:141–143

Gauthier-Clerc M, Manguin S, Le Bohec C, Gendner JP, Le Maho Y (2003) Comparison of behaviour, body mass, haematocrit level, site fidelity and survival between infested and non-infested king penguin *Aptenodytes patagonicus* by ticks *Ixodes uriae*. Polar Biol 26:379–382

Georgiev BB, Vasileva GP, Chipev NH, Dimitrova ZM (1996) Cestodes of seabirds at Livingston Island, South Shetlands. Bulg Antarct Res Life Sci 1:111–127

Gonzalez-Acuña D, Hernandez J, Moreno L, Herrmann B, Palma R et al (2013) Health evaluation of wild gentoo penguins (*Pygoscelis papua*) in the Antarctic Peninsula. Polar Biol 36:1749–1760

Gothe R, Kunze K, Hoogstraal H (1979) The mechanisms of pathogenicity in the tick paralysis. J Med Entomol 16:357–369

Hoberg EP (1986) Aspects of ecology and biogeography of Acanthocephala in Antarctic seabirds. Ann Parasit Hum Comp 61:199–214

Hoberg EP (1987) *Tetrabothrius shinni* sp. nov. (Eucestoda) from *Phalacrocorax atriceps bransfieldensis* (Pelecaniformes) in Antarctica with comments on morphological variation, host-parasite biogeography, and evolution. Can J Zool 65:2969–2975

Hoberg EP (1996) Faunal diversity among avian parasite assemblages: the interaction of history, ecology and biogeography in marine systems. Bull Scand Soc Parasitol 6:65–89

Hoberg EP (2005) Marine birds and their helminth parasites. In: Rohde K (ed) Marine parasitology, (Chapter 10, Economic, environmental and medical importance). CSIRO, Sydney, pp 414–421

Hunter PE (1970) Acarina: Mesostigmata: free-living mites of South Georgia and Heard Island. Pacific Insects Monograph 23:43–70

Holloway HL Jr, Bier W (1967) Notes on the host specificity of *Corynosoma hamanni* (Linstow, 1892). Bull Wildl Dis Assoc 3:76–78

Ippen R, Odening K, Henne D (1981) Cestode Parorchites zederi and sarcosporidian Sarcocystis spp. Infections in penguins of the South Shetland Islands. Erkr Zootiere 22:203–210

Johnston TH (1937) Australian Antarctic Expedition 1911–1914. Scientific Reports. Series C, Zoology and Botany, vol X, part 4. Cestoda, p 77

Johnston TH, Mawson PM (1945) Parasitic nematodes. B.A.N.Z.A.R.E. Reports, Series B, vol. V, part 2, pp 73–160

Kagei N, Asano K, Kihata M (1978) On the examination against the parasites of antarctic krill, *Euphausia superba*. Sci Rep Whales Res Inst 30:311–313

Kerry KR, Riddle MJ (2009) Health of Antarctic wildlife: a challenge for science and policy. Springer, Berlin

Kleinertz S, Christmann S, Silva LMR, Hirzmann J, Hermosilla C, Taubert A (2014) Gastrointestinal parasite fauna of Emperor Penguins (*Aptenodytes forsteri*) at the Atka Bay. Antarct Parasitol Res 113:4133–4139. doi:10.1007/s00436-014-4085-4

Kloser H, Plotz J, Palm H, Bartsch A, Hubold G (1992) Adjustment of anisakid nematode life cycles to the high Antarctic food web as shown by *Contracaecum radiatum* and *C. osculatum* in the Weddell Sea. Antarct Sci 4:171–178

Laskowski Z, Zdzitowiecki K (2005) The helminth fauna of some notothenioid fishes collected from the shelf of Argentine Islands, west Antarctica. Pol Polar Res 26:315–324

Laskowski Z, Korczak-Abshire M, Zdzitowiecki K (2012) Changes in acanthocephalan infection of the Antarctic fish Notothenia coriiceps in Admiralty Bay, King George Island, over 29 years. Pol Polar Res 33:99–108

Leiper RT, Atkinson EL (1914) Helminthes of the British Antarctic Expedition, 1910–13. P.Z.S., pp 222–226

Lescröel A, Ridoux V, Bost C-A (2004) Spatial and temporal variation in the diet of the gentoo penguin (*Pygoscelis papua*) at Kerguelen Islands. Polar Biol 27:206–216

Mangin S, Gauthier-Clerc M, Frenot Y, Gendner JP, Le Maho Y (2003) Ticks Ixodes uriae and the breeding performance of a colonial seabird king penguin Aptenodytes patagonicus. J Avian Biol 34:30–34

Markowski S (1952) The Cestodes of seals from the antarctica – vol 1 num 7. Published by Bulletin of the British Museum (Natural History) Zoology

Martín MA, Ortiz JM, Seva J, Vidal V, Valera F, Benzal J, Cuervo J, de la Cruz C, Belliure J, Martínez AM, Diaz JI, Motas M, Jerez S, D'Amico VL, Barbosa A (2016) Mode of attachment and pathology caused by *Parorchites zederi* in three species of penguins: *Pygoscelis papua, Pygoscelis adeliae*, and *Pygoscelis antarctica in* Antarctica Journal of Wildlife Diseases, 52: 568–575. DOI: 10.7589/2015-07-200

Mawson PM (1953) Parasitic nematoda collected by the Australian National Antarctic Research Expedition: Heard Island and Macquarie Island 1948–1951. Parasitology 43:291–297

McCoy KD, Beis P, Barbosa A, Cuervo JJ, Fraser WR, Gonzalez-Solis J, Jourdain E, Poisbleau M, Quillfeldt P, Leger E, Dietrich M (2013) Population genetic structure and colonisation of the western Antarctic Peninsula by the seabird tick *Ixodes uriae*. Mar Ecol Prog Ser 459:109–120

Meggitt FJ (1924) The cestodes of mammals. London, p 282

de Meeus T, Renaud F (2002) Parasites within the new phylogeny of eukariotes. Trends Parasitol 18:247–251

Meredith MP, King JC (2005) Rapid ocean climate change at the WAP. Geophys Res Lett 32:L19604

Miyazaki I (1991) An illustrated book of helminthic zoonoses. Southeast Asian Medical Information Center (International Medical Foundation of Japan) Nihon Kokusai Iryōdan

Moller AP (1997) Parasitism and the evolution of host life history. In: Clayton DH, Moore J (eds) Host-parasite evolution. General principles and avian moldels. Oxford University press, New York. pp 105–127
Montero E, Gonzalez LM, Chaparro A, Benzal J, Bertellotti M, Masero JA, Colominas-Ciuró R, Vidal V, Barbosa A (2016) First record of Babesia in Antarctic penguins. Ticks Tick Borne Dis 7(3):498–501
Montes-Hugo M, Doney SC, Ducklow HW, Fraser W, Martinson D, Stammerjohn SE, Schofield O (2009) Recent changes in phytoplankton communities associated with rapid regional climate change along the western Antarctic peninsula. Science 323:1470–1473
Morand S, Deter J (2009) Parasitism and regulation of the host population. In: Thomas F, Guégan JF, Renaud F (eds) Ecology and evolution of parasitism. Oxford University Press, Oxford, pp 83–104
Morgan IR, Westbury HA, Caple IW, Campbell J (1981) A survey of virus infection in sub-antarctic penguins on Macquarie Island, Southern Ocean. Aust Vet J 57:333–335
Murray MD, Vestjens WJM (1967) Studies on the ectoparasites of seals and penguins. Aust J Zool 15:715–725
Murray MD, Palma RL, Pilgrim RLD (1991) Ectoparasites of Australian, New Zealand and Antarctic birds. Appendix I. In: Marchant S, Higgins PJ (eds) Handbook of Australian, New Zealand and Antarctic birds, vol I, part A. Oxford University Press, Melbourne
Muzaffar SB, Jones IL (2004) Parasites and diseases of the auks (Alcidae) of the world and their ecology. Mar Ornithol 32:121–146
Olsen B, Duffy DC, Jaenson TGT, Gylfe A, Bonnedahl J, Berström S (1995) Transhemispheric exchange of Lyme disease spirochetes by seabirds. J Clin Microbiol 33:3270–3274
Palacios MJ, Valera F, Barbosa A (2012) Experimental assessment of the effects of gastrointestinal parasites on offspring quality in chinstrap penguins (*Pygoscelis antarctica*). Parasitology 139:819–824
Palma RL, Horning DS (2002) The lice (Insecta:Phthiraptera) from Macquarie island. ANARE Res Notes 105:1–27
Perrot-Minnot M-J, Cézilly F (2009) Parasites and behaviour. In: Thomas F, Guégan J-F, Renaud F (eds) Ecology and evolution of parasitism. Oxford University Press, Oxford, pp 49–67
Poulin R, Morand S (2004) Parasite biodiversity. Smithsonian Books, Washington, DC, p 216
Price PW (1980) Evolutionary biology of parasites. Monogr Popul Biol 15:1–237
Prudhoe S (1969) Cestodes from fish, birds and whales. BANZARE Rep Ser B VIII (Part 9)
Pütz K (1995) The post-moult diet of Emperor Penguins (*Aptenodytes forsteri*) in the eastern Weddell Sea. Antarct Polar Biol 15:457–463
Rocka A (2003) Cestodes of the Antarctic fishes. Polar Res 24:261–276
Rocka A (2004) Nematodes of the Antarctic fishes. Polar Res 25:135–152
Schmidt H (1965) *Tetrameres (G.) wetzeli* sp. n. (Nematoda, Spirurida), eine neue Tetrameresart aus dem Felsenpinguin, *Eudyptes (=Catarrhactes) chrysocome* Forst (Aves, Sphenisciformes). Z f Parasitenkunde 26:71–81
Schramm F, Gauthier-Clerc M, Fournier JC, McCoy KD, Barthel C, Postic D, Handrich Y, Le Maho Y, Jaulhac B (2014) First detection of *Borrelia burgdorferi* sensu lato DNA in king penguins (*Aptenodytes patagonicus halli*). Ticks Tick Borne Dis 5:939–942. doi:10.1016/j.ttbdis.2014.07.013
Schultz A, Petersen SL (2003) Absence of haematozoa in breeding Macaroni Eudyptes chrysolophus and Rockhopper E. chrysocome Penguins at Marion Island. *African Journal of Marine Science* 25:499–502
Siers S, Merkel JF, Bataille A, Vargas FH, Parker PG (2010) Ecological correlates of microfilarial prevalence in endangered Galapagos birds. J Parasitol 96:259–272
Sutherst RW (2001) The vulnerability of animal and human health to parasites under global change. Int J Parasitol 31:933–948
Stammerjohn S, Martinson D, Smith R, Yuan X, Rind DH (2008) Trends in Antarctic annual sea ice retreat and advance and their relation to El Niño-Southern Oscillation and Southern Annular Mode variability. J Geophys Res 113:C03S90

Tragardh (1908) The Acari of the Swedish South Polar Expedition. Wissensch. Ergebn. Schwed. Südpolar Expedition 5:1–34

Trivelpiece WZ, Hinke JT, Miller AK, Reiss CS, Trivelpiece SG, Watters GM (2011) Variability in krill biomass links harvesting and climate warming to penguin population changes in Antarctica. Proc Natl Acad Sci U S A 108:7625–7628

Vidal V, Ortiz J, Diaz JI et al (2012) Gastrointestinal parasites in chinstrap penguins from Deception Island, South Shetlands, Antarctica. Parasitol Res 111:723–727. doi:10.1007/s00436-012-2892-z1

Vidal V et al (2016) Morphological, molecular and phylogenetic analyses of the spirurid nematode *Stegophorus macronectes* (Johnston & Mawson, 1942). J Helminthol. doi:10.1017/S0022149X15000218

Williams TD (1995) The penguins. Spheniscidae. (Birds Families of the Word, No 2). Oxford University Press, p 328

Wilson N (1967) Acarina: Mesostigmata: Halarachnidae, Rhynonisidae of South Georgia, Heard and Kerguelen. Pacific Insect Monogr 23:71–77

Windsor DA (1998) Controversies in parasitology. Most of the species on Earth are parasites. Int J Parasitol 28:1939–1941

Yabsley MJ, Parsons NJ, Horne EC, Shock BC, Purdee M (2012) Novel relapsing fever *Borrelia* detected in African penguins (*Spheniscus demersus*) admitted to two rehabilitation centers in South Africa. Parasitol Res 110:1125–1130

Zdzitowiecki K (1991) Synopses of the Antarctic benthos koenigstein koeltz scientific books. Antarctic Acanthocephala, Koenigstein, p 116

Zdzitowiecki K, Drózdz J (1980) Redescription of *Stegophorus macronectes* (Johnston et Mawson, 1942) and description of *Stegophorus arctowskii* sp. n. (Nematoda, Spirurida) from birds of South Shetlands (the Antarctic). Acta Parasitol 26:205–212

Chapter 10
Lice on Seals in the Antarctic Waters and Lice in Temperate Climates

Birgit Mehlhorn and Heinz Mehlhorn

10.1 Introduction

Lice (Phthiraptera) are ectoparasites, which stroll on the surface of their warm-blooded hosts. The members of the suborder *Anoplura* suck blood, while the so-called biting lice (*Mallophaga*) feed on skin particles and/or hair of their hosts. The species of the skin feeding mallophages parasitize practically exclusively at terrestrial animals, while among the bloodsucking species of the Anoplura also semi-aquatic species exist, which parasitize permanently at marine mammalian animals.

It is known since 200 years that the bloodsucking lice may occur on the skin – especially along the flippers – of marine mammals in cold or even polar regions (von Olfers 1816). Comparing the facts known at this time and looking at the material obtained during several German South Polar expeditions, Enderlein described the new genus, *Antarctophthirius*, in 1906 with the type species *Antarctophthirius ogmorhini* (Enderlein 1906). This taxonomic work includes four genera (*Proechinophthirus, Echinophthirus*, *Lepidophthirus,* and *Antarctophthirus*) within the family Echinophthiriidae. Leonardi et al. (2014) published a survey on the recent status of body lice of such aquatic (often Antarctic) biotopes in our days. These authors listed 13 lice species (belonging to 5 genera) within the single family Echinophthiriidae.

The genus *Antarctophthirius* contains up to now seven recognized species:

- *A. callorhini* on fur seals
- *A. microchir* on sea lions

B. Mehlhorn • H. Mehlhorn (✉)
Institute for Parasitology, Heinrich-Heine-University,
Düsseldorf, Germany
e-mail: mehlhorn@uni-duesseldorf.de

S. Klimpel et al. (eds.), *Biodiversity and Evolution of Parasitic Life in the Southern Ocean*, Parasitology Research Monographs, Vol. 9,
DOI 10.1007/978-3-319-46343-8_10

- *A. trichechi* on walruses
- *A. lobodontis* on Antarctic true seals
- *A. ogmorhini* on Antarctic true seals
- *A. mawsoni* on Antarctic true seals
- *A. carlini* on Weddell seals

The other six species belong to the following four genera:

- *Echinophthirius* (1 species: *E. horridus* on Antarctic true seals)
- *Lepidophthirius* (*L. macrorhini* and another species on Antarctic true seals)
- *Latagophthirius* (1 species on river otters)
- *Proechinophthirius* (*P. fluctus* and another species on fur seals and sea lions)

However, the exact host specificity of these species is not completely elucidated. For example, *A. ogmorhini* is found on the leopard seal (*Hydrurga leptonyx*) and on the Weddell seal (*Leptonychotes weddelli*), while *A. callorhini*, *A. trichechi*, *A. lobodontis,* and *A. mawsoni* are considered as host specific. However, the available data material is rather scarce due to the fact that it can only be obtained during rather short expeditions.

Anyway, all these bloodsucking insects have to survive the influences of high-graded saltwaters, very low temperatures as well as high pressures, since their hosts are divers catching their food often in deeper zones (up to 450 m) of their water biotopes (Plötz et al. 2001).

10.2 Morphology of *Antarctophthirius ogmorhini*

The morphology of the until now described seal lice species is rather similar, so that the following features obtained from studies on *A. ogmorhini* will cover the available sound facts of the whole group (without neglecting species specificities).

10.2.1 General Aspects

1. All stages have a moderately swollen (rounded) hind body (Fig. 10.1) with visible borders of the segments. The abdomen of males appears more ovoid than spherical.
2. Their eyeless, conical head is longer than wide (Fig. 10.2).
3. The antennae of the adults have five segments which appear marbled in light microscopy by broad, dense annuli (Fig. 10.2).
4. The forelegs are smaller and more slender than the middle and hind legs and are equipped with claws being different from those of the other legs (Figs. 10.1 and 10.2). The claws of the foreleg appear needle-like, while those of the other legs are strong and bended.

Fig. 10.1 Prof. Dr. Mehlhorn and a veterinarian colleague from the German Dallmann Summer Research Station on King George Island (Antarctica) checking elephant seals (*Mirounga leonina*) for lice

Fig. 10.2 Light micrograph of a female louse of the species *Antarctophthirius ogmorhini* attached at a hair

5. The quadratic thorax of all stages is closely connected to the abdomen (Fig. 10.1).
6. The pseudopenis of males appears v-shaped.
7. Females have patches of genital setae which are arranged convergently.
8. The dorsal and ventral surfaces of the abdomen are covered by differently shaped scales which apparently trapped bubbles of air around the body.
9. The dorsal and ventral surfaces are covered with strands of stout spines of different lengths which appear in different arrangements.
10. The intersegmental regions of the thorax and the abdomen are insignificantly invaginated compared, for example, to lice from terrestrial mammals (e.g., pigs and humans).
11. The outer margin of the abdomen does not form deep invaginations along the border of segments, but is rounded, giving rise to a more or less spherical appearance of the whole abdomen especially in females.
12. The females glue their eggs onto the hair of the seal with the operculum pointing towards the tip of the shaft.
13. The glue is so tenacious that it cannot be dissolved without disrupting the hair, although it covers only one fourth of the egg.
14. The ovoid eggs (Fig. 10.3) reach a length of about 0.4–0.5 μm and thus are large compared to the size of the females (Table 10.1).
15. The egg operculum (cover) has in contrast to human head lice only a single, rather large opening (stigma) being situated in the center of the cover (Fig. 10.3). This is in contrast to the nits of human head lice, where several small openings occur at special place of the operculum.

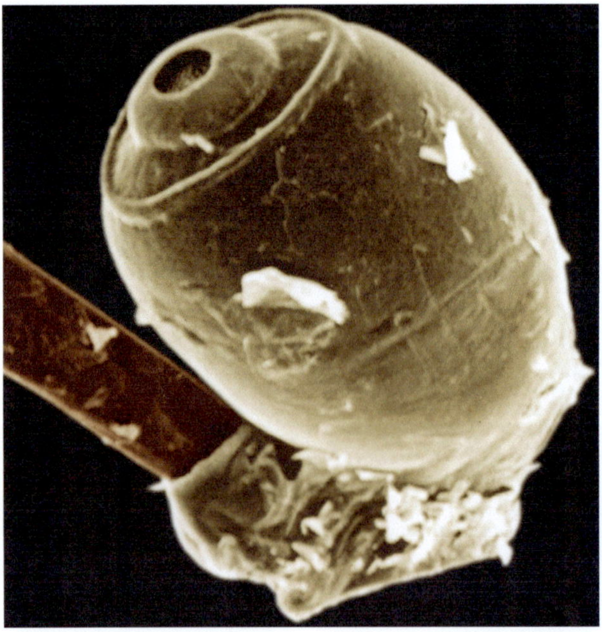

Fig. 10.3 Scanning electron micrograph of an egg of *Antarctophthirius ogmorhini*. Note the central opening on the upper surface of the operculum

Table 10.1 Body measurements (length) of some "aquatic" lice

Species	Males (mm)	Females (mm)	Body shape
Antarctophthirius carlinii	2.29 (±0.23)	2.77 (±0.52)	Ovoid-spherical
Antarctophthirius microchir	2.48 × 1.26	2.78 × 1.64	Ovoid
Antarctophthirius ogmorhini	2.0 × 1.2	2.2 × 1.3	Ovoid

10.2.2 Peculiarities of *Antarctophthirius ogmorhini*

1. The diameter of the cuticle is rather thick compared to sucking lice from terrestrial hosts and reaches about 1/6–1/10 of the whole diameter in the abdominal region of the body.
2. In general, it has at least double the width on the dorsum compared to the ventral part of the body.
3. The cuticle of the head is thinner than that in other portions of the body.
4. At the segmental borders of the abdomen, the cuticle is rather smooth and the plates are connected by intersegmental membranes.
5. Along the inner side of each of the three thoracic segments, a thick ridge is formed which is used as an anchor-point for strong muscle strands. These ridges are also visible from outside (Fig. 10.2) and run to a central point in the metathorax, where a depression can be seen when seen from above.
6. The thorax and the abdomen are closely covered by small scales which are arranged in a tile-like manner. The scales of the dorsal side of the body appear like the leaves of a European lime tree on the dorsal surface, while those on the ventral side of the body have an arrowhead shape on the ventral surface. In both cases, however, there was some air-filled space between the scales and the solid layers of the cuticle. This space is apparently filled by air bubbles during diving.
7. The segmental plates of the thorax and the abdomen, as well as the head, are spotted with regular rows of short, arrowhead-like, solid spines which are formed by the cuticle. These spines, which are directed obliquely to the posterior end of the louse, are shorter and broader in the head region and along the mid-thorax and abdomen, while they were longer and more pointed on the lateral sides of the body. These spines are apparently used to envelope the louse in a thick layer of the seal's sebum and thus provide another additional means of protection against low temperatures.
8. The dorsal hind border of the head as well as the dorsal surface of the thoracic segments has symmetrically arranged, long, cuticular hairs. These are found in a semicircular arrangement on the head, but occur only at the margin of the thorax (mostly in groups of four).
9. At the posterior end of the lice – especially around the genital openings – smooth hairs are found. These are considerably shorter than the thoracic hairs but longer and smoother than the body spines.
10. Similar, rather short hairs can be seen along the five segments of the relatively thick antennae, which appear striated due to alternating dense and white bands when studied by help of light microscopy. Other fine hairs can be found on the segments of the legs.

10.2.3 Comparisons to Other Lice

1. All genera of Antarctic lice of pinnipedia except for *Echinophthirius* have forelegs, which are smaller than the middle and hind ones. A similar phenomenon is seen in the human crab louse *Phthirus pubis* (Figs. 10.4 and 10.5), while human head and body lice (*Pediculus humanus capitis, P. h. corporis*) have legs all of the same size and shape (Figs. 10.6, 10.7, and 10.8).
2. The cover (operculum) of the lice eggs of other species is different and seems species specific, too (Figs. 10.3, 10.4, 10.5, 10.9, and 10.10).

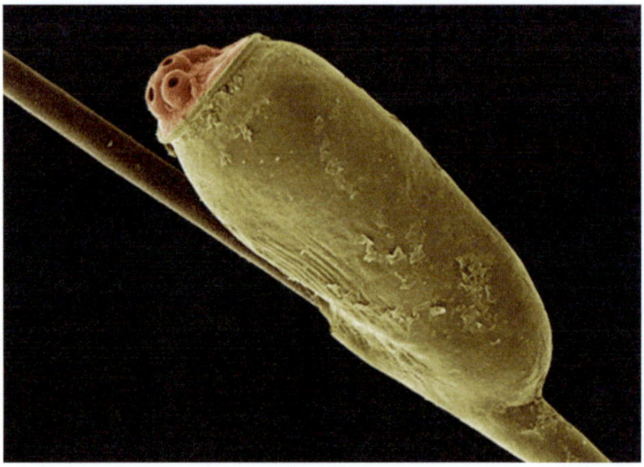

Fig. 10.4 Scanning electron micrograph of an egg of the human head louse *Pediculus humanus capitis*

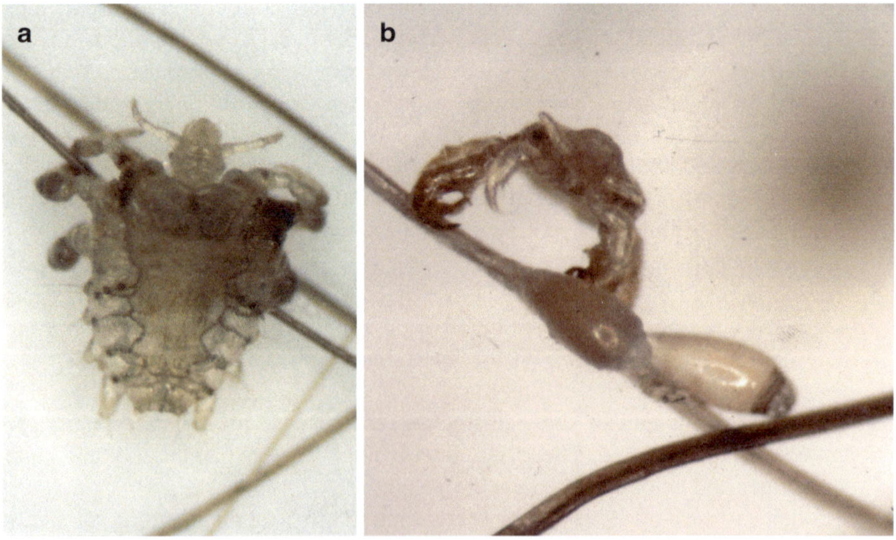

Fig. 10.5 Light micrographs of an adult human crab (pubic) louse (**a**) *Phthirus pubis* and the egg (**b**)

Fig. 10.6 Light micrograph of a human head louse

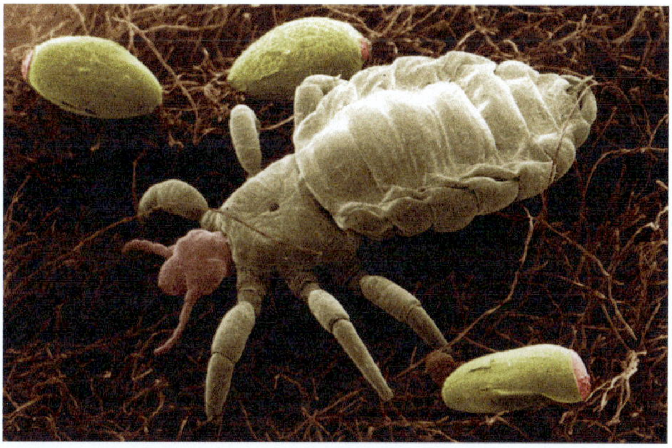

Fig. 10.7 Scanning electron micrograph of a human body louse (*Pediculus humanus corporis*) and its eggs on clothes

Fig. 10.8 Scanning electron micrograph of a pig louse (*Haematopinus suis*). Note the prolonged head

Fig. 10.9 Scanning electron micrograph of two eggs of the pig louse *Haematopinus suis*. One egg is empty (the larva has hatched and thus the cover is lacking)

Fig. 10.10 Light micrograph of eggs of the pig louse *Haematopinus suis*. Note their whitish appearance

3. All species studied so far have a characteristic body shape. While the closely related human body lice and head lice have, for example, a rather slender abdomen with marginal striations at the segmental borders, the hind body – especially in species of the genus *Antarctophthirus* – appears more ovoid to spherical with rounded margins, although the segmental borders are visible even at low magnification. The abdominal and thoracal portions of the pubic louse (*Phthirus pubis*) appear fused and thus appear unique (Fig. 10.5).
4. The outer surface of lice of the family Echinophthiriidae is absolutely unique and apparently represents an adaptation to the cold temperatures in the biotopes of their hosts.

   (a) The body surface is covered with regular rows of *stout spines* of a species-specific length. Those of *A. ogmorhini* are medium-sized in comparison to *Echinophthirius horridus*, *Lepidophthirus macrorhini,* or *A. trichechi* (Murray 1976; Scherf 1963). The spines on the ventral surface of the body and on the outer body margin are considerably larger and thicker than those on the mid-body and thorax. The main function of these spines became evident in our scanning electron micrographs. They are apparently used to fix a thick layer of the seal's sebum to their body surface. This sebum layer would offer protection against low water temperatures. The contact of the host's body surface with the spines of the lice probably induces an increased production of sebum.
   (b) In addition to these stiff body spines, there is *longer hair* seen on the dorsal surface of the louse's body. These thoracic hairs – probably representing sensillae (setae) – are species specific. Thus, *A. ogmorhini* has groups of four while *A. trichechi* (Scherf 1963) has only two on each side. The function of these longer hairs, however, is unknown.

5. Another prominent characteristic of the surface of the Antarctic lice (except for *Echinophthirius*) is the presence of numerous *small scales* which cover the abdomen and thus are produced by the rather thick cuticle. These scales, which may cover some air-filled space, are postulated to function in the same way as the plastron found in other insects, and apparently trap bubbles of air when the louse (together with its host) is immersed in water. These structures would therefore increase the oxygen uptake of the lice via the body surface, when direct contact via the stigmata is impossible (Hinton 1976; Murray 1976).
6. The fact that the dorsal cuticle of specimens of the family Echinophthiriidae is considerably thicker (reaching up to 1/6 of the whole diameter) than that of the ventral cuticle may also be explained as an adaptation to the cold environment, since the dorsal surface interfaces directly with the cold water, while the ventral surface, with the thinner cuticle, is attached to the warmer surface of the seal's body. In contrast to these Antarctic lice, the lice of terrestrial animals are much thinner and the dorsal and ventral cuticle plates are connected by rather thin membranes. This helps to regulate the body temperature.

7. Thus, considering items 4–6, the body surface of these lice from cold waters has three peculiarities that do not occur in lice from temperate climates, and guarantee that they can maintain a suitable body temperature. In addition, the surface scales, which apparently trap air bubbles, may help the lice to survive the rather long (30 min) and deep (up to 450 m) diving periods of the seal (Plötz et al. 2001).

10.3 Transmission of Agents of Diseases

Since some louse species change the hosts (apparently during body contacts when resting close together at the shore), agents of diseases may be transmitted. While it is well known that body lice transmit (via the oral-fecal route) the agents of the classic "spotted fever" induced by *Rickettsia prowazekii* (Mehlhorn 2011), the knowledge of transmissions of Antarctic lice is scarce. However, the paper of Linn et al. (2001) showing the transmission of α-viruses by the seal louse indicates that there is a large unknown background in the transmission story.

References and Further reading

Abdel-Ghaffar F et al (2010) Comparative in-vitro test on the efficacy and safety of 13 anti-head-lice products. Parasitol Res 106:423–429

Abdel-Ghaffar F et al (2012) Efficacy of a single treatment of head lice with a neem seed extract: an in vitro study on nits and motile stages. Parasitol Res 110:277–280

Al-Quraishy S et al (2015) Head louse control by suffocation due to blocking their oxygen uptake. Parasitol Res 114:3105–3110

Aznar FJ et al (2009) Population dynamics of *Antarctophthirius microchir* in pups from South American sea lion (*Otaria flavescens*) in Northern Patagonia. Parasitology 136:293–303

Drali R et al (2014) *Bartonella quintana* in body lice from scalp hair of homeless persons, France. Emerg Infect Dis 20:907–908

Drali R et al (2015) A new clade of African body and head lice infected by *Bartonella quintana* and *Yersinia pestis*. Am J Trop Med Hyg 93:990–993

Durden A, Musser M (1994) Sucking lice. Bull Am Mus Nat Hist 218:3–31

Enderlein G (1904) Die Gattung Echinophthiriidae. Zool Anz 28:136–142

Enderlein G (1906) Läusestadien.V. Schuppen als sekundäre Atmungsorgane, sowie über eine neue antarktische Echinophthiriiden-Gattung. Zool Anz 29:659–665

Freund L (1928a) Anoplura pinnipediorum. In: Grimpe G, Wagler E (eds) Tierwelt der Nord- und Ostsee, vol 4, part XI. Geest und Portig, Leipzig, pp 1–36

Freund L (1928b) Anoplura. In: Brohmer P (ed) Die tierwelt Mitteleuropas, vol 4, part IX. Quelle und Meyer, Leipzig, pp 1–26

Hase A (1931) Läuse. In: Schulze K (ed) Biologie der Tiere Deutschlands, part 30. Borntraeger, Berlin, pp 1–58

Hinton HE (1976) Respiratory adaptations. In: Cheng L (ed) Marine insects. North-Holland, Amsterdam

Kim KC (1972) Louse populations of the Northern fur seal (*Callorhinus ursinus*). Am J Vet Res 33:2027–2036

Kim KC (1985) Evolution and host association of Anoplura. In: Kim KC (ed) Coevolution of parasitic arthropods and mammals. Wiley, New York, pp 197–229

Leonardi MS, Lazzari CR (2014) Uncovering deep mysteries: the underwater life of an amphibious louse. J Insect Physiol 71:164–169

Leonardi MS et al (2009) Redescription of *Antarctophthirus microchir*, Trouessart and Neumman 1888 (Anoplura: Echinophthiriidae) from South American sea lion, *Otaria flavescens*, from Patagonia, Argentina. J Parasitol 95:1086–1092

Leonardi MS et al (2012a) Scanning electron microscopy of *Antarctophthirus microchir* (Anoplura: Echinophthiriidae): studying morphological adaptations to aquatic life. Micron 43:929–936

Leonardi MS et al (2012b) Life begins when the sea lion is ashore: microhabitat use by a louse living on a diving mammal host. Bull Entomol Res 102:444–452

Leonardi MS et al (2013) Lousy mums: patterns of vertical transmission of an amphibian louse. Parasitol Res 112:3315–3323

Leonardi MS et al (2014) *Antarctophthirius carlinii*: a new species from the Weddell seal *Leptonychotes weddelli*. Parasitol Res 113:3947–3951

Linn ML et al (2001) Arbovirus of marine mammals: a new alpha-virus isolated from the elephant seal louse (*Lepidophthirus macrorhini*). J Virol 75:4103–4109

Marcus AD et al (2015) Health assessment of free-ranging endangered Australian sea lion (*Neophora cinera*) pups: effects of haematophaggous parasites on haematological parameters. Comp Biochem Physiol A Mol Integr Physiol 184:132–143

Martini E (1946) Lehrbuch der medizinischen Entomologie, 3rd edn. Fischer, Jena

Mehlhorn H (2011) Head lice and their control. A long lasting story. In: Preedy VR (ed) Handbook of hair in health and disease. Wageningen Academic Publishers, The Netherlands, pp 355–386

Mehlhorn H (ed) (2016) Encyclopedia of parasitology, vols 1, 2, 3; 4th edn. Springer, New York/Berlin/Heidelberg

Mehlhorn B et al (2003) Light and scanning electron microscopical studies on *Antarctophthirius ogmorhini* lice from the Antarctic seal *Leptonychotes weddelli*. Parasitol Res 88:651–660

Miller FH Jr (1971) Scanning electron microscopy of *Echinophthirius horridus* (vonOlfers), *Antarctophthirus callorhini* (Osborn) and *Proechinophthirus fluctus* (Ferris) with emphasis on the antennal structures (Anoplura: Echinophthiriidae). J Parasitol 57:668–674

Murray MD (1967) Ectoparasites of Antarctic seals and birds. Japanese Antarctic Research Expedition JARE, Scientific Reports 1, pp 185–191

Murray MD (1976) Insect parasites of marine birds and mammals. In: Cheng JC (ed) Marine insects. North Holland, Amsterdam

Plötz J, Bornemann H, Knust R, Schröder A, Bester M (2001) Foraging behaviour of Weddell seals, and its ecological implications. Polar Biol 24:901–909

Sangaré AK et al (2014) Detection of *Bartonella quintana* in African body and head lice. Am J Trop Med Hyg 91:294–301

Scherf H (1963) Ein Beitrag zur Kenntnis zweier Pinnipedierläuse (*Antarctophthirus trichechi* Boheman und *Echinophthirius horridus* Olfers). Parasitol Res 23:16–44

Semmler M et al (2010) Repellency against head lice (*Pediculus humanus capitis*). Parasitol Res 106:729–731

Thompson PM et al (1998) Prevalence and intensity of the ectoparasites *Echinophthirius horridus* on harbour seals (*Phoca vitulina*): effects of host age and inter-annual variability in host food availability. Parasitology 117:393–403

Trouessart E, Neumann G (1888) Le pou de L'Otarie. Le Naturaliste 10:80–82

Von Olfers A (1816) De vegetativis et animatis corporibus animalis reperiundis commentarius, part 1 (cited by Durden and Musser 1994)

Werner R, Campagna C (1995) Diving behaviour of lactating Southern sea lions (*Otaria flavescens*) in Patagonia. Can J Zool 73:1975–1982

Index

S. Klimpel et al. (eds.), *Biodiversity and Evolution of Parasitic Life in the Southern Ocean*, Parasitology Research Monographs, Vol. 9,
DOI 10.1007/978-3-319-46343-8

D

M

R

MIX
Papier aus verantwortungsvollen Quellen
Paper from responsible sources
FSC® C105338

If you have any concerns about our products,
you can contact us on
ProductSafety@springernature.com

In case Publisher is established outside the EU,
the EU authorized representative is:
Springer Nature Customer Service Center GmbH
Europaplatz 3, 69115 Heidelberg, Germany

Printed by Libri Plureos GmbH
in Hamburg, Germany